Computational and Mathematical Techniques for Heat Transfer and Fluid Dynamics

Computational and Mathematical Techniques for Heat Transfer and Fluid Dynamics offers an in-depth exploration of modern modeling, simulation, and optimization strategies for solving complex problems in thermal and fluid systems. This edited volume presents a diverse collection of chapters that blend classical theories with advanced computational methods and emerging technologies to address a wide spectrum of thermal-fluid phenomena.

Covering foundational concepts as well as cutting-edge research, the book introduces readers to the fundamental principles of heat transfer and fluid mechanics, followed by in-depth discussions on non-Newtonian fluid models, nanofluid and hybrid nanofluid behaviors, magnetic and chemical interactions, and micropolar and porous media flows. A distinctive feature of this volume is the integration of advanced numerical techniques—including similarity transformations, finite difference solvers, special functions, and fractional calculus—with optimization strategies and machine learning tools.

Special emphasis is placed on biomedical flows, multiphase systems, and the application of artificial intelligence in modeling and prediction. Readers will also find case studies related to energy systems, polymer processing, and chemical reactors, making this book highly relevant to both academic researchers and industry professionals.

This book:

- Introduces essential concepts of heat transfer and fluid dynamics using practical modeling approaches.
- Explores nanofluid, hybrid nanofluid, and non-Newtonian fluid flow problems under multiple physical effects.
- Integrates modern computational tools such as MATLAB solvers, statistical optimization, and machine learning models.
- Discusses the application of fractional calculus and exact analytical techniques for complex flow and thermal scenarios and presents real-world applications in microfluidics, biomedical systems, renewable energy, and advanced cooling technologies.
- Offers detailed insights into optimization and control strategies for enhancing thermal-fluid system performance.

This book is an essential resource for graduate students, researchers, engineers, and professionals in mechanical, chemical, aerospace, and applied mathematics fields who seek to deepen their understanding of computational heat transfer and fluid dynamics in both traditional and emerging applications.

Advances in Manufacturing, Design and Computational Intelligence Techniques

Series Editor-Ashwani Kumar- Senior Lecturer, Mechanical Engineering, at Technical Education Department, Uttar Pradesh, Kanpur, India

The book series editor is inviting edited, reference and text book proposal submission in the book series. The main objective of this book series is to provide researchers a platform to present state of the art innovations, research related to advanced materials applications, cutting edge manufacturing techniques, innovative design and computational intelligence methods used for solving nonlinear problems of engineering. The series includes a comprehensive range of topics and its application in engineering areas such as additive manufacturing, nanomanufacturing, micromachining, biodegradable composites, material synthesis and processing, energy materials, polymers and soft matter, nonlinear dynamics, dynamics of complex systems, MEMS, green and sustainable technologies, vibration control, AI in power station, analog-digital hybrid modulation, advancement in inverter technology, adaptive piezoelectric energy harvesting circuit, contactless energy transfer system, energy efficient motors, bioinformatics, computer aided inspection planning, hybrid electrical vehicle, autonomous vehicle, object identification, machine intelligence, deep learning, control-robotics-automation, knowledge based simulation, biomedical imaging, image processing and visualization. This book series compiled all aspects of manufacturing, design and computational intelligence techniques from fundamental principles to current advanced concepts.

Heat Exchanger Technologies for Sustainable Renewable Energy Systems
Mukesh Kumar Awasthi, Ashwani Kumar, Nitesh Dutt, Sivasakthivel Thangavel

Green Hydrogen Production: Storage, Transportation and Applications
Ashwani Kumar, Sivasakthivel Thangavel and Gaurav Dwivedi

Computational and Mathematical Techniques for Heat Transfer and Fluid Dynamics
Satya Ranjan Mishra, Apul Narayan Dev, Alok Kumar Pandey, and Mukesh Kumar Awasthi

www.routledge.com/Advances-in-Manufacturing-Design-and-Computational-Intelligence-Techniques/book-series/CRCAIMDCIT?publishedFilter=alltitles&pd=published,forthcoming&pg=1&pp=12&so=pub&view=list?publishedFilter=alltitles&pd=published,forthcoming&pg=1&pp=12&so=pub&view=list

Computational and Mathematical Techniques for Heat Transfer and Fluid Dynamics

Edited by
Satya Ranjan Mishra, Apul Narayan Dev,
Alok Kumar Pandey, and
Mukesh Kumar Awasthi

CRC Press
Taylor & Francis Group
Boca Raton London New York

CRC Press is an imprint of the
Taylor & Francis Group, an informa business

Designed cover image: Shutterstock

First edition published 2026
by CRC Press
2385 NW Executive Center Drive, Suite 320, Boca Raton FL 33431

and by CRC Press
4 Park Square, Milton Park, Abingdon, Oxon, OX14 4RN

CRC Press is an imprint of Taylor & Francis Group, LLC

ISBN: 9781041077749 (hbk)
ISBN: 9781041077787 (pbk)
ISBN: 9781003642169 (ebk)

DOI: 10.1201/9781003642169

Typeset in Sabon
by Newgen Publishing UK

Contents

Aim and scope *vii*
Preface *ix*
About the editors *xi*
List of contributors *xv*
Acknowledgments *xix*

1 Introduction to heat transfer and fluid dynamics 1
MUKESH KUMAR AWASTHI, HIMANSHU UPRETI, AND ALOK KUMAR PANDEY

2 Rheological models for non-Newtonian fluids 18
A.J.D. NANTHAKUMAR

3 Analysis of buoyancy force, Brownian motion and thermophoretic effect in assisting and opposing flow of Prandtl fluid in presence of Joule heating and viscous dissipation 37
PRASANTA PARIDA, KHARABELA SWAIN, AND HIBAH ISLAHI

4 Magnetic field effects on Casson–Maxwell nanofluid flow over a stretching surface in porous media 59
SEEMA TINKER, OM PRAKASH, AND PIYUSH KUMAR GUPTA

5 Hybrid nanofluid heat transfer: Williamson–Maxwell interactions 78
SEEMA TINKER, OM PRAKASH, AND PIYUSH KUMAR GUPTA

6 Enhanced thermal performance of magnetized Maxwell hybrid nanofluid transportation over a radiative stretching sheet with exponential heat source 93
SUBHAJIT PANDA, FOLARIN OLUWASEUN, TITILAYO M. AGBAJE, S. BAAG, AND P. K. PATTNAIK

7 Ree–Eyring hybrid nanofluid flow bounded by double-revolving disks with Darcy–Forchheimer medium and variable thermal conductivity: heat transfer analysis 107
SK ENAMUL AND SURENDER ONTELA

8 Illustration of *n*th order chemical reaction on second-grade nanofluid flow past a Riga sensor plate with the influence of viscous dissipation 137
ABHISHEK SHARMA AND RAM PRAKASH SHARMA

9 Thermo-bioconvective transport in a lid-driven porous cavity with oxytactic microorganisms under magnetic field and Joule heating effects 154
SAHIN AHMED, ANIL NANGKAR, AND BIKASH DAS

10 Heat and mass transfer effects on a stagnation flow considering slip and radiation impact 177
T. PRADHAN, S. JENA, AND B. K. PAIKARAY

11 Mathematical model of immiscible Newtonian fluids flow through an anisotropic porous channel 190
AMIT KUMAR AND KRISHNA PRASAD MADASU

12 Exact solutions of micropolar stagnation point slip flow with heat transfer: Special functions categories 211
WAQAR KHAN USAFZAI AND EMAD HASSAN ALY

13 Application of Jacobi algorithm to physical problems 229
PRASANNA BANSODE

Index *247*

Aim and scope

The book *Computational and Mathematical Techniques for Heat Transfer and Fluid Dynamics* aims to provide a comprehensive understanding of the latest analytical, numerical, and computational methods applied in the field of thermal and fluid sciences. With growing industrial demands and the emergence of complex flow and heat transfer phenomena, this volume addresses the need for advanced modeling techniques, hybrid computational approaches, and intelligent algorithms to tackle real-world engineering challenges. It serves as a valuable resource for graduate students, researchers, and engineers working in mechanical, chemical, and aerospace engineering, as well as applied mathematics and computational science.

The scope of the book begins with foundational knowledge in Chapter 1, which lays the groundwork by discussing essential principles of heat transfer mechanisms and fluid dynamics, focusing on flow regimes, fluid properties, and the derivation of conservation laws essential for mathematical modeling. In Chapter 2, readers are introduced to rheological models that describe the behavior of non-Newtonian fluids, alongside a discussion of how machine learning tools can refine the accuracy and speed of simulations involving such complex fluids.

Chapter 3 presents an in-depth study of nanofluid dynamics by incorporating multiple physical effects such as buoyancy, Brownian motion, and thermophoresis, solved using similarity transformations and numerical tools like MATLAB's bvp4c. This is extended in Chapter 4, which examines magnetic and thermal effects in a Casson–Maxwell nanofluid system through porous media, using bvp5c-based simulations to illustrate the impact of magnetohydrodynamics on thermal performance.

Further exploration of hybrid nanofluids is undertaken in Chapter 5, which analyzes the combined thermal effects of magnetic fields, thermophoresis, and radiation on stretched surfaces. Chapter 6 continues this theme by modeling a magnetized hybrid nanofluid containing graphene oxide and silver nanoparticles over a radiatively heated surface, highlighting the use of Runge–Kutta methods and shooting techniques for improved thermal predictions.

Chapter 7 addresses complex flows between rotating disks, focusing on Ree–Eyring hybrid nanofluids in porous media and studying the interplay between rotational forces, magnetic fields, and nanoparticle concentrations. Chapter 8 discusses second-grade nanofluid behavior over a Riga plate under chemical reactions and viscous dissipation, integrating computational results with optimization using response surface methodology (RSM) to improve flow and thermal control.

Chapter 9 expands into the domain of biological flows by examining the transport of oxytactic microorganisms in a magnetized lid-driven porous cavity, providing valuable insights into heat and oxygen transfer in biomedical and microfluidic systems. In Chapter 10, stagnation-point flow with slip and radiation effects is analyzed, offering dual-solution profiles that reflect the complexity of industrial cooling and materials processing environments.

A mathematical model for immiscible Newtonian fluid flow through anisotropic porous channels is detailed in Chapter 11, with particular attention to the roles of anisotropy and viscosity contrast in petroleum and multiphase systems. Chapter 12 offers exact analytical solutions for micropolar stagnation-point flows with slip and heat transfer using special functions to describe dual solutions relevant to industrial coating technologies. Chapter 13 introduces the Jacobi algorithm for solving eigenvalue problems, illustrating its application to quantum systems such as the wave and Schrödinger equations and showcasing its computational efficiency.

Overall, this book integrates classical theory with cutting-edge computational and data-driven techniques, delivering a holistic treatment of the modeling, analysis, and optimization of heat transfer and fluid dynamics problems across scientific and engineering disciplines.

Satya Ranjan Mishra, Apul Narayan Dev,
Alok Kumar Pandey, and Mukesh Kumar Awasthi

Preface

The intricate interplay between heat transfer and fluid dynamics is fundamental to countless natural phenomena and engineering applications, ranging from energy systems and aerospace engineering to biomedical devices and environmental management. With the increasing complexity of real-world systems and the growing demand for energy-efficient and sustainable technologies, the role of advanced computational and mathematical tools has never been more critical. This book, *Computational and Mathematical Techniques for Heat Transfer and Fluid Dynamics*, offers a comprehensive exploration of modern analytical, numerical, and data-driven approaches to tackle current challenges in the field. The text aims to serve as a valuable reference for researchers, engineers, and graduate students involved in thermal sciences and fluid mechanics.

Beginning with Chapter 1, the book lays the foundational concepts of heat transfer and fluid dynamics by presenting an overview of conduction, convection, radiation, and the governing conservation laws, thereby setting the stage for the advanced discussions that follow. Chapter 2 introduces rheological models for non-Newtonian fluids and discusses how machine learning can refine simulations involving complex fluid behaviors encountered in various industries. In Chapter 3, the effects of buoyancy, Brownian motion, and thermophoresis on non-Newtonian nanofluid flow are examined using similarity transformations and MATLAB-based solutions.

Chapter 4 delves into magnetohydrodynamic nanofluid dynamics, analyzing the role of magnetic fields, thermophoresis, and thermal radiation in a porous media setup through numerical techniques. Continuing the exploration of nanofluid behavior, Chapter 5 investigates hybrid nanofluid heat transfer and assesses how different physical phenomena impact heat transfer enhancement over a stretched surface. Chapter 6 presents an in-depth study of graphene oxide–silver hybrid nanofluids subjected to exponential heating, showcasing the potential of hybrid materials for advanced thermal applications using numerical schemes.

Chapter 7 extends the study to Ree–Eyring hybrid nanofluid flows between rotating disks, examining the combined influence of magnetic fields, nanoparticle loading, and porous media on thermal behavior. Chapter 8 focuses on chemically reactive flows of second-grade nanofluids near Riga plates, employing numerical techniques and statistical optimization to uncover the interaction between chemical reactions and fluid properties. In Chapter 9, thermo-bioconvective transport is modeled within a porous cavity filled with oxytactic microorganisms, highlighting the effects of magnetic fields and Joule heating on microfluidic and biomedical systems.

Chapter 10 discusses stagnation-point flow and addresses the role of slip and radiation on heat and mass transfer, revealing the presence of multiple solution branches relevant to cooling and manufacturing technologies. In Chapter 11, a model for immiscible Newtonian fluids in anisotropic porous channels is developed, offering insights into flow phenomena critical to petroleum engineering and multiphase systems. Chapter 12 provides exact solutions for micropolar slip flows using special functions, serving as a valuable reference for analytical studies in coating and material processing industries.

Chapter 13 applies the Jacobi algorithm to solve eigenvalue problems in wave and quantum mechanics, illustrating the power of classical iterative techniques in solving modern physical systems. Together, these chapters represent a synthesis of traditional principles and cutting-edge innovations, aiming to bridge theoretical insights with industrial and technological advancements. We hope this book not only deepens the understanding of thermal-fluid phenomena but also inspires novel research and development across multiple disciplines.

EDITORS

Satya Ranjan Mishra, Apul Narayan Dev, Alok Kumar Pandey, and Mukesh Kumar Awasthi

About the editors

Satya Ranjan Mishra is currently working as a professor in the Department of Mathematics in Siksha O Anusandhan Deemed to be University. He has 19 years of teaching experience at both the postgraduate and under graduate levels. He completed his PhD from Siksha O Anusandhan in 2013 and since then has actively engaged in his research work. His area of interest is heat transfer, magnetohydrodynamics, and porous media within the broad area of fluid dynamics. He has published nearly 250 papers in the international journals of repute, and all are either SCI or Scopus indexed that can be viewed in various databases like Scopus, Research Gate, and Google Scholar. With a huge citation of his work, he took a position of top 2% world scientists by Stanford University, USA, for four consecutive years – 2020–2023. He has also guided 13 research scholars, and six more students are working under his guidance. Dr. Mishra has delivered several lectures in various conferences and faculty development program organized by Rajasthan Technical University in 2019 on the topic "Introduction to MATLAB", FDP organized by Poornima Group of Institutions, Rajasthan, on the topic "Numerical solutions using scientific tool MAPLE" in 2020 (online), Summer Instructional School organized by NIT Arunachachal Pradesh on the topic "Linear algebra" in the year 2020. He has edited a book series published by Springer Nature in 2020 as a leading editor and also editor/associate editor in several reputed international journals. Dr. Mishra's forthcoming book *Learning Numerical Method using MATLAB* is going to be published soon.

Apul Narayan Dev is currently working as an associate professor and coordinator at the Centre for Data Science in Siksha O Anusandhan Deemed to be University, Bhubaneswar. He has 12 years of teaching experience at both post graduate and under graduate levels. He obtained his M.Phil and Ph.D. from Gauhati University in 2011 and 2016, respectively. His area of interest is basic study of theoretical plasmas, degenerate and non-degenerate plasmas, mathematical method, fluid flow, etc. with the broad area of plasmas and fluid dynamics. He has published nearly 45 papers in reputed international journals, and all are either SCI or Scopus indexed that can be viewed in various databases like Scopus, Research Gate, and Google Scholar. His book *Shock Wave Phenomena in Dusty Plasma* is available on Amazon. He has also guided one research scholar, and four more students are working under his guidance. He has organized two international conferences (AMSE-2019, AMSE-2022) and two international webinars (Maple-2020, Mathematica-2020). He also edited a *Scopus index Conference Proceedings* published by AIP Publishing in 2022 as a leading editor.

Alok Kumar Pandey is an assistant professor in the Department of Mathematics in Graphic Era (Deemed to be University), India. He obtained his Ph.D. from Govind Ballabh Pant University of Agriculture and Technology, Pantnagar, Uttarakhand, India. His area of research is computational fluid dynamics. He has published more than 60 research articles in international journals. He was included in the world's top 2% scientists 2023 list (by Stanford University). He was awarded the Publons Peer Review Award in 2018. He is a certified reviewer of more than 100 international journals. He is currently the editor for *Open Physics* and *Journal of Engineering Researcher* and Lecturer, Associate Editor for the *Journal of Advanced Research in Numerical Heat Transfer* and *Journal of Advanced Research in Micro and Nano Engineering*. He serves as a member of the editorial board for journals such as *Teknomekanik* and *SCIREA Journal of Mathematics*. His scientific metrics according to Google Scholar show h-index=30, citations=2126, and i10-index=43.

Mukesh Kumar Awasthi has done his Ph.D. titled "Viscous Correction for the Potential Flow Analysis of Capillary and Kelvin-Helmholtz instability". He is working as an assistant professor in the Department of Mathematics in Babasaheb Bhimrao Ambedkar University, Lucknow. Dr. Awasthi is specialized in the mathematical modeling of flow problems. He has taught courses on fluid mechanics, discrete mathematics, partial differential equations, abstract algebra, mathematical methods, and measure theory to postgraduate students. He has acquired excellent knowledge in the mathematical modeling of flow problems, and he can solve these problems analytically as well as numerically. He has a good grasp of the subjects like viscous potential flow, electro-hydrodynamics, magnetohydrodynamics, heat, and mass transfer. He has excellent communication skills and leadership qualities. He is self-motivated and responds to suggestions in a more convincing manner. Dr. Awasthi has qualified National Eligibility Test (NET) conducted on all India level in the year 2008 by the Council of Scientific and Industrial Research (CSIR) and obtained Junior Research Fellowship (JRF) and Senior Research Fellowship (SRF) for performing research. He has published 135 plus research publications (journal articles/books/book chapters/conference articles) in many national and international journals and conferences with reputed publishers. He has published 15 books. He is also a series editor of *Artificial Intelligence and Machine Learning for Intelligent Engineering Systems* published by CRC Press. He has attended many symposia, workshops, and conferences in mathematics as well as fluid mechanics. He has got the "Research Awards" consecutively four times from 2013–2016 by the University of Petroleum and Energy Studies, Dehradun, India. He has also received the start-up research fund for his project "Nonlinear Study of the Interface in Multilayer Fluid System" from the University Grants Commission, New Delhi. He is also listed among the top 2% of influential researchers in the world, as prepared by Stanford University based on Scopus data for the years 2022 and 2023.

Contributors

Titilayo M. Agbaje
School of Computer Science and Applied Mathematics, University of the Witwatersrand, Johannesburg, South Africa
Center for Sustainable Smart Cities, Department of Mathematical and Physical Sciences, Central University of Technology, Bloemfontein, South Africa

Sahin Ahmed
Department of Mathematics, Rajiv Gandhi University, Rono Hills, Doimukh, Arunachal Pradesh, India

Emad Hassan Aly
Department of Mathematics, Faculty of Education, Ain Shams University, Roxy, Cairo, Egypt

Mukesh Kumar Awasthi
Department of Mathematics, Babasaheb Bhimarao Ambedkar University, Lucknow, India

S. Baag
Department of Physics, College of Basic Science and Humanities, OUAT, Bhubaneswar, Odisha, India

Prasanna Bansode
Northwestern University, United States

Bikash Das
Department of Mathematics, Rajiv Gandhi University, Rono Hills, Doimukh, Arunachal Pradesh, India

Sk Enamul
Department of Mathematics, National Institute of Technology Mizoram, Aizawl, India

Piyush Kumar Gupta
Department of Mathematics, JECRC University, Jaipur, India

Hibah Islahi
Institute of Applied Sciences, Mangalayatan University, Aligarh, India

S. Jena
Department of Mathematics, Centurion University of Technology and Management, Odisha, India

Amit Kumar
Department of Mathematics, National Institute of Technology, Raipur, Chhattisgarh, India

Krishna Prasad Madasu
Department of Mathematics, National Institute of Technology, Raipur, Chhattisgarh, India

Anil Nangkar
Department of Mathematics, Rajiv Gandhi University, Rono Hills, Doimukh, Arunachal Pradesh, India

A.J.D. Nanthakumar
Vehicle Dynamics Laboratory, Department of Automobile Engineering, SRM Institute of Science and Technology, Kattankulathur, Tamil Nadu, India

Folarin Oluwaseun
School of Mathematics, Statistics, and Computer Science, University of the KwaZulu-Natal, Private Bag X01, Scottsville, Pietermaritzburg, South Africa

Surender Ontela
Department of Mathematics, National Institute of Technology Mizoram, Aizawl, India, Department of Mathematics, National Institute of Technology Kurukshetra, Haryana, India

B. K. Paikaray
Department of Computer Science and Engineering, Siksha 'O' Anusandhan (Deemed to be) University, Odisha, India

Subhajit Panda
Centre for Data Science, Department of Computer Science and Engineering, Siksha 'O' Anusandhan Deemed to be University, Bhubaneswar, Odisha, India

Alok Kumar Pandey
Department of Mathematics, Graphic Era (Deemed to be University), Dehradun, Uttarakhand, India

Prasanta Parida
Institute of Applied Sciences, Mangalayatan University, Aligarh, India

P. K. Pattnaik
Department of Mathematics, Odisha University of Technology and Research, Bhubaneswar, Odisha, India

T. Pradhan
Department of Mathematics, Centurion University of Technology and Management, Odisha, India

Om Prakash
Department of Mathematics, JECRC University, Jaipur, India

Abhishek Sharma
Department of Basic and Applied Science, National Institute of Technology Arunachal Pradesh, Jote, Papum Pare District, Arunachal Pradesh, India

Ram Prakash Sharma
Department of Mechanical Engineering, National Institute of Technology Arunachal Pradesh, Jote, Papum Pare District, Arunachal Pradesh, India

Kharabela Swain
Department of Mathematics, GIFT Autonomous, Bhubaneswar, India

Seema Tinker
Department of Mathematics, JECRC University, Jaipur, India

Himanshu Upreti
School of Engineering and Technology, BML Munjal University, Gurugram, Haryana, India

Waqar Khan Usafzai
School of Mathematics and Physics, Nanjing Institute of Technology, Nanjing, China

Acknowledgments

The successful completion of this book, *Computational and Mathematical Techniques for Heat Transfer and Fluid Dynamics*, has been made possible through the invaluable contributions and collaborative efforts of many individuals and institutions.

First and foremost, we extend our deepest gratitude to the contributing authors. Their scholarly dedication, technical expertise, and insightful research have significantly shaped this book into a comprehensive and forward-looking resource. Each chapter reflects the authors' commitment to advancing knowledge in the fields of heat transfer and fluid dynamics through the integration of computational and mathematical methodologies.

We are equally thankful to our academic peers and professional collaborators for their continuous encouragement, constructive feedback, and intellectual support during the development of this book. The detailed reviews and thoughtful suggestions provided by the peer reviewers have been instrumental in refining the quality and clarity of the content.

We sincerely appreciate the support of our respective institutions, which provided a stimulating environment and the necessary resources to carry out this work. Their encouragement and infrastructure have played a vital role in facilitating the research and collaboration behind this project.

Our heartfelt thanks go to the publishing team for their expert guidance throughout the editorial and production process. Their professionalism and dedication ensured a seamless transformation of our manuscript into its final published form.

Finally, we are profoundly grateful to our families and friends for their unwavering support, patience, and understanding. Their encouragement and sacrifices behind the scenes have been a source of strength throughout the journey of completing this book.

This book represents a collective intellectual endeavor, and we are truly appreciative of all who have contributed to its realization. We hope it serves as a valuable reference for researchers, students, and professionals seeking to explore and apply advanced computational and mathematical techniques in heat transfer and fluid dynamics.

EDITORS

Satya Ranjan Mishra, Apul Narayan Dev, Alok Kumar Pandey, Mukesh Kumar Awasthi

Chapter 1

Introduction to heat transfer and fluid Dynamics

Mukesh Kumar Awasthi, Himanshu Upreti, and Alok Kumar Pandey

1.1 OVERVIEW OF HEAT TRANSFER MECHANISMS

Heat transfer is the science of governing energy exchange due to temperature differences, occurring via three primary modes: conduction, convection, and radiation(see Figure 1.1). These mechanisms often coexist in practical systems, necessitating combined analysis. Understanding their constitutive equations and interactions is critical for engineering applications ranging from electronics cooling to climate control.

1.1.1 Conduction

Conduction is heat transfer through a stationary medium (solid or static fluid) driven by a temperature gradient. Governed by Fourier's Law, the heat flux $q''(\mathrm{W/m^2})$ is proportional to the negative temperature gradient as

$$q'' = -k\frac{dT}{dx} \tag{1.1}$$

where $k(\mathrm{W/m \cdot K})$ is thermal conductivity, and $\frac{dT}{dx}$ is the spatial temperature variation [1]. For transient conditions, the heat diffusion equation applies

$$\rho c_p \frac{\partial T}{\partial t} = \nabla \cdot (k\nabla T) + \dot{q}_{gen} \tag{1.2}$$

Here, $\rho(\mathrm{kg/m^3})$ is density, $c_p\,(\mathrm{J/kg \cdot K})$ is specific heat, $q_{gen}\,(\mathrm{W/m^3})$ is internal heat generation, and t is time. Materials with high k, such as metals, facilitate rapid conduction, while insulators like polymers suppress it. Steady-state solutions simplify to $\nabla^2 T = 0$, pivotal in designing heat sinks and building insulation [2].

DOI: 10.1201/9781003642169-1

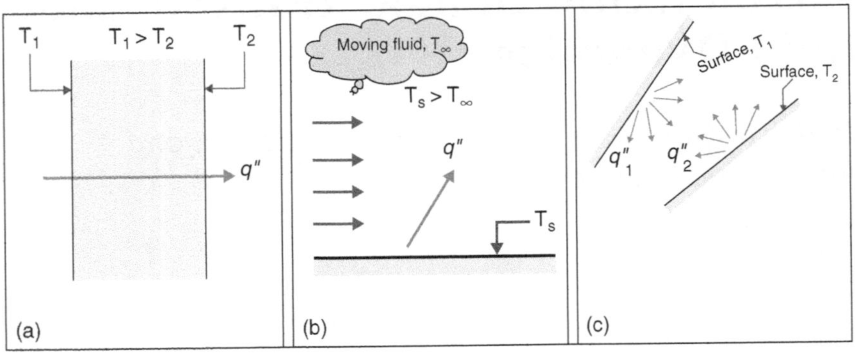

Figure 1.1 Modes of heat transfer: (a) conduction, (b) convection, and (c) radiation.

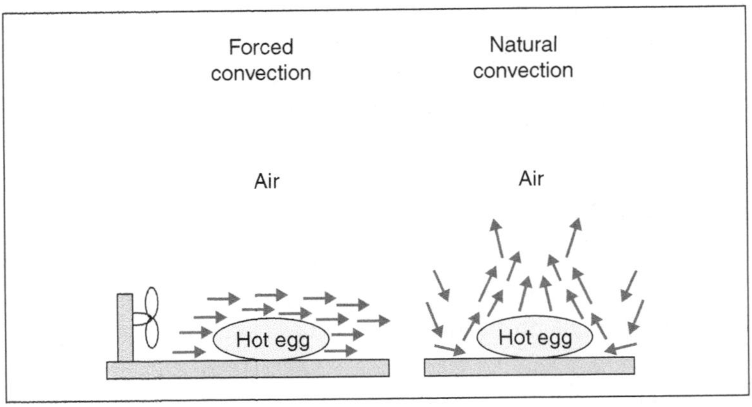

Figure 1.2 Forced and natural convection.

1.1.2 Convection

Convection involves heat transfer between a surface and a moving fluid, described by Newton's law of cooling:

$$q'' = h\left(T_s - T_\infty\right) \tag{1.3}$$

where $h\left(\mathrm{W/m^2} \cdot K\right)$ is the convective coefficient, T_s is the surface temperature, and T_∞ is the fluid temperature. Convection is categorized as *natural* (buoyancy-driven) or *forced* (externally driven), a pictorial representation is shown in Figure 1.2. The dimensionless Nusselt number ($Nu = hL/k_f$, where L is characteristic length and k_f is fluid conductivity) correlates h

to fluid properties and flow regime [1]. Solving convection problems often requires coupling momentum equations (Navier–Stokes) with energy conservation, complicating analytical solutions. Empirical correlations (e.g., $Nu = C\,\mathrm{Re}^m\,\mathrm{Pr}^n$ for forced flow) are widely used in heating, ventilation, and air conditioning(HVAC) and turbine cooling [2].

1.1.3 Radiation

Radiation is energy transfer via electromagnetic waves, requiring no medium. The Stefan–Boltzmann law quantifies emissive power:

$$q'' = \varepsilon\sigma A\left(T^4 - T_{sur}^4\right) \tag{1.4}$$

where ε is emissivity $(0 \le \varepsilon \le 1)$, $\sigma = 5.67 \times 10^{-8}\ \mathrm{W/m^2} \cdot K^4$ is the Stefan–Boltzmann constant, A is surface area, and T_{sur} is surroundings temperature [3]. Unlike conduction/convection, radiation depends on the fourth power of temperature, dominating at high temperatures (e.g., furnaces, stars). Real surfaces emit less than a blackbody $(\varepsilon = 1)$, with spectral dependence described by Planck's distribution. Applications include solar collectors and spacecraft thermal management, where radiation is the sole mode in vacuums.

1.1.4 Combined heat transfer modes

Real systems often involve multiple modes. For instance, a building wall loses heat via conduction through layers, convection to air, and radiation to surroundings. The thermal resistance networksimplifies analysis:

$$Q = \frac{\Delta T}{\sum R_{th}}, \quad R_{th} = \frac{L}{kA}\text{(conduction)}, \quad R_{th} = \frac{1}{hA}\ \text{(convection)}$$

Radiation resistance is nonlinear but can be linearized as $R_{rad} = \frac{1}{h_r A}$, where $h_r = 4\varepsilon\sigma T_{avg}^3$. The overall heat transfer coefficient $U\left(\mathrm{W/m^2} \cdot K\right)$ for a composite system is:

$$\frac{1}{U} = \frac{1}{h_i} + \sum \frac{L_j}{k_j} + \frac{1}{h_o + h_r} \tag{1.5}$$

where h_i and h_o are internal/external convective coefficients [1]. Combined modes are critical in electronics (conduction-convection), solar panels (radiation-convection), and aerospace (all three modes). Advanced modeling often employs computational tools (e.g., CFD) to resolve coupled phenomena.

1.2 BASIC CONCEPTS IN FLUID DYNAMICS

1.2.1 Classification of fluids and flow types

Fluids are broadly classified based on their response to shear stress. Newtonian fluids, such as water and air, obey Newton's law of viscosity, which states that the shear stress τ is proportional to the strain rate

$$\tau = \mu \frac{du}{dy} \tag{1.6}$$

where μ is dynamic viscosity and $\frac{du}{dy}$ is the velocity gradient perpendicular to flow direction [4]. In contrast, non-Newtonian fluids exhibit nonlinear stress–strain relationships, Figure 1.3 shows stress–strain relationship for different fluid models. For example, power-law fluids follow:

$$\tau = K\left(\frac{du}{dy}\right)^n \tag{1.7}$$

where K is the consistency index and n is the flow behavior index [5]. Examples include blood ($n < 1$, shear-thinning) and cornstarch suspensions ($n > 1$, shear-thickening).

Flows are categorized by their spatial and temporal characteristics. Steady flows (time-invariant) vs. unsteady flows (time-dependent), and

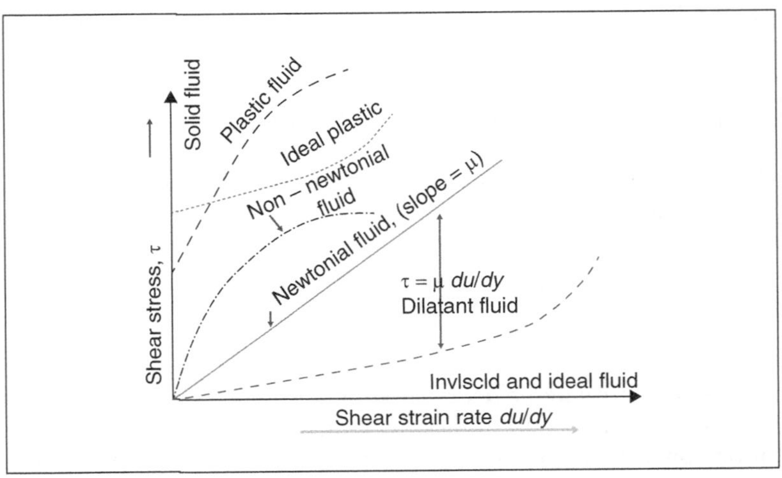

Figure 1.3 A stress–strain plot for non-Newtonian fluids.

incompressible flows (constant density, $\nabla \cdot V = 0$) vs. compressible flows (density varies, relevant at high Mach numbers). These classifications guide the choice of governing equations, such as the Euler equations for inviscid flows or the Navier–Stokes equations for viscous flows.

1.2.2 Continuum hypothesis

The continuum hypothesis assumes fluids are infinitely divisible, neglecting molecular discreteness. This approximation holds when the Knudsen number (K_n), defined as the ratio of molecular mean free path λ to characteristic length L, satisfies $K_n \ll 1$;

$$K_n = \frac{\lambda}{L} \tag{1.8}$$

For $K_n < 0.01$, continuum mechanics is valid, as in most engineering applications [6]. However, in rarefied gases $(K_n > 0.1)$ or microfluidic systems, molecular effects dominate, necessitating kinetic theory or direct simulation Monte Carlo methods. The hypothesis underpins field variables like density $\rho(x,t)$, treated as smoothly varying functions.

1.2.3 Properties of Fluids: Density, Viscosity, etc.

Density (ρ), defined as mass per unit volume $(\rho = m/V)$, influences inertial forces and buoyancy. Viscosity (μ) quantifies resistance to shear deformation, with kinematic viscosity given by $\nu = \mu/\rho$. Temperature dependence is modeled via Sutherland's law for gases or Arrhenius-type equations for liquids [7].

Additional properties include surface tension (σ), governing capillary effects, and bulk modulus (K), measuring compressibility:

$$K = -V\frac{dP}{dV} \tag{1.9}$$

These properties dictate phenomena from droplet formation to pressure wave propagation. For example, the Young–Laplace equation relates surface tension to pressure jump across curved interfaces:

$$\Delta P = \sigma\left(\frac{1}{R_1} + \frac{1}{R_2}\right) \tag{1.10}$$

where R_1, R_2 are principal radii of curvature [8].

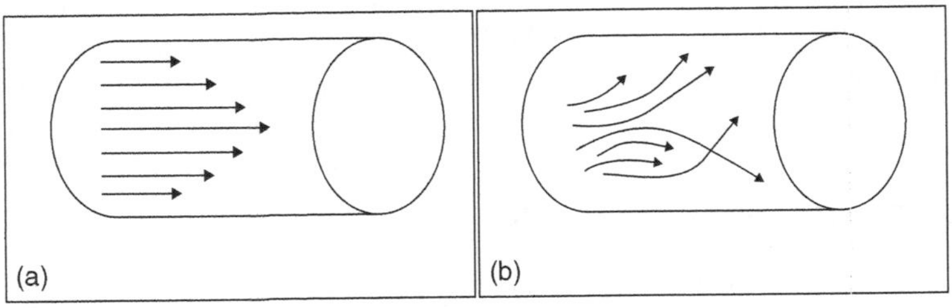

Figure 1.4 Schematic representation of (a) laminar, (b) turbulent flow.

1.2.4 Flow Regimes: Laminar vs Turbulent

Flow regimes are distinguished by the Reynolds number (Re), a dimensionless ratio of inertial to viscous forces:

$$\mathrm{Re} = \frac{\rho V L}{\mu} \tag{1.11}$$

For pipe flow, laminar flow (Re < 2300) features smooth, layered motion described by the Hagen–Poiseuille equation:

$$\Delta P = \frac{128 \mu L Q}{\pi D^4} \tag{1.12}$$

where Q is flow rate and D is diameter. Turbulent flow (Re > 4000) is chaotic, with eddies enhancing mixing and momentum transfer, a pictorial representation of laminar and turbulent flow is shown in Figure 1.4.The transition is marked by instability, as demonstrated by Reynolds' dye experiments [9].

Turbulence modeling often employs time-averaged Navier–Stokes equations (RANS) or direct numerical simulation (DNS). Key features include the energy cascade, where kinetic energy transfers from large to small eddies until dissipated by viscosity. Understanding these regimes is critical for designing pipelines, aircraft, and HVAC systems, where pressure losses and shear stresses vary dramatically with Re.

1.3 GOVERNING EQUATIONS

1.3.1 Conservation of mass (continuity equation)

The principle of mass conservation states that mass within a fluid system is neither created nor destroyed. For a control volume, the rate of change of mass equals the net flux across its boundaries. Expressed mathematically, this yields the continuity equation,

$$\frac{\partial \rho}{\partial t} + \nabla \cdot (\rho \boldsymbol{u}) = 0 \tag{1.13}$$

where ρ is the fluid density, $\boldsymbol{u}$ is the velocity vector, and t is time [10]. The first term represents local density changes, while the second term accounts for convective mass flux.

For incompressible flows (constant density), the equation simplifies to:

$$\nabla \cdot \boldsymbol{u} = 0 \tag{1.14}$$

This implies velocity divergence is zero, reflecting volume preservation. The continuity equation is foundational in fluid dynamics, constraining solutions to physically realistic flows [11].

1.3.2 Conservation of Momentum (Navier–Stokes Equations)

Newton's second law applied to fluid motion governs momentum conservation. The net force on a fluid element equals its rate of momentum change. The Navier–Stokes equations in vector form are:

$$\rho\left(\frac{\partial \rho}{\partial t} + (\boldsymbol{u} \cdot \nabla)\boldsymbol{u}\right) = -\nabla p + \nabla \cdot \boldsymbol{\tau} + \rho \boldsymbol{g} \tag{1.15}$$

where p is pressure, $\boldsymbol{\tau}$ is the viscous stress tensor, and $\boldsymbol{g}$ is gravitational acceleration [8]. The left-hand side represents inertial forces, while the right-hand side includes pressure, viscous, and body forces.

For Newtonian fluids, the stress tensor $\boldsymbol{\tau}$ relates linearly to the strain rate via the constitutive equation:

$$\boldsymbol{\tau} = \mu\left((\nabla \boldsymbol{u}) + (\nabla \boldsymbol{u})^T\right) - \frac{2}{3}\mu(\nabla \cdot \boldsymbol{u})\boldsymbol{I} \tag{1.16}$$

where μ is dynamic viscosity and $\boldsymbol{I}$ is the identity tensor. For incompressible flows, this simplifies to $\nabla \cdot \boldsymbol{u} = \mu \nabla^2 \boldsymbol{u}$, reducing the momentum equation to:

$$\rho\left(\frac{\partial \rho}{\partial t} + (\boldsymbol{u} \cdot \nabla)\boldsymbol{u}\right) = -\nabla p + \mu \nabla^2 \boldsymbol{u} + \rho \boldsymbol{g} \tag{1.17}$$

These equations are nonlinear and challenging to solve, reflecting turbulence and complex flow phenomena [12]

1.3.3 Conservation of energy

The first law of thermodynamics dictates energy conservation, balancing internal energy changes with work and heat transfer. The energy equation for a fluid is:

$$\rho\left(\frac{\partial e}{\partial t}+(\boldsymbol{u}\cdot\nabla)e\right)=-p\nabla\cdot\boldsymbol{u}+\nabla\cdot(k\nabla T)+\Phi \tag{1.18}$$

where e is specific internal energy, k is thermal conductivity, T is temperature, and Φ is viscous dissipation [2]. The term $-p\nabla\cdot\boldsymbol{u}$ represents compressive work, while $\nabla\cdot(k\nabla T)$ describes conductive heat transfer via Fourier's law:

$$q=-k\nabla T \tag{1.19}$$

Viscous dissipation Φ, the conversion of kinetic energy to heat due to friction, is given by:

$$\Phi=\tau:\nabla\boldsymbol{u} \tag{1.20}$$

For ideal gases or incompressible liquids, the equation is often recast in terms of enthalpy $h=e+p/\rho$ or temperature, facilitating practical applications like combustion or heat exchanger analysis [13].

1.4 BOUNDARY CONDITIONS AND INITIAL CONDITIONS

1.4.1 Types of boundary conditions

Boundary conditions (BCs) are critical for defining the behavior of a system governed by partial differential equations (PDEs). They constrain the solution at the domain boundaries, ensuring physical realism and mathematical uniqueness. Common types include Dirichlet, Neumann, Robin, periodic, and mixed conditions, each applicable to distinct physical scenarios.

Dirichlet boundary conditions specify the value of a dependent variable on the boundary. For example, in heat transfer, fixing the temperature at a surface Γ_D is expressed as:

$$T(\boldsymbol{x},t)=T_0(\boldsymbol{x},t)\ for\ \boldsymbol{x}\in\Gamma_D \tag{1.21}$$

where T_0 is the prescribed temperature [2]. These BC models scenarios are like a constant-temperature wall or a fixed displacement in structural mechanics.

Neumann boundary conditions prescribe the gradient of the variable, often representing fluxes. In heat conduction, a specified heat flux q'' at boundary Γ_N is:

$$-k\frac{\partial T}{\partial n} = q''(\boldsymbol{x},t) \text{ for } \boldsymbol{x} \in \Gamma_N \tag{1.22}$$

where k is thermal conductivity and $\frac{\partial T}{\partial n}$ is the normal derivative [4]. A zero-flux condition $(q'' = 0)$ models insulated surfaces.

Robin (Mixed) boundary conditions combine the variable and its gradient, common in convective heat transfer:

$$-k\frac{\partial T}{\partial n} = h\left[T(\boldsymbol{x},t) - T_\infty\right] \text{ for } \boldsymbol{x} \in \Gamma_R \tag{1.23}$$

where h is the convective coefficient and T_∞ is the ambient temperature [2]. This balances conduction and convection at the boundary.

Periodic boundary conditions enforce symmetry or repetitiveness, such as in wave propagation or cyclic systems:

$$u(\boldsymbol{x},t) = u(\boldsymbol{x} + \boldsymbol{L},t) \text{ for } \boldsymbol{x} \in \Gamma_P \tag{1.24}$$

where $\boldsymbol{L}$ is the periodicity length [14]. These are used in spectral methods and turbulence modeling.

Mixed BCs involve different types on disjoint boundary segments. For example, a domain might have Dirichlet conditions on Γ_D and Neumann on Γ_N, requiring partitioned implementation in numerical solvers [15]. BCs must be carefully formulated to avoid over specification, which can render the problem ill-posed.

The choice of BCs directly impacts solution uniqueness. For elliptic equations (e.g., Laplace's equation), a well-posed problem requires one BC type per boundary segment. In numerical methods, BCs are integrated into discretized equations, often via penalty methods or Lagrange multipliers [14]. Misapplication can lead to unphysical oscillations or divergence.

1.4.2 Initial conditions in transient analysis

Transient problems require initial conditions (ICs) to define the system's state at $t = 0$. ICs are essential for time-marching solutions, as they provide the starting point for evolution governed by PDEs. For a scalar field $u(\boldsymbol{x},t)$, the IC is:

$$u(\boldsymbol{x},t) = u_0(\boldsymbol{x}) \text{ for } \boldsymbol{x} \in \Omega \tag{1.25}$$

where Ω is the spatial domain [16]. In fluid dynamics, this could represent an initial velocity field; in heat transfer, a temperature distribution.

ICs influence the temporal evolution of the system. For example, in solving the transient heat equation:

$$\rho c_p \frac{\partial T}{\partial t} = \nabla \cdot (k \nabla T) \tag{1.26}$$

The initial temperature $T_0(\boldsymbol{x})$ determines how heat diffuses over time. Discrepancies between ICs and BCs at $t = 0$ (e.g., a cold region adjacent to a suddenly heated wall) introduce sharp gradients, necessitating adaptive time-stepping in numerical schemes [17].

In nonlinear systems, ICs can affect stability and convergence. For instance, in turbulent flow simulations, small perturbations in the initial velocity field may lead to divergent trajectories due to chaos [18]. Sensitivity to ICs underscores the need for accurate measurements or probabilistic approaches in uncertainty quantification.

Practical challenges arise in specifying ICs for real-world systems. Experimental data may be sparse, requiring interpolation or data assimilation [19]. In scenarios where ICs are unknown, steady-state solutions or heuristic approximations often serve as proxies, though these risks misrepresent transient dynamics.

Numerical stability hinges on IC consistency with BCs. A discontinuous jump between $u_0(\boldsymbol{x})$ and boundary values at $t = 0$ can cause oscillations, necessitating regularization or implicit time integration [16]. For example, in computational fluid dynamics, abrupt pressure changes may require smoothed ICs to avoid solver divergence.

1.5 RELEVANCE IN ENGINEERING AND INDUSTRY

The principles of heat transfer and fluid dynamics form the backbone of modern engineering systems, enabling innovations in energy efficiency, thermal management, and industrial processes. This section explores their critical applications in heat exchangers, fluid flow in pipelines, thermal management in electronics, and HVAC systems, emphasizing constitutive equations that govern these systems.

1.5.1 Applications in heat exchangers

Heat exchangers are ubiquitous in industries such as power generation, chemical processing, and refrigeration, facilitating energy transfer between fluids. Their design relies on fundamental heat transfer principles, including conduction, convection, and the log mean temperature difference (LMTD) method. The overall heat transfer rate Q is governed by:

$$Q = U \cdot A \cdot \Delta T_{\text{lm}} \tag{1.27}$$

where U is the overall heat transfer coefficient, A is the surface area, and ΔT_{lm} is the LMTD, calculated as:

$$\Delta T_{\text{lm}} = \left(\frac{\Delta T_1 - \Delta T_2}{\ln(\Delta T_1 / \Delta T_2)} \right) \tag{1.28}$$

Here, ΔT_1 and ΔT_2 are the temperature differences at the inlet and outlet [2]. For complex geometries, the effectiveness-NTU method is employed, where effectiveness $\epsilon\epsilon$ depends on the number of transfer units (NTU) and the heat capacity rate ratio [20]. Constitutive equations like Fourier's law and Newton's law of cooling underpin the thermal resistance network in heat exchanger design [21]. Recent advancements focus on enhancing U through nanostructured surfaces and computational fluid dynamics (CFD) optimization [22].

1.5.2 Fluid flow in pipelines

The transport of fluids through pipelines is critical in oil and gas, water distribution, and chemical industries. Pressure drops ΔP due to viscous effects is predicted using the Darcy–Weisbach equation:

$$\Delta P = f \cdot \frac{L}{D} \cdot \frac{\rho v^2}{2} \tag{1.29}$$

where f is the Darcy friction factor, L and D are pipe length and diameter, ρ is fluid density, and v is velocity [7]. The friction factor f is determined via the Colebrook–White equation for turbulent flow:

$$\frac{1}{\sqrt{f}} = -2 \log \left(\frac{\varepsilon / D}{3.7} + \frac{2.51}{\text{Re}\sqrt{f}} \right) \tag{1.30}$$

where ε is pipe roughness and $\text{Re} = \rho v D / \mu$ is the Reynolds number [23]. Laminar flow $(\text{Re} < 2300)$ follows $f = 64/\text{Re}$. Pipeline design also incorporates Bernoulli's equation for energy conservation:

$$\frac{P_1}{\rho g} + \frac{v_1^2}{2g} + z_1 = \frac{P_2}{\rho g} + \frac{v_2^2}{2g} + z_2 + h_{loss}$$

Optimization of pipeline networks minimizes pumping costs and ensures safe operation under varying flow regimes [24].

1.5.3 Thermal management in electronics

With the rise of high-power electronics, thermal management has become vital to prevent overheating and ensure reliability. Fourier's law governs heat conduction in microprocessors, while convective cooling via heat sinks follows:

$$q'' = h\left(T_s - T_\infty\right) \tag{1.31}$$

where $h\left(\mathrm{W/m^2} \cdot K\right)$ is the convective coefficient, T_s is the surface temperature, and T_∞ is the fluid temperature [25]. Advanced systems employ phase-change materials (PCMs) and heat pipes, leveraging latent heat $Q = mL$ (where m is mass and L is latent heat of fusion). Thermal resistance networks, analogous to electrical circuits, model heat flow:

$$R_{total} = R_{cond} + R_{conv} = \frac{L}{kA} + \frac{1}{hA} \tag{1.32}$$

CFD tools like ANSYS Fluent optimize heat sink designs, while nanomaterials enhance thermal conductivity [26]. Challenges include managing hotspots in miniaturized devices and integrating cooling solutions in compact form factors [27].

1.5.4 HVAC and energy systems

HVAC systems regulate indoor climates and energy efficiency using principles of psychometrics and thermodynamics. The cooling load Q_{cool} is calculated via:

$$Q_{cool} = \dot{m} \cdot \left(h_{in} - h_{out}\right) \tag{1.33}$$

where $\dot{m}$ is air mass flow rate and h is specific enthalpy. The refrigeration cycle's coefficient of performance (COP) is:

$$COP = \frac{Q_L}{W_{in}} \tag{1.34}$$

where Q_L is heat removed and W_{in} is work input [28]. Constitutive equations like the ideal gas law ($PV = nRT$) and humidity ratio $\omega = m_v / m_a$ underpin psychrometric charts for air conditioning design [29]. Modern systems integrate variable refrigerant flow (VRF) and geothermal heat pumps to reduce energy consumption. Smart HVAC controls, driven by IoT sensors, dynamically adjust loads based on occupancy and weather data [30].

1.6 CHALLENGES IN TRADITIONAL MODELING APPROACHES

1.6.1 Analytical limitations

Traditional analytical approaches in heat transfer and fluid dynamics rely on solving PDEs derived from fundamental conservation laws, such as the Navier–Stokes equations for fluid flow or Fourier's law for heat conduction. These methods excel in idealized scenarios where assumptions like steady-state conditions, linear material properties, and simple geometries hold true. For instance, analytical solutions for laminar flow in a straight pipe or one-dimensional heat conduction in a homogeneous slab are straightforward and yield precise results. However, these simplifications often diverge from real-world applications, where systems exhibit nonlinearities, transient behavior, and complex interactions.

A critical limitation arises when governing equations become intractable due to nonlinear terms. For example, turbulent flow, temperature-dependent material properties, or radiative heat transfer introduce nonlinearities that defy closed-form solutions. Analytical methods struggle to address phenomena like buoyancy-driven convection or phase-change processes (e.g., melting or boiling), where coupled heat and mass transfer create feedback loops. Even in cases where approximate solutions exist, such as perturbation techniques, their accuracy diminishes rapidly with increasing complexity.

Additionally, analytical models often fail to account for multi-dimensional or transient effects. While separation of variables or integral transforms can resolve time-dependent or spatially varying problems in simple domains, they become unwieldy in systems with irregular boundaries or coupled physics. For instance, predicting heat dissipation in electronic devices with intricate geometries or analyzing unsteady aerodynamic forces on a vibrating structure demands computational tools beyond analytical reach. These limitations underscore the gap between textbook problems and engineering realities, necessitating alternative approaches.

1.6.2 Complex geometries and turbulence

The inability of analytical methods to handle complex geometries is a major challenge in traditional modeling. Real-world applications—such as heat exchangers with finned surfaces, vascular networks in biomedical devices, or turbine blades in jet engines—involve irregular shapes that defy simple coordinate systems. Analytical solutions often require domain transformations or symmetry assumptions that distort physical accuracy. For example, cylindrical or spherical coordinate systems may simplify equations for specific geometries but fail for arbitrarily shaped components, limiting their utility in multidisciplinary design processes.

Turbulence further compounds these challenges. As a chaotic, multiscale phenomenon, turbulence involves eddies spanning orders of magnitude in size and energy, making it resistant to deterministic modeling. Traditional Reynolds-averaged Navier–Stokes (RANS) equations attempt to statistically model turbulent fluctuations but rely heavily on empirical closure models, which introduce uncertainties. These models often inadequately capture flow separation, vortex shedding, or boundary layer transitions, leading to errors in drag prediction, heat transfer rates, or mixing efficiency. In industries like automotive or aerospace, such inaccuracies can result in suboptimal designs or costly experimental iterations.

Combined geometrical and turbulent complexities create a "perfect storm" for traditional approaches. For instance, simulating airflow over a vehicle involves not only the car's curved surfaces but also unsteady wake dynamics and heat generation from the engine. Similarly, modeling blood flow through stenotic arteries requires resolving both the geometry of plaque deposits and the transition to turbulent flow under pulsatile conditions. Experimental validation in such cases is resource-intensive, further highlighting the need for advanced computational frameworks.

1.6.3 Importance of computational techniques

Computational fluid dynamics (CFD) and numerical heat transfer (NHT) have emerged as indispensable tools to overcome the shortcomings of analytical methods. By discretizing governing equations over a computational mesh, techniques like the finite volume method (FVM) or finite element method (FEM) can handle complex geometries through flexible grid generation. For example, unstructured meshes conform to irregular shapes, enabling simulations of heat sinks with micro channels or aerodynamic profiles of wind turbine blades. These methods also accommodate nonlinearities and transient terms through iterative solvers and time-stepping algorithms, bridging the gap between theory and practice.

Advanced turbulence modeling within computational frameworks has revolutionized predictive capabilities. Large eddy simulation (LES) and DNS resolve turbulent structures with varying degrees of fidelity, offering insights into phenomena like combustion instabilities or atmospheric boundary layers. Hybrid approaches, such as detached eddy simulation

(DES), balance accuracy and computational cost, making them viable for industrial applications. Similarly, conjugate heat transfer models integrate fluid flow and thermal analysis, enabling coupled simulations of cooling systems in nuclear reactors or electronic packaging.

The rise of high-performance computing (HPC) has further amplified the impact of numerical techniques. Parallel processing allows simulations with billions of grid points, capturing multiscale physics in reasonable timeframes. However, challenges persist in model validation, grid sensitivity, and computational expense. Recent advancements in machine learning and reduced-order modeling promise to address these issues by accelerating simulations and optimizing designs. As computational power grows, these tools will continue to democratize access to high-fidelity modeling, transforming how engineers tackle the intertwined challenges of heat transfer and fluid dynamics.

While traditional analytical methods provide foundational insights, their limitations in addressing nonlinearities, complex geometries, and turbulence underscore the necessity of computational techniques. These tools not only enhance predictive accuracy but also enable innovation in fields ranging from renewable energy to biomedical engineering, marking a paradigm shift in modern engineering analysis.

1.7 CONCLUSION

This chapter has provided a foundational understanding of heat transfer and fluid dynamics—two essential pillars of thermal and fluid engineering. It has explored the primary mechanisms of heat transfer: conduction, convection, and radiation, highlighting their governing laws and practical relevance. Additionally, it introduced the classification of fluids and flow types, elaborated on the continuum hypothesis, and discussed critical fluid properties influencing behavior under different conditions. The chapter also presented the distinction between laminar and turbulent flow regimes, underscoring the importance of Reynolds number and flow stability. Finally, the governing equations derived from conservation laws offer the analytical basis for modeling complex thermal-fluid systems. This theoretical groundwork is vital for advanced study and practical application in fields ranging from aerospace and energy systems to biomedical and environmental engineering.

REFERENCES

1. Incropera, F. P., DeWitt, D. P., Bergman, T. L., &Lavine, A. S., *Principles of Heat and Mass Transfer* (7th ed.). Wiley (2013).
2. Bergman, T. L., Lavine, A. S., Incropera, F. P., &DeWitt, D. P., *Fundamentals of Heat and Mass Transfer* (7th ed.). Wiley (2011).

3. Modest, M. F, *Radiative Heat Transfer* (3rd ed.). Academic Press (2013).
4. Bird, R. B., Stewart, W. E., &Lightfoot, E. N., *Transport Phenomena*. Wiley (2007).
5. Chhabra, R. P., &Richardson, J. F., *Non-Newtonian Flow and Applied Rheology*. Elsevier (2008).
6. Kundu, P. K., Cohen, I. M., &Dowling, D. R., *Fluid Mechanics*. Academic Press (2016).
7. White, F. M., *Fluid Mechanics*. McGraw-Hill (2016).
8. Batchelor, G. K., *An Introduction to Fluid Dynamics*. Cambridge University Press (2000).
9. Munson, B. R., Okiishi, T. H., Huebsch, W. W., &Rothmayer, A. P., *Fluid Mechanics*. Wiley (2020).
10. Anderson, J. D., *Computational Fluid Dynamics: The Basics with Applications*. McGraw-Hill (1995).
11. White, F. M., *Viscous Fluid Flow*. McGraw-Hill (2006).
12. Landau, L. D., &Lifshitz, E. M. *Fluid Mechanics*. Pergamon Press (1987).
13. Kays, W. M., *Convective Heat and Mass Transfer* (4th edition). McGraw-Hill (2004).
14. Hughes, T. J. R., *The Finite Element Method*. Dover (2000).
15. Zienkiewicz, O. C., Taylor, R. L., &Zhu, J. Z., *The Finite Element Method: Its Basis and Fundamentals*. Elsevier (2005).
16. Patankar, S. V., *Numerical Heat Transfer and Fluid Flow*. CRC Press (1980).
17. LeVeque, R. J., *Finite Difference Methods for Ordinary and Partial Differential Equations*. SIAM (2007).
18. Ferziger, J. H., Perić, M., &Street, R. L. *Computational Methods for Fluid Dynamics*. Springer (2002).
19. Kalnay, E. *Atmospheric Modeling, Data Assimilation, and Predictability*. CambridgeUniversity Press (2003).
20. Shah, R. K. and Sekulić,D. P.*Fundamentals of Heat Exchanger Design*, Wiley, (2003).
21. Lienhard IV, J. H. and Lienhard V, J. H., *A Heat Transfer Textbook*, 5th ed., Phlogiston Press(2020).
22. Kandlikar, S. G., Colin, S., Peles, Y., Garimella, S., Pease, R. F., Brandner, J. J., & Tuckerman, D. B. (2013). Heat transfer in microchannels—2012 status and research needs. *Journal of Heat Transfer*, 135(9), 091001.
23. Colebrook, C. F., & White, C. M. (1937). The reduction of carrying capacity of pipes with age. *Journal of the Institution of Civil Engineers*, 7(1), 99–118.
24. Çengel, Y. A.and Cimbala,J. M. (2018) *Fluid Mechanics: Fundamentals and Applications*, 4th ed., McGraw-Hill.
25. Bar-Cohen, A., Arik, M., & Ohadi, M. (2006). Direct liquid cooling of high flux micro and nano electronic components. *Proceedings of the IEEE*, 94(8), 1549–1570.
26. Chung, D. D. L. (2001) Cement-matrix composites for thermal engineering *Applied Thermal Engineering*, 21(16), 1607–1619.

27. Mills, A. F. *Heat Transfer*, 2nd ed., Prentice Hall(1999).
28. Moran, M. J. et al., *Principles of Engineering Thermodynamics*, 9th ed., Wiley(2018).
29. Fox, R. W.et al., *Introduction to Fluid Mechanics*, 8th ed., Wiley(2011).
30. Klein, S. A., & Hall, A. (2015). Emergent constraints for cloud feedbacks. Current climate change reports, 1(4), 276–287.

Chapter 2

Rheological models for non-Newtonian fluids

A.J.D. Nanthakumar

2.1 INTRODUCTION

2.1.1 States of matter

Matter exists in three primary states: solid state, liquid state, and gaseous state. The solid state maintains a fixed volume and a defining shape, both of which are influenced by temperature and external forces. The liquid state takes the shape of the container and the gas state expands and occupies the entire volume of the container. In the solid phase, the distinction between molecules and atoms becomes less distinct. At equilibrium, atoms vibrate around fixed spatial positions. In the liquid phase, intermolecular forces are not strong enough to confine molecules to fixed equilibrium positions but are sufficient to prevent excessive separation. In the gaseous phase, intermolecular distances are large enough that these forces become negligible. Gas molecules move rapidly relative to one another and interact primarily through elastic collisions. Consequently, they disperse throughout the container, and the pressure exerted on the container walls arises from molecular collisions. The definition of fluid includes the liquid and the gas, which at rest can transmit a pressure force normal to the bounding surface whereas when tangential forces come into play, they result in the relation motion among the fluidic layers. Such tangential forces create the shear stresses inside the fluid domain. The study of fluid mechanics deals with the macro-mechanical analysis of the behavior of liquids and gases.

2.1.2 Study of fluid mechanics

Many scientists have contributed to the study of fluids and their characteristics and the domain itself has evolved in a concurrent manner. Archimedes (285–212 BC) formulated the principles of buoyancy and floating bodies. Leonardo da Vinci (1452–1519) established the initial equation of continuity, whereas Isaac Newton (1642–1727) proposed the law of linear viscous stresses. Leonhard Euler (1707–1783) derived Bernoulli's equation

DOI: 10.1201/9781003642169-2

by resolving the fundamental equations. L. M. H. Navier (1785–1836) established the fundamental differential equations governing viscous flow. Jean Louis Poiseuille (1799–1869) performed the initial studies on laminar flow in tubes, whereas Osborne Reynolds (1842–1912) elucidated the transition to turbulence. The boundary layer theory was proposed by Ludwig Prandtl in the nineteenth century developed boundary layer theory, which forecasts flow separation, whereas Theodore von Karman (1881–1963) significantly advanced aerodynamics and turbulence theory.

2.1.3 Newtonian fluids

In engineering terms, the differentiation between a solid and a fluid is based on their capacity to withstand applied shear (tangential) stress, which is responsible for the change in shape of materials. The solid state of matter is capable of withstanding shear stress by deforming to a certain degree, but the fluid state experiences continuous deformation under any applied shear stress, regardless of its amount. The solid state of matter exhibits the proportionality of stress with respect to strain whereas the fluid state exhibits the property of stress with respect to strain rate. The deformation of a solid stops at a strain angle representative of material strength but fluid undergoes continuous deformation representing constant strain rate. Fluids possess the characteristic of viscosity, a representation of internal resistance of fluids against motion. The deformation rate is proportional to gradient of flow velocity. Empirical findings reveal that for most of the fluids, the gradient of velocity depends upon the shear stress applied.

$$\tau \propto \frac{du}{dy} \tag{2.1}$$

Fluids that demonstrate a linear correlation between the rate of deformation and shear stress are classified as Newtonian fluids, which was first articulated by Sir Isaac Newton. Typical examples of Newtonian fluids are water, air, petrol, and oils. For the one-dimensional flow of fluids under shearing conditions, the state of shear stress follows a linear relation with that of rate of shear:

$$\tau = \mu \frac{du}{dy} \tag{2.2}$$

where μ is termed as viscosity coefficient or the dynamic viscosity coefficient. For practical calculations, the ratio between dynamic viscosity and density (coefficient of kinematic viscosity)is generally used (υ). Fluids that do not follow the viscosity law of Newton are classified as non-Newtonian fluids. The basic analysis in fluid mechanics arise from the assumption of fluids being ideal, which is assumed to have zero viscosity. The theoretical analysis of ideal fluids can help in gaining insight into the real fluid behavior.

2.1.4 Non-Newtonian fluids

Non-Newtonian fluids are distinguished from Newtonian fluids by their nonlinear relationship between shear stress and shear rate, resulting in a viscosity that varies depending on the applied stress or shear rate. These fluids are categorized based on their specific flow behaviors.

A non-Newtonian fluid is characterized by a flow curve (shear stress vs. shear rate) that is either nonlinear or does not pass through the origin. The apparent viscosity, calculated as the ratio of shear stress and shear rate, varies with changes in shear conditions, even at a constant temperature and pressure. Rather, it is contingent upon flow variables, including geometry, shear rate, and occasionally the fluid's motion history. These fluids can be categorized into three distinct classifications:

1. Time-independent fluids: Often referred to as purely viscous, inelastic, or generalized Newtonian fluids, are characterized by a shear rate at any location being exclusively dictated by the instantaneous shear stress at that location.
2. Time-dependent fluids: They are intricate fluids whose connection between shear stress and shear rate is influenced by both the time length of shearing and their flow motion history.
3. Viscoplastic fluids: Materials exhibiting properties of fluids as well as elastic solids, showing partial elastic recovery after deformation.

Out of the above types, the most widely prevalent type is the non-Newtonian fluids that are time-independent.

This chapter exclusively addresses this category of non-Newtonian fluids. Figure 2.1 illustrates the correlation between shear stress and rate of shear for time-independent non-Newtonian fluid categories. Pseudoplastic fluids with yield require a certain shear stress (yield stress) to start the flow of the fluid and demonstrate shear-thinning behavior thereafter, as observed in substances such as toothpaste and ketchup. Bingham plastic fluids possess a yield stress and exhibit a constant viscosity post-yielding, as demonstrated by substances such as mud and mayonnaise. Pseudoplastic fluids, like blood and paint, exhibit shear-thinning characteristics devoid of yield stress, wherein viscosity diminishes as shear rate escalates. Dilatant fluids demonstrate shear-thickening behavior, characterized by an increase in viscosity

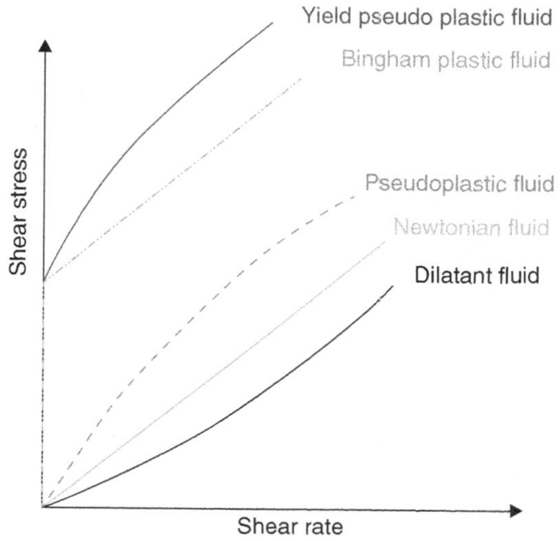

Figure 2.1 Time-independent non-Newtonian fluids.

with rising shear rate, as seen in cornstarch-water mixtures and certain suspensions.

2.2 LITERATURE REVIEW

2.2.1 Non-Newtonian fluids

Non-Newtonian fluids encompass diverse materials, such as polymeric liquids, biological fluids, and suspensions, that cannot be described by the classical Newtonian model (Rajagopal, 1993). Certain materials flow only after a critical shear stress is reached, and complex models like rate-type fluids are required to describe their behavior. Except for simple flow fields, their analysis is highly challenging and often relies on numerical solutions. Chhabra (2010) examined the widespread occurrence of non-Newtonian fluid behavior in various natural and technological applications. The various types of non-Newtonian behavior, including shear-thinning behavior, shear-thickening characteristics, viscoplastic nature, time-dependent property, and viscoelastic behaviors, were described using experimental data on real materials. Methods to engineer these properties by manipulating microstructure and their impact on process modeling and analysis were also discussed. (Irgens, 2014) presented a concise and thorough exploration of non-Newtonian fluids, focusing on their behavior, mechanical properties, and governing principles. Key topics include deformation kinematics, viscometric and extensional flows, material functions, generalized Newtonian fluids, and advanced fluid models.

A mathematical framework using fractional-order impedance models was proposed (Ionescuet al., 2020) to characterize fluid mechanics properties from practical datasets belonging to the frequency domain. Four non-Newtonian fluids were analyzed using the method of nonlinear least squares, genetic algorithms, and optimization using particle swarm technique. The compact model effectively captured viscoelasticity, supporting real-time identification for prediction and control applications. Hydrodynamic lubrication was analyzed by (Veltkampet al., 2023) for shear-thinning and viscoelastic polymer solutions, revealing that elasticity had minimal impact on friction, while shear-thinning significantly affected lubricating layer thickness and friction dependence on velocity. A developed hydrodynamic model incorporating shear-thinning aligned well with experimental data but underestimated layer thickness by a factor of 1.5. Non-Newtonian fluid flow in porous media (Eberhard et al., 2019) involved spatially varying viscosity due to diverse shear rates, typically described using an effective viscosity at the Darcy scale. The study developed a semi-analytical model using Carreau fluid rheology, avoiding empirical correction factors and classical capillary bundle assumptions, and validatedit against flow cell experiments, showing strong agreement and providing insights into upscaling non-Newtonian flows.

Haoet al. (2020) developed an intelligent fluidic system integrating two non-Newtonian rheological properties mediated by an autocatalytic enzyme reaction. By engineering associative polyelectrolytes, shear thickening and thinning behaviors were achieved at different pH levels, with tunable non-Newtonian states driven by a chemical-fueled process. Malekiet al. (2019) analyzed heat transfer and fluid flow of a pseudoplastic non-Newtonian nanofluid over a permeable surface with injection and suction. The results showed differing heat transfer behaviors based on injection, suction, and nanoparticle types, with non-Newtonian nanofluids outperforming Newtonian ones in injection cases on impermeable plates. Non-Newtonian fluids exhibit flow properties distinct from Newtonian fluids, with viscosity dependent on shear rate or history and cannot be described by a constant viscosity. This chapter introduces common rheological models and solutions for non-Newtonian fluid flow in circular pipes, annular, and rectangular ducts. Non-Newtonian fluids exhibit flow properties distinct from Newtonian fluids, with viscosity dependent on shear rate or history, and cannot be described by a constant viscosity. Nguyen and Nguyen (2012) discussed common rheological models and solutions for non-Newtonian fluid flow in circular pipes, annular, and rectangular ducts. Bohme (2012) employed continuum mechanics concepts to analyze non-Newtonian fluids, emphasizing the derivation of constitutive equations based on flow kinematics. It progresses from basic principles and simple flows to more complex motions, requiring no external references.

2.2.2 Rheological modeling

Sun et al. (2024) reviewed constitutive models for non-Newtonian fluids, including time-independent, time-dependent, and viscoelastic fluids, while systematically comparing traditional integer-order models with newly proposed fractional derivative equations. The results indicated that, under reasonable assumptions, space fractional derivative models can replicate classical constitutive models. The advantages and limitations of these models were discussed to guide their selection for industrial and biomedical applications, highlighting their practical relevance and usability. Sunet al. (2018) developed a space fractional-order constitutive equation (FCE) using a Caputo-type fractional velocity gradient to describe non-Newtonian fluid flow driven by spatially nonlocal velocities. The FCE model encompassed existing non-Newtonian models as special cases by adjusting the fractional index and effectively captured shear stress growth for varying velocity gradients. Further analysis provided insights into velocity profiles, frictional head loss, and Reynolds numbers, offering tools to extend standard models for complex fluid dynamics. Benyahia (2024) extended traditional computational fluid dynamics (CFD) models for gas-solids fluidization to non-Newtonian fluids by incorporating a shear-thinning power-law viscosity and an updated drag correlation based on direct numerical simulations. Validation against experimental data for fluidized beds of steel and glass particles demonstrated improved accuracy across a wide range of porosity and Reynolds numbers compared to prior semi-empirical models. (Kornaeva et.al., 2022). Variational approaches combined with artificial neural networks were explored for modeling non-Newtonian fluid flows, using physics-based functionals as loss functions to approximate solutions with boundary conditions. The method demonstrated up to 4 percent error in velocity predictions compared to analytical and numerical solutions, validating its applicability for complex fluid systems.

The study conducted by Kaveh and Malcherek (2024) extended the cross-flow curve model to incorporate particle size and concentration, capturing shear-thinning, Newtonian, and shear-thickening behaviors. It provided insights into suspensions and complex fluid-solid interactions. Ershadnia et al. (2020) investigated non-Newtonian fluid flow in rotating annular media, developing a highly accurate predictive model and validating it along-side CFD simulations against 730 experimental cases. The findings revealed distinct pressure loss dynamics influenced by shear-thinning effects, inertial forces, and key system parameters. A mathematical model for Taylor flow hydrodynamics in micro channels with combinations of Newtonian and non-Newtonian fluids was developed by Abiev (2022), providing velocity profiles and pressure gradients for various phases. The results explained previously observed phenomena and highlighted the need for experimental validation in cases involving non-Newtonian fluids. Non-Newtonian fluids used in hydraulic fracturing and other applications complicate flow prediction in porous media, requiring advanced modeling approaches. L. Yang & Du (2020) combined column experiments with a pore-scale lattice Boltzmann model (LBM) incorporating cross rheology, achieving good agreement with experimental results and providing a tool for scaling non-Newtonian fluid dynamics.

Mahmoudabadbozchelou et al. (2022) introduced non-Newtonian physics-informed neural networks (nn-PINNs) to solve coupled PDEs for complex fluid modeling, bypassing mesh generation through automatic differentiation. Tested on various constitutive models and flow protocols, nn-PINNs accurately captured spatiotemporal dynamics and resolved unknown boundary conditions using sparse velocity data. Droplet formation in a flow-focusing device was modeled by Chenet al. (2020) using Open FOAM for power-law fluids, showing that higher power-law index (nn) or consistency coefficient (KK) reduced droplet size and spacing but increased formation frequency and velocity. A capillary number method and scaling law were developed to support microfluidic device design.

The literature review highlighted the importance of investigating the rheological modeling of non-Newtonian fluids because of its varied applications and intricate behaviors. Non-Newtonian fluids have distinct properties, including shear-thinning, shear-thickening, and viscoelasticity, which are inadequately represented by traditional Newtonian models. Advanced mathematical and computational frameworks, such as fractional-order models, machine learning methodologies, and specialized computational fluid dynamics techniques, are crucial for tackling this complexity. These models serve as essential instruments for the analysis and prediction of fluid behavior in industrial and biological contexts, yielding insights into flow dynamics, stress distributions, and the optimization of processes involving non-Newtonian fluids. The persistent focus on experimental validation and the creation of precise predictive models underscores the essential function of rheological modeling in connecting theoretical comprehension with practical application.

2.3 RHEOLOGICAL MODELS FOR NON-NEWTONIAN FLUIDS

Rheological models are fundamental in fluid mechanics as they define fluid responses to applied forces, enabling the prediction of flow behavior under various conditions. Many real-world fluids, such as blood, polymers, slurries, and food products, exhibit non-Newtonian characteristics, where viscosity changes with stress or strain rate. Models like the power-law, Bingham law, and the relationship of Herschel–Bulkley model are instrumental in characterizing these complex fluids, aiding industries such as oil and gas, pharmaceuticals, and food processing. They enhance computational fluid dynamics (CFD) models for precise engineering predictions and contribute to optimizing industrial processes, improving material performance, and understanding biological and environmental phenomena like blood flow in medicine and lava movement in geology. By integrating these models, scientists and engineers advance process efficiency, material development, and disaster management, making them indispensable in modern fluid mechanics.

2.3.1 Shear thinning fluids

Shear thinning behavior, otherwise called pseudo plasticity, is characterized by the apparent viscosity parameter. For such fluids, apparent viscosity decreases with increasing shear rate. A practical inside into our everyday life reveals sheer thinning recruitments for many natural phenomena as well as scientific applications. Human blood is a shear thinning fluid which enables it to flow through capillaries easily. Food items like sauces, dairy products, chocolates need to thin during shear so that they can be easily consumed. Chemical paints that remain thick tend to shear thin while applying to reduce human labor and energy involved. Drilling fluids used in petroleum engineering are shear thinning fluids. Many rheological models have been proposed in the literature to expressions from the fundamental ideas of statistical mechanics. Few of the models are discussed here:

1 Power law model:
The relationship between shear stress and shear strain is given by

$$\tau = K[\dot{\gamma}]^n \tag{2.3}$$

The apparent viscosity expression is given by

$$\mu = K[\dot{\gamma}]^{n-1} \tag{2.4}$$

The constant K represents fluid consistency coefficient under and n represents flow behavior index. When $n< 1$, fluid exhibits shear thinning behavior. When $n=1$, fluid exhibits Newtonian behavior. When $n>1$, fluid exhibits shear thickening behavior. The main limitations of the power law model is its applicable range of shear rates for which K and n hold good. The dimensions of the consistence index correlate with the index of flow behavior for such fluids. The power law model is widely deployed due to its simplicity.

Abbas et. al. (2022) analyzed power-law nanofluid flow over a variable Riga stretching sheet with variable thermal conductivity and viscosity under heat generation effects. Numerical results revealed velocity increases with modified Hartmann number and variable viscosity, while temperature and concentration profiles varied significantly with thermophore sis, Brownian motion, and other parameters under suction and injection conditions. Praveshet al. (2022) examined mixed convection heat transfer around an inline array of heated cylinders in non-Newtonian power-law fluids, revealing that Nusselt numbers increase with higher Reynolds, Prandtl, and Richardson numbers and shear-thinning behavior but decrease for shear-thickening fluids. Volume fractions significantly influenced thermal characteristics, with shear-thinning fluids showing unsteady behavior at larger fractions.

2 Carreau model

The power law model is not reliable at extremely high or low shear rates. To include the behavior of the fluids at very high and very low shear rates, Carreau model that includes limiting values at high shear rates and low shearing conditions is deployed. The Carreau model is given by

$$\frac{\mu - u_{\infty}}{\mu_0 - \mu_{\infty}} = \left[1 + (\lambda\dot{\gamma})^2\right]^{(n-1)/2} \tag{2.5}$$

where n and λ are curve fitting coefficients. The Carreau expression reduces toNewtonian model when $n = 1$ or $\lambda = 0$ or both. The Carreau model is a four-parameter model.

The study conducted by Bowers and Miller (2023) extended a microscale-averaged modeling approach to Carreau fluids, predicting flow in isotropic and anisotropic porous media using rheological properties, a Newtonian resistance tensor, and a pore morphology-based length-scale tensor. The method eliminates reliance on experimental GNF flows and provides accurate macro scale predictions. Daunaiset al. (2023) proposed a high-order finite element method for shear-thinning flows utilizing power-law and Carreau rheological models, validated through simulations of flow past a sphere across various Reynolds numbers. Results highlighted nonlinear drag behavior.

3 Cross-viscosity model:

The cross-viscosity model also contains four parameters and its expression is given by

$$\frac{\mu - u_{\infty}}{\mu_0 - \mu_{\infty}} = \frac{1}{1 + k\dot{\gamma}^n} \tag{2.6}$$

where n and k are two fitting parameters, μ and μ_0 are apparent viscosity values at low share rate and high share rate respectively. The cross model reduces to Newtonian model when $k \rightarrow 0$ and power law model when $\mu \ll \mu_0$ and $\mu \gg \mu_{\infty}$. Initially the value of n was proposed as 2/3, which had been treated as a model parameter of late.

Zamanet al. (2023) analyzed blood flow as a cross fluid in a curved, stenosed channel, incorporating entropy generation due to dissipation and magneto hydrodynamic effects. The results showed that the Bejan number and entropy generation parameter are significantly influenced by curvature and magnetic parameters, with flow patterns visualized via streamlines. Al-Kouzet al. (2021) investigated the heat transfer and entropy production in a magnetized cross fluid within a microchannel, considering Joule heating, external magnetism, and Newton's heating. The results showed that the power-law index reduced entropy, velocity, and temperature.

4 Ellis's fluid model:

It is a three-parameter model meant to model a non-Newtonian fluid that deviates from the power law at very low shear rates. The Ellis model differs from other models wherein the earlier model takes the inverse form of other models. The Ellis model is given by

$$\mu = \frac{1}{1 + \left(\tau/\tau_{1/2}\right)^{\alpha - 1}} \tag{2.7}$$

where μ_0 is zero shear viscosity, α and $\tau_{1/2}$ are variable model coefficients. The parameter α represents degree of shear thinning with higher the values of α the greater is the extent of shear thinning and $\tau_{1/2}$is the shear stress value at which apparent stress has dropped to half of its zero-shear value.

Alam et al. (2021) examined the steady-state flow of viscoelastic Ellis fluid on a vertical cylinder for drainage and lift phenomena, deriving closed-form solutions and analyzing film thickness, flow rate, and velocity profiles. Saleem et al. (2023) analyzed nanofluid flow and heat transfer in a divergent wavy channel using the Ellis fluid model, considering electro-osmotic effects and thermos physical properties of blood and gold. Results showed convection increases with nanofluid concentration.

2.3.2 Viscoplastic fluids

Viscoplastic fluid, as the term indicates, exhibits both viscous as well as plastic behavior. The transition from the plastic behavior to the viscous behavior is characterized by the parameter yield stress. Until the applied stress reaches the yield point value, the material behaves elastically or plastically moving as a single rigid body of mass. Once the external stress crosses the yield point value the state of matter starts exhibiting viscous characteristic thereby displaying shear stress versus shear rate relationship instead of stress versus strain relationship. The flow curve formed after crossing the yield stress can be linear or nonlinear and without crossing the origin. It can be understood that a viscoplastic fluid need not possess a free surface when stored in a container. This has made possible the utilization of such fluids for varied engineering applications that need a phase transition characteristic for its operation. The assumption of the initial solid phase characteristics emanates from the understanding that the material possesses a three-dimensional internal structure, which is broken down when yield stress value is overcome by the stress that is applied. Many of the viscoplastic materials also exhibit reversibility of such phase change, thereby enabling the material to regain the yield stress after being disturbed. Many of the present-day engineered fluids such as magnetorheological fluid and electrorhelogical fluid belong to this category.

Many empirical relationships have been found in the literature either as a curve fitting expression or as an expression pertaining to a specific fluid for a specific application. An engineering relation depending upon fundamental concepts is to be formulated yet. Three types of models have been widely accepted to represent viscoplastic behavior in their formulation.

1 Bingham model:
The main parameter of the Bingham model is the yield stress parameter that distinguishes such fluids from other Newtonian and non-Newtonian fluids. Thus the model for Bingham fluid includes the parameter as a sum term along with shear stress strain expression. The expression for Bingham's model is given by

$$\tau = \tau_0 + \mu\dot{\gamma} \tag{2.8}$$

In reality, fluids need not show distinct yield stress limitation rather a yield stress like behavior within a small range. The Bingham model thus forms here basic skeleton over which further complex models can be formulated. Bingham model is a two-parameter model.

Duanet al. (2023) explored high-performance heat exchangers for coal water slurry, analyzing flow and heat transfer mechanisms using the bi-viscous Bingham model. Farinaet al. (2024) examined two-dimensional viscoplastic fluid flows in symmetric channels using the Bingham model and its regularizations, leveraging the lubrication approximation for analytical solutions. The results showed regularized flows align with the Bingham model in plane channels but deviate significantly in curved geometries. Yang et al. (2022) proposed a column diffusion mechanism for Bingham fluid, incorporating the displacement effect of grout on groundwater, validated through grouting experiments. Numerical simulations using Comsol Multiphysics showed that the theoretical diffusion radius aligns better with experimental results compared to traditional Bingham fluid models, aiding practical grouting engineering.

2 Herschel–Bulkley model:

Bingam model assumes linearity in fluid flow once the limiting shear stress is crossed. in practical applications, fluids with yield stress also exhibit non-linearity during flow. Such fluid flows can be expressed through the Herschel–Bulkley model. It is a three-parameter model and the expression is given by

$$\tau = \tau_0 + K(\dot{\gamma}) for\, |\tau| > |\tau_0| \tag{2.9}$$

$$\dot{\gamma} = 0 \quad for \quad |\tau| < |\tau_0|$$

The dimensionality characteristics of the constant K is determined by the values of n. The model provides a better representation of fluids flowing with yield stress characteristic than that of Bingham model. The model reduces to Bingham model for $n = 1$.

Magnon and Cayeux (2021) proposed a method for calibrating the Herschel–Bulkley rheological model using pressure gradient measurements at varying bulk velocities combined with physics-based equations and nonlinear optimization. Experimental validation showed improved accuracy in estimating Herschel–Bulkley parameters for shear-thinning polymeric fluids, especially at low flow rates, with minimal computational resources required. Majeed et al. (2024) investigated the hydrodynamic forces on a barrier in a channel using the Herschel–Bulkley model for viscoplastic materials, considering yield stress and power-law index. A finite element method (FEM) approach was used to solve the nonlinear equations, validating results for special cases. Hao and Zhou (2022) prepared a nano-scale magnetic fluid (MF) and analyzed its rheological properties revealing that yield stress, viscosity coefficient, and shear-thinning behavior increase with magnetic field intensity but decrease with rising temperature.

3 Casson model:
Casson model is a two-parameter model that has been widely employed for steady shear-stress shear rate behavior of biological fluids. It is also well suited for foodstuff materials and biological materials such as blood, particulate suspensions etc. The expression for the Casson model is given by

$$|\tau|^{1/2} = |\tau_0|^{1/2} + \left|\frac{\mu}{\dot{\gamma}}\right| \; for \; |\tau| > |\tau_0| \tag{2.10}$$

$$\dot{\gamma} = 0 \quad for \quad |\tau| < |\tau_0|$$

Divyaet al.(2021) analyzed the peristaltic motion of a Casson fluid in a non-uniform channel under a radial magnetic field, considering variable viscosity and thermal conductivity. Using perturbation methods, the model revealed the effects of variable viscosity on fluid velocity near channel walls and bolus size, offering insights into blood flow through small vessels. Nadeem et al. (2023) analyzed the steady flow of a non-Newtonian Casson nanofluid with water-based aluminum oxide nanoparticles over a stretchable surface of variable thickness. Using different viscosity models, including Brinkman and Einstein, it revealed the deteriorating effect of Casson fluid.

2.3.3 Shear thickening or dilatant fluids

Shear-thickening, or dilatant fluids, resemble pseudoplastic fluids without exhibiting yield stress. In dilatant fluids, the apparent viscosity rises as rate of shear increases. Such a behavior is widely observed in suspensions, colloidal dispersions, and polymer solutions. The reason for the increase in shear stress with shear rate is due to the fact that when shear rate is low, the fluid present lubricates the motion of particles. As the sheer rate is high, the material moves and expands with no such fluid to occupy the empty space among the particles resulting in solid versus solid interactions. This results in increased friction and thereby high shear stresses. Thus, the apparent viscosity is increased. For such type of fluids, the viscosity shear rate relations are depicted in double algorithmic scale for a given range of shear rate. For such cases of flow, the power law is employed to illustrate the behavior of such fluid flows. The dilatant fluids are represented by the power law wherein the index of power is greater than one. The expression for such dilatant fluids is given by

$$\mu = K\dot{\gamma}^{n-1} \tag{2.11}$$

Such dilatant fluid behavior can be found in protective gear applications industrial slurries and biomedical fluid flows. The nonlinear rheology of such fluids is attributed to the combined interplay among the hydrodynamic characteristics and frictional behavior. Liuet al. (2023) reviewed the applications of shear thickening fluids (STFs), highlighting their energy absorption and dissipation properties in impact scenarios and discussed STF mechanisms, impact resistance of impregnated composites, recent advancements in dampers and shock absorbers. Shendeet al. (2021) proposed an empirical equation, based on Doolittle's free volume theory, to model the viscosity of shear-thickening fluids across various regimes and conditions. Validated against experimental data, the model predicted rheological behavior under varying physico-chemical factors and revealed bimodal pore-scale distributions in 2D porous medium simulations. Garg and Prasad (2024) derived analytical formulas for velocity profiles and flow rates of yield stress shear thinning/thickening fluids using the Herschel–Bulkley model in rigid and deformable channels. The results showed that wall deformability increases flow rate and channel height, while higher yield stress reduces velocity, yield surface height, and flow rate, with shear thickening further decreasing flow rate. Table 2.1 shows the model equations, fluid flow characteristics, and applications of the fluid models.

2.4 CONCLUSION

Scientists and engineers face critical challenges while designing systems involving non-Newtonian fluids. It has been found in the research literature that conventional models including power law, Bingham law, Herschel–Bulkley expression of fluid flow and Casson expression of fluid flow can be considered as fundamental paradigms to analyze non-Newtonian fluid behavior. Such models are capable of explaining non-Newtonian behavior such as shear-thinning nature, shear thickening nature, and flow under the influence of yield stress. These models can thus be utilized for applications ranging from industrial processing to fluid flow in biological systems. The recent developments in rheological modeling can provide accurate predictions of non-Newtonian fluid flow. To achieve the desired accuracy, a holistic approach of integrating fractional calculus, machine learning and computational fluid dynamics is quintessential. Research studies that home utilized Carreon model, cross model, and Ellis model have successfully explained the transitional nature of shear dependent viscosities. Many of the so conducted numerical simulations and successive optimization exercises have been validated through experimental observations. This has increased the reliability of such fluid models. The advent of machine learning technique has further improvised the model accuracy. Of such machine learning techniques, physics-based models have gained prominence due to their proximate output in alignment with physical laws and relationships.

Table 2.1 Comparison of non-Newtonian fluid modes

Fluid model	*Model equation*	*Fluid flow characteristics*	*Applications*
Power law model	$\tau = K[\dot{\gamma}]^n$	Describes shear-thinning(\(n<1\)) and shear-thickening (\(n>1\)) fluids. Viscosity varies as a power function of shear rate.	Polymer solutions, slurries, food processing, biofluids.
Carreau model	$\frac{\mu - u_\infty}{\mu_0 - \mu_\infty} = \left[1 + (\lambda\dot{\gamma})^2\right]^{(n-1)/2}$	Exhibits both Newtonian behavior at low rate of shearing conditions and power-law characteristics at high rate of shearing conditions.	Polymer melts, biological fluids, paints, lubricants.
Cross-viscosity-model	$\frac{\mu - u_\infty}{\mu_0 - \mu_\infty} = \frac{1}{1 + k\dot{\gamma}^n}$	Smooth transition from Newtonian to non-Newtonian behavior. Handles limiting viscosity at both low and high shear rates. Four parameters.	Industrial coatings, emulsions, suspensions.
Ellis model	$\mu = \frac{1}{1 + \left(\tau/\tau_{1/2}\right)^{\alpha-1}}$	Primarily used for shear-thinning fluids. Modifies power law at low shear rates. Introduces a limiting viscosity.	Cosmetics, pharmaceuticals, protein solutions.
Bingham model	$\tau = \tau_0 + \mu\dot{\gamma}$	Yield stress model with linear viscous behavior beyond the yield point; behaves like a solid below yield stress.	Drilling muds, toothpaste, mayonnaise, suspensions.
Herschel–Bulkley model	$\tau = \tau_0 + K(\dot{\gamma})\ for\ \lvert\tau\rvert > \lvert\tau_0\rvert$ $\dot{\gamma} = 0 \quad for \quad \lvert\tau\rvert < \lvert\tau_0\rvert$	Generalizes Bingham and power-law models; includes yield stress and nonlinear flow behavior.	Drilling fluids, food products, biological fluids.

Table 2.1 (Cont.)

Fluid model	*Model equation*	*Fluid flow characteristics*	*Applications*
Casson model	$\lvert\tau\rvert^{\frac{1}{2}} = \lvert\tau_0\rvert^{\frac{1}{2}} + \left\lvert\frac{\mu}{\dot{\gamma}}\right\rvert$ *for* $\lvert\tau\rvert > \lvert\tau_0\rvert$ $\dot{\gamma} = 0$ *for* $\lvert\tau\rvert < \lvert\tau_0\rvert$	Describes yield stress fluids with shear-thinning properties; primarily used for blood and suspensions.	Blood flow modeling, chocolate, ink formulations.
Dilatant fluids	$\mu = K\dot{\gamma}^{n-1}$	Shear-thickening behavior where viscosity increases with shear rate, often due to particle interactions in suspensions.	Cornstarch–water mixtures (oobleck), protective gear, industrial slurries, food suspensions.

The advantages of using such rheological models can be applied to domains wherever non-Newtonian fluids are present. Few of the many successful implementations include blood flow modeling through Casson model, cardiovascular modeling through Herschel–Bulkley model, and magnetorheological fluid modeling through the elementary Bingham model. Especially the food and pharmaceutical industries rely on non-Newtonian rheological models to a great extent since most of the fluids in the biomedical domain exhibit non-Newtonian characteristics. There are also areas that still pose challenges for the current rheological modeling paradigms that include multi-phone flow and flow through porous media. Such domains require hybrid-modeling approach integrating molecular dynamics and microstructural effects. The domain of smart fluids has necessitated the inclusion of the effect of external stimuli such as magnetic fields and electric fields. Rheological modeling of non-Newtonian fluids is a complex as well as a rich field of research scope. The next steps in rheological modeling can be the refining of the constitutive equations, enhancing the quality of data obtained from experiments and proper exploitation of the computational advancements including machine learning and artificial intelligence. These activities can bridge the gap between the theoretical predictions and experimental observations.

REFERENCES

Abbas, N., Rehman, K. U., Shatanawi, W., &Abodayeh, K. (2022). Mathematical model of temperature-dependent flow of power-law nanofluid over a variable stretching riga sheet. *Waves in Random and Complex Media*,1–18.

Abiev, R. S. (2022). Mathematical model of two-phase Taylor flow hydrodynamics for four combinations of non-Newtonian and Newtonian fluids in microchannels. *Chemical Engineering Science*, 247, 116930.

Alam, M. K., Memon, K., Siddiqui, A., Shah, S., Farooq, M., Ayaz, M., ...Ahmad, H. (2021). Modelling and analysis of high shear viscoelastic Ellis thin liquid film phenomena. *Physica Scripta*, 96 (5), 055201.

Al-Kouz, W., Reddy, C. S., Alqarni, M., &Mahanthesh, B. (2021). Spectral quasi-linearization and irreversibility analysis of magnetized cross fluid flow through a microchannel with two different heat sources and newton boundary conditions. *The European Physical Journal Plus*, 136 (6), 645.

Benyahia, S. (2024). Simulation of heavy particles fluidized with non-Newtonian fluids. *Powder Technology*, 433, 119261.

Bohme, G. (2012). *Non-Newtonian Fluid Mechanics*. Elsevier.

Bowers, C. A., &Miller, C. T. (2023). Modelling flow of Carreau fluids in porous media. *Physical Review E*, 108 (6), 065106.

Chen, Q., Li, J., Song, Y., Christopher, D. M., &Li, X. (2020). Modelling of Newtonian droplet formation in power-law non-Newtonian fluids in a flow-focusing device. *Heat and Mass Transfer*, 56, 2711–2723.

Chhabra, R. P. (2010). Non-Newtonian fluids: an introduction. *Rheology of complex fluids*,3–34.

Daunais, C.-A., Barbeau, L., &Blais, B. (2023). An extensive study of shear thinning flow around a spherical particle for power-law and Carreau fluids. *Journal of Non-Newtonian Fluid Mechanics*, 311,104951.

Divya, B., Manjunatha, G., Rajashekhar, C., Vaidya, H., &Prasad, K. (2021). Analysis of temperature dependent properties of a peristaltic MHD flow in a non-uniform channel: A Casson fluid model. *Ain Shams Engineering Journal*, 12 (2), 2181–2191.

Duan, X., Xiao, J., Zhou, A., Wang, S., &Wen, J. (2023). Numerical investigation on flow and heat transfer performance of non-Newtonian Bingham fluid in novel spiral wound tube heat exchangers.*Case Studies in Thermal Engineering*, 49, 103352.

Eberhard, U., Seybold, H. J., Floriancic, M., Bertsch, P., Jimenez-Martınez, J., Andrade Jr, J. S., &Holzner, M. (2019). Determination of the effective viscosity of non-Newtonian fluids flowing through porous media. *Frontiers in Physics*, 7, 71.

Ershadnia, R., Amooie, M. A., Shams, R., Hajirezaie, S., Liu, Y., Jamshidi, S., &Soltanian, M. R. (2020). Non-Newtonian fluid flow dynamics in rotating annular media: Physics-based and data-driven modeling. *Journal of Petroleum Science and Engineering*, 185, 106641.

Farina, A., Fusi, L., Vergori, L., &Zanetti, E. (2024). Viscoplastic flows in channels with small aspect ratio: Bingham versus regularised models. *International Journal of Engineering Science,*199, 104074.

Garg, A., &Prasad, P. (2024). Yield-stress shear thinning and shear thickening fluid flows in deformable channels. *Physica Scripta*, 99(3), 035240.

Hao, L., &Zhou, J. (2022). Experimental study on the rheological properties of mf lubricant based on the Herschel Bulkley model. *Journal of Superconductivity and Novel Magnetism*, 35(11), 3333–3346.

Hao, X., Yang, K., Wang, H., Peng, F., &Yang, H. (2020). Biocatalytic feedback-controlled non-Newtonian fluids. *AngewandteChemie International Edition*, 59(11), 4314–4319.

Ionescu, C.-M., Birs, I. R., Copot, D., Muresan, C., &Caponetto, R. (2020). Mathematical modelling with experimental validation of viscoelastic properties in non-Newtonian fluids. *Philosophical Transactions of the Royal Society A*, 378(2172), 20190284.

Irgens, F. (2014). *Rheology and Non-Newtonian Fluids*(Vol. 1). Springer.

Kaveh, K., &Malcherek, A. (2024). Enhancing non-Newtonian fluid modeling: a novel extension of the cross flow curve model. *Journal of Hydro-environment Research*, 56, 17–27.

Kornaeva, E., Kornaev, A., Fetisov, A., Stebakov, I., &Ibragimov, B. (2022). Physics-based loss and machine learning approach in application to non-Newtonian fluids flow modelling. In 2022IEEECongress on Evolutionary Computation (CEC) (pp. 1–8).

Liu, H., Fu, K., Cui, X., Zhu, H., &Yang, B. (2023). Shear thickening fluid and its application in impact protection: a review. *Polymers*, 15(10), 2238.

Magnon, E., &Cayeux, E. (2021). Precise method to estimate the Herschel Bulkley parameters from pipe rheometer measurements. *Fluids*, 6 (4), 157.

Mahmoudabadbozchelou, M., Karniadakis, G. E., &Jamali, S. (2022). nn-pinns: Non-Newtonian physics-informed neural networks for complex fluid modelling. *Soft Matter*, 18(1), 172–185.

Majeed, A. H., Siddique, I., Mehmood, A., Ghazwani, H. A., Manzoor, S., &Ahmad, S. (2024). Finite element simulations of Herschel–Bulkley visco-plastic materials over a cylinder: Drag and lift correlation analysis. *Modern Physics Letters B*, 38(26), 2450248.

Maleki, H., Safaei, M. R., Alrashed, A. A., &Kasaeian, A. (2019). Flow and heat trans fer in non-newtonian nanofluids over porous surfaces. *Journal of Thermal Analysis and Calorimetry*, 135, 1655–1666.

Nadeem, S., Ishtiaq, B., Ben Hamida, M. B., Almutairi, S., Ghazwani, H. A., Eldin, S. M., &Al-Shafay, A. (2023). Reynolds nano fluid model for casson fluid flow conveying exponential nanoparticles through a slandering sheet. *Scientific Reports*, 13(1), 1953.

Nguyen, Q.-H., &Nguyen, N.-D. (2012). Incompressible non-Newtonian fluid flows. *Continuum Mechanics—Progress in Fundamentals and Engineering Applications*, 1,47–72.

Pravesh, R., Dhiman, A., &Bharti, R. (2022). Thermal features of mixed convection from an inline periodic array of circular cylinders in non-Newtonian power-law fluids. *Case Studies in Thermal Engineering*, 36, 102175.

Rajagopal, K. R. (1993). Mechanics of non-Newtonian fluids. In*Recent Developments in Theoretical Fluid Mechanics*(pp. 129–162). Chapman and Hall/CRC.

Saleem, S., Hussain, F., Irfan, M., Siddique, I., Nazeer, M., &Eldin, S. (2023). Theoretical investigation of heat transfer analysis in Ellis nanofluid flow through the divergent channel. *Case Studies in Thermal Engineering*, 48, 103140.

Shende, T., Niasar, V. J., &Babaei, M. (2021). An empirical equation for shear viscosity of shear thickening fluids. *Journal of Molecular Liquids*, 325, 115220.

Sun, H., Jiang, Y., Zhang, Y., &Jiang, L. (2024). A review of constitutive models for non-Newtonian fluids. *Fractional Calculus and Applied Analysis*, 1–44.

Sun, H., Zhang, Y., Wei, S., Zhu, J., &Chen, W. (2018). A space fractional constitutive equation model for non-Newtonian fluid flow. *Communications in Nonlinear Science and Numerical Simulation*, 62,409–417.

Veltkamp, B., Jagielka, J., Velikov, K., &Bonn, D. (2023). Lubrication with non-newtonian fluids. *Physical Review Applied*, 19(1), 014056.

Yang, L., &Du, K. (2020). A comprehensive review on the natural, forced, and mixed convection of non-Newtonian fluids (nanofluids) inside different cavities. *Journal of Thermal Analysis and Calorimetry*, 140, 2033–2054.

Yang, Z., Zhang, D., Li, C., Zhang, Z., Zhu, Y., Yang, Y (2022). Column penetration and diffusion mechanism of Bingham fluid considering displacement effect. *Applied Sciences*, 12 (11), 5362.

Zaman, A., Khan, A. A., &Kousar, N. (2023). Simulation of magneto-hydrodynamics effects on cross fluid (blood) model with entropy generations in multiple stenosed (aneurysm) curved channel. *Chinese Journal of Physics*, 85, 616–628.

Chapter 3

Analysis of buoyancy force, Brownian motion and thermophoretic effect in assisting and opposing flow of Prandtl fluid in presence of Joule heating and viscous dissipation

Prasanta Parida, Kharabela Swain, and Hibah Islahi

3.1 INTRODUCTION

Usually, metal oxide, metal, or carbon-based nanoparticles are mixed with a base fluid, like ethylene glycol, water, or oil, to create Newtonian nanofluid (NF). Compared to their base/traditional fluid, they show an improved thermal, optical, and electrical property, which makes them interesting materials for a range of uses. Choi (1995) initially proposed the idea of a NF, demonstrating how adding nanoparticles to a fluid can dramatically increase its heat conductivity. The boundary layer flow past a stretched plate in a viscous fluid is studied by Siddappa and Abel (1985). The study found that the velocity and shear stress distributions depend on the properties of the non-Newtonian fluid, such as its consistency and flow index. The results highlighted the impact of the stretching rate and the rheological properties of the fluid on the flow characteristics. This work provides foundational insights into the dynamics of non-Newtonian fluid flows, which are relevant in various industrial processes like polymer extrusion and coating. The parameters of an electrically conducting fluid's heat transmission across a stretched sheet with varying wall temperatures and internal heat generation are examined by Vajravelu and Rollins (1992). They found that the applied magnetic field retards the fluid motion and alters the temperature profile so that a thicker thermal boundary layer (TBL) is formed. In both forced and free convection flows, Wang and Mujumdar (2007) focused on the fluid flow and heat transmission properties of NF. They analyzed the mechanisms responsible for the enhanced thermal conductivity, including Brownian motion and the formation of a nanoparticle layer near the fluid's surface. Additionally, the authors compared experimental results and theoretical models to provide insights into the practical applications of NFs in cooling and heat transfer systems. The inadequacy of macroscopic

DOI: 10.1201/9781003642169-3

analysis in describing heat transfer in NFs is examined by Keblinskiet al. (2002). They also offer fresh explanations for the fluids' higher thermal conductivity, which has been shown in experiments. Swainet al. (2021) studied stagnation-point flow of a magnetized NF with Brownian motion and thermophoresis effects. They found that increasing magnetic intensity enhanced the Lorentz force, which reduced fluid velocity but increased TBL thickness. Variable thermal conductivity significantly influenced the heat transfer rate, while opposite effects on nanoparticle concentration by Brownian motion and thermophoresis. Biswal et al. (2023) analyzed the radiative magneto-hydrodynamic (MHD) flow of a non-Newtonian fluid past a nonlinearly elongating sheet. The findings reveal that magnetic fields and nonlinear stretching significantly impact velocity, temperature, and concentration profiles, while Soret and Dufour effects enhance the coupled heat and mass transfer processes. The research offers valuable insights into optimizing thermal and mass transfer in engineering systems involving non-Newtonian fluids. Ghennai and Bessaïh (2021)explored heat transfer analysis of hybrid Newtonian nanofluid (HNF) within a cubical cavity. Using numerical simulations based on the finite volume method, they analyzed the impacts of nanoparticle concentration, type, and distribution on thermal performance. Their results showed that the use of hybrid NFs significantly improves heat transfer compared to conventional single-nanoparticle fluids. Additionally, increasing nanoparticle volume fraction enhances the average Nusselt number, indicating better thermal conductivity, although it also increases fluid viscosity. The study highlights the potential of hybrid NFs in improving the efficiency of thermal management systems like cooling devices and energy storage units Santosh and Parida (2023) studied magneto-hydrodynamic (MHD) viscous NF flow over a nonlinearly stretching sheet. The results showed that an applied magnetic field and heat source parameters significantly affect the velocity and temperature profiles of the NF. Specifically, a stronger magnetic field tends to slow down the fluid motion while enhancing TBL, and the heat source intensifies fluid temperature. This work provides insights into the design of industrial and engineering processes involving thermal control using NFs. Jena et al. (2024) studied the three-dimensional magneto-hydrodynamic (MHD) rotating flow of a radiative NF over a stretched sheet, incorporating the effects of homogeneous–heterogeneous chemical reactions. Their mathematical model considers the impact of magnetic fields, fluid rotation, thermal radiation, and simultaneous bulk and surface chemical reactions, which are important for advanced manufacturing and chemical processes. They found that magnetic fields and rotational effects significantly influence the velocity and temperature distributions, while the presence of chemical reactions alters the nanoparticle concentration profiles. The study provides valuable insights into controlling heat and mass transfer in rotating systems involving reactive NFs. Furthermore, because of their distinct and durable qualities,

NFs have drawn a lot of interest and are being used in a variety of industries, including manufacturing, microelectronics, energy, and biomedicine.

Joule heating and viscous dissipation play essential roles in various engineering and scientific applications. In electronic cooling systems, Joule heating generates heat in conductive materials, requiring efficient thermal management solutions to prevent overheating of components. In magnetohydrodynamics (MHD), Joule heating and viscous dissipation are critical for controlling temperature in systems like fusion reactors and electromagnetic pumps. In power generation systems, these effects influence the heat transfer in components like turbines and heat exchangers, ensuring efficient energy conversion. Material processing techniques, such as laser welding and additive manufacturing, rely on Joule heating and viscous dissipation to control material properties. In environmental systems, such as geothermal energy and ground source heat pumps, these effects enhance energy efficiency. Additionally, biomedical applications like hyperthermia treatments and enhanced oil recovery (EOR) use Joule heating to manipulate temperatures for therapeutic or extraction purposes. Overall, Joule heating and viscous dissipation are critical in optimizing performance, energy efficiency, and heat management across multiple industries and scientific fields.

In addition to familiar objects like toothpaste, blood, paints, emulsions, oil, and polymer sheets, non-Newtonian fluids are used in a variety of engineering, industrial, and technical operations. There are many branches within the class of non-Newtonian fluids based on their physical properties. Shear thinning fluids are one of the most significant branches within this group. In these fluids, the apparent viscosity decreases with applied shear force. Many fluid models are proposed to explain the characteristics of shear thinning fluids and the Prandtl fluid model is one that can adequately explain the phenomenon of shear thinning. Kuik (2004) studied the Prandtl fluid flow behavior at infinity. To introduce a characteristic length scale, he initially derived the flow nearby an exponential spiral of finite length. In the limit where the spiral's length approaches infinity, Prandtl's original equations are recovered, but with modifications that incorporate the influence of the flow conditions at infinity. Khan et al. (2017) examined the effects of H-H chemical reactions on the magneto-hydrodynamic (MHD) flow of a Prandtl fluid over a widening surface. Their model captures the complex interface between magnetic fields, non-Newtonian fluid behavior, and simultaneous bulk–surface chemical reactions. The study revealed that the magnetic field retards the fluid motion, thickening the momentum boundary layer, while chemical reactions significantly affect the species concentration profiles. The findings are useful for designing chemical and industrial processes where fluid flow and reaction dynamics are interconnected. Ganesh Kumar et al. (2019) considered Joule heating to study the melting effects on flow of Prandtl fluid over an expanded sheet and found that velocity profiles are directly proportional to the Prandtl parameter. The results indicated that the magnetic field and Joule heating significantly influence the velocity and temperature fields, with Joule heating enhancing the TBL thickness and melting rate. The research provides important insights for applications involving electrically conducting non-Newtonian fluids and melting processes in materials engineering. Eidet al. (2021) studied heat transfer on the 3-D flow of Prandtl NF over a convectively heated porous sheet in presence of non-linear radiation and high-order chemical reaction. The higher-order chemical reactions significantly impact the concentration profiles, while the flow and thermal fields are affected by nanoparticle effects and stretching rates. Asad (2023) examined entropy generation on the mixed convective MHD flow of Prandtl fluid past a nonlinear stretched sheet. Zafar et al. (2023) investigated the irreversibility flow of Prandtl NF over an elongating sheet in the medium of Darcy–Forchheimer. The activation energy and chemical reactions significantly affect the concentration profiles of the NF, while the radiative and Darcy–Forchheimer effects influence the velocity and temperature distributions.

The buoyant force produced by temperature changes inside a fluid is referred to as thermal buoyancy. A heated fluid expands and loses density in comparison to the surrounding, colder fluid. Consequently, the density differential between the heated fluid and the surrounding fluid causes the heated fluid to feel buoyancy, or an upward push. The interplay of heat and mass transfer along with buoyancy effects is crucial in a wide range of practical applications to enhance the efficiency and effectiveness of various thermal and chemical processes like solar collectors, nuclear reactor coolants, chemical process industries, environmental engineering, etc. Raju and Sandeep (2016) demonstrated the assisting opposing flow non-Newtonian fluid past a cone with Lotentz force. The study revealed that the magnetic field significantly influences the velocity and temperature profiles, with the effects being more pronounced in assisting flows. The radiative heat transfer and the Casson fluid's yield stress were shown to alter the TBL, which impacts the heat dissipation rate. Habib et al. (2021)presented a mathematical model that incorporated the effects of buoyancy-induced forces (Stephen blowing) and the interaction between magnetic fields and electrically conductive fluids (Lorentz force) on the flow and heat transfer behavior of Prandtl NFs. By employing the Keller box approach, a numerical solution was obtained for the nonlinear governing equations, which revealed that both the Stephen blowing effect and Lorentz force significantly alter the velocity and temperature profiles of the fluid. The similar observations are mentioned by Yaseen et al. (2021) considering hybrid NF flow past a permeable moving surface. (Ramanjini et al. (2023) examined the effects of first-order slip, and chemical reaction on unsteady Prandtl nanofluid flow through stretching sheet due to mixed convection. Recently, many researchers Shaw et al., 2022; Kavya and Nagendramma, 2023; Nagaraja et al., 2023; Khan et al., 2023) explained the importance of assisting and opposing flow of NF and hybrid NF by taking different flow geometry. The objectives of the study are:

- To investigate the impact of buoyancy forces on the behavior of Prandtl fluid in both assisting and opposing flow conditions.
- To analyze the role of Brownian motion and thermophoretic effects in enhancing or impeding heat and mass transfer in the Prandtl fluid flow.
- To study the influence of Joule heating on the temperature distribution and thermal boundary layer characteristics of the fluid.
- To examine how viscous dissipation affects the energy balance and thermal performance of Prandtl fluid under varying flow regimes.
- To develop a mathematical model incorporating buoyancy, Brownian motion, thermophoresis, Joule heating, and viscous dissipation for Prandtl fluid flow over a stretching surface.
- To provide insights and potential applications of the findings in thermal management systems, materials processing, and energy systems involving non-Newtonian fluids.

3.2 RESEARCH QUESTIONS

The following scientific research issues are addressed in this study:

- What is the effect of Lorentz force, fluid parameter, elastic parameter and porosity parameter on the velocity profile?
- How heat is transported in presence of joule heating, viscous dissipation, and Brownian motion?
- What is the effect on nanoparticle concentration under the influence of the chemical reaction parameter, Schmidt number, thermophoresis, and Brownian motion?
- How skin friction coefficient, Nusselt number, and Sherwood number change with the effect of the fluid parameter, elastic parameter, Eckert number, and Biot number?

3.3 MATHEMATICAL FORMULATION

The steady MHD incompressible flow of an electrically conducting NF past a two-dimensional stretching sheet as illustrated in Figure 3.1. The thermophoresis and Brownian motion are taken into account in the energy equation. Further, the energy equation is empowered by joule heating, thermal radiation, and viscous dissipation. The nanoparticles and the base fluid are maintained in thermal equilibrium. A constant magnetic field of specified strength is applied perpendicular to the surface. The induced

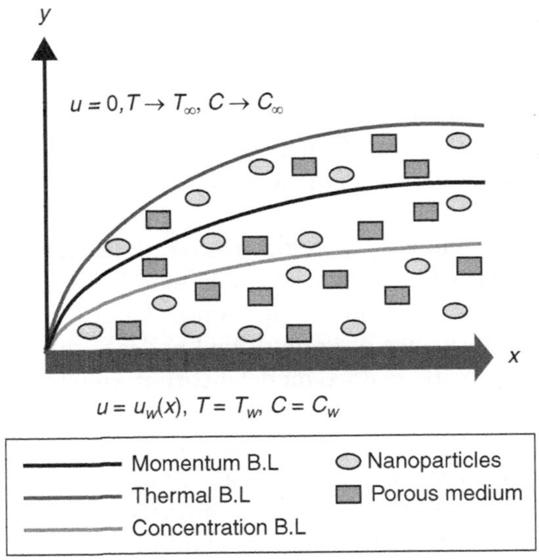

Figure 3.1 Physical model of the problem.

magnetic field is assumed to be negligible due to low Reynolds number and thus is not taken into account.The stretching sheet is placed along the x-axis and the y-axis is normal to the flow direction. The sheet is stretched with the velocity $u_w(x) = ax$ with $a > 0$ is constant. The temperature and concentration of the stretching surface are taken T_w, and C_w while the ambient temperature and concentration are taken T_∞ and C_∞, respectively.

The Cauchy stress tensor for the Prandtl fluid is given by

$$S = \frac{A\sin^{-1}\left\{\frac{1}{c}\left[\left(\frac{\partial u_i}{\partial x_j}\right)^2 + \left(\frac{\partial u_j}{\partial x_i}\right)^2\right]^{\frac{1}{2}}\right\}}{\left[\left(\frac{\partial u_i}{\partial x_j}\right)^2 + \left(\frac{\partial u_j}{\partial x_i}\right)^2\right]^{\frac{1}{2}}} A_1 \tag{3.1}$$

where A and care the material parameters and A_1 is the first Rivlin–Erickson tensor.

Under the above assumptions, the governing equations and corresponding boundary conditions are given by

$$\frac{\partial u}{\partial x}+\frac{\partial v}{\partial y}=0, \tag{3.2}$$

$$u\frac{\partial u}{\partial x}+v\frac{\partial u}{\partial y}=\upsilon\frac{A}{c}\left\{1+\frac{1}{2c^2}\left(\frac{\partial u}{\partial y}\right)^2\right\}\frac{\partial^2 u}{\partial y^2}-\frac{\sigma B_0}{\rho}u$$
$$-\frac{\upsilon}{k^*}u+g\beta_1\left(T-T_\infty\right), \tag{3.3}$$

$$u\frac{\partial T}{\partial x}+v\frac{\partial T}{\partial y}=\left(\alpha+\frac{16\sigma^* T_\infty^3}{3\rho c_p k^*}\right)\frac{\partial^2 T}{\partial y^2}++\frac{\sigma B_0}{\rho c_p}u^2$$
$$+\tau\left\{D_B\frac{\partial C}{\partial y}\frac{\partial T}{\partial y}+\frac{D_T}{T_\infty}\left(\frac{\partial T}{\partial y}\right)^2\right\}+\frac{\mu}{\rho c_p}\left(\frac{\partial u}{\partial y}\right)^2, \tag{3.4}$$

$$u\frac{\partial C}{\partial x}+v\frac{\partial C}{\partial y}=D_B\frac{\partial^2 C}{\partial y^2}+\frac{D_T}{T_\infty}\frac{\partial^2 T}{\partial y^2}-k_c\left(C-C_\infty\right)^n. \tag{3.5}$$

$$\left.\begin{array}{l} at\, y=0:\ u=u_w,\ \mathrm{v}=0,-k\dfrac{\partial T}{\partial y}=h\left(T_w-T\right),\ C=C_w, \\ as\, y\to\infty:\ u\to 0, T\to T_\infty, C\to C_\infty. \end{array}\right\} \tag{3.6}$$

Applying the similarity transformations in the following form,

$$u=axf'(\eta), v=-\sqrt{a\upsilon}f, \theta(\eta)=\frac{T-T_\infty}{T_w-T_\infty},$$
$$\eta=\sqrt{\frac{a}{\upsilon}}y, \phi(\eta)=\frac{C-C_\infty}{C_w-C_\infty}. \tag{3.7}$$

Equations (3.2)–(3.6) become

$$\alpha\left(1+\beta f''^2\right)f'''+ff''-f'^2-Mf'-k_pf'+\lambda\theta=0 \tag{3.8}$$

$$\frac{1}{\Pr}\left(1+\frac{4}{3}Rd\right)\theta+f\theta'+Nb\theta'\phi'+Nt\theta'^2-MEcf'^2+Ecf''^2=0 \tag{3.9}$$

$$\phi''+\frac{Nt}{Nb}\theta''+Scf\phi'-k_rSc\phi^n=0 \tag{3.10}$$

$$\left.\begin{array}{l} f(0)=0, f'(0)=1, \theta'(0)=-\gamma\left[1-\theta(0)\right], \phi(0)=1, \\ f'(\infty)=0, \theta(\infty)=0, \phi(\infty)=0. \end{array}\right\} \tag{3.11}$$

Here $\alpha = \frac{A}{c}$ (fluid parameter), $\beta = \frac{a^2 \,\mathrm{Re}_x}{2c^2}$ (elastic parameter), $\gamma = \frac{h}{k}\sqrt{\frac{\upsilon}{a}}$ (Biot number), $M = \frac{\sigma B_0}{a\rho}$ (magnetic parameter), $k_p = \frac{\upsilon}{ak^*}$ (porosity parameter), $\lambda = \frac{g\beta}{a^2 x}\left(T_w - T_\infty\right)$ (thermal buoyancy parameter), $R = \frac{4\sigma^* T_\infty^3}{kk^*}$ (radiation parameter), $\Pr = \frac{\upsilon}{\alpha}$ (Prandtl number), $Nb = \frac{(\rho c)_p D_B}{(\rho c)_f \upsilon}\left(C_w - C_\infty\right)$ (Brownian diffusion parameter), $Nt = \frac{(\rho c)_p D_T}{(\rho c)_f T_\infty \upsilon}\left(T_w - T_\infty\right)$ (thermophoresis parameter), $Ec = \frac{u_w^2}{c_p (T_w - T_\infty)}$ (Eckert number), $Sc = \frac{\upsilon}{D_B}$ (Schmidt number), $k_r = \frac{k_c}{a}$ (chemical reaction parameter), and $\mathrm{Re}_x = \frac{u_w x}{\upsilon}$ (local Reynolds number).

The physical parameters such as the skin friction coefficient, local Nusselt number, and local Sherwood number are respectively defined as

$$C_f = \frac{2\tau_w}{\rho u_w^2}, Nu_x = \frac{xq_w}{k(T_w - T_\infty)}, \text{ and } Sh_x = \frac{xq_m}{D_B(C_w - C_\infty)} \tag{3.12}$$

Here, $\tau_w = \mu\left(\frac{\partial u}{\partial y}\right)_{y=0}$ (shear stress), $q_w = -k\left(\frac{\partial T}{\partial y}\right)_{y=0} + \left(q_r\right)_{y=0}$ (heat flux), and $q_m = -D_B\left(\frac{\partial C}{\partial y}\right)_{y=0}$ (mass flux).

Finally, we get $\sqrt{\mathrm{Re}_x} C_f = f''(0)$, $\sqrt{\mathrm{Re}_x} Nu_x = -\left(1 + \frac{4}{3}R\right)\theta'(0)$, $\sqrt{\mathrm{Re}_x} Sh_x = -\phi'(0)$ (3.13)

3.4 RESULTS AND DISCUSSION

The system of nonlinear coupled ODEs along with the boundary conditions (3.8–3.11) is numerically solved using the Runge-Kutta (RK) fourth-order method along with the shooting technique, implemented through MATLAB's bvp4c solver. The solution's convergence, stability, and accuracy have been verified using a step size of 10^{-2} and an error tolerance of 10^{-5}. To validate the numerical code, the results for the skin friction coefficient were computed for different values of relevant parameters, as presented in

Table 3.1 Validation of the skin friction coefficient

M	*(Habib et al.(2021)*	*Ramanjiniet al.(2023)*	*Present study*
0	−1.8428	−1.8478	−1.8465
1	−1.0100	−1.0012	−1.0126
5	−0.5213	−0.5205	−0.5218
10	−0.4150	−0.4156	−0.4162

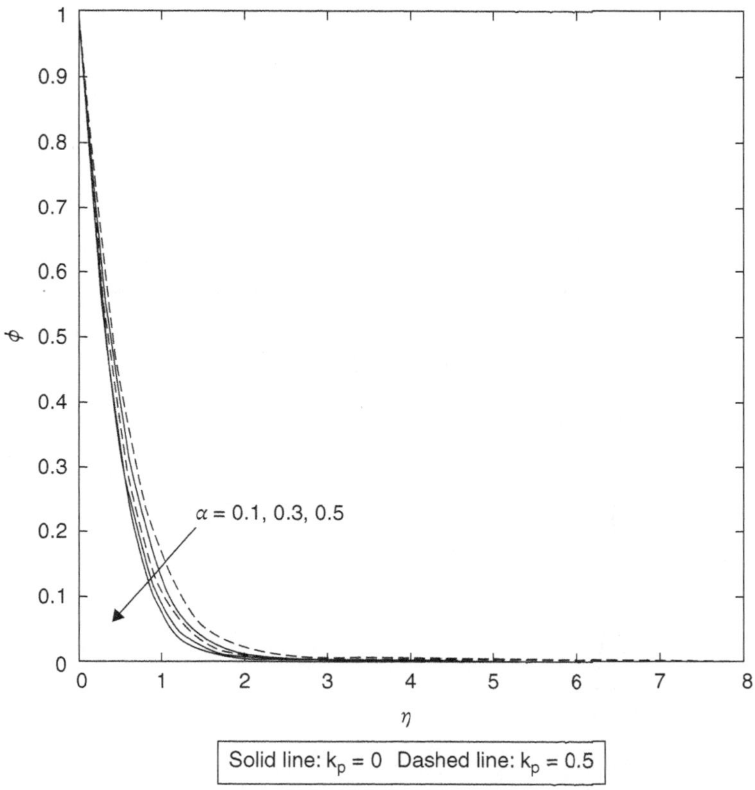

Figure 3.2 Effects of α and k_p on $\phi(\eta)$.

Table 3.1. These results show excellent agreement with those previously reported by Habib et al. (2021)and Ramanjini et al. (2023).

The effects of porosity and fluid parameters on NF are shown in Figures 3.2–3.4. Porosity and fluid parameters strongly influence the concentration, temperature, and velocity profiles in porous medium flows. Higher porosity generally means the medium offers less resistance to fluid motion, resulting in higher velocity profiles. Conversely, lower porosity restricts the flow,

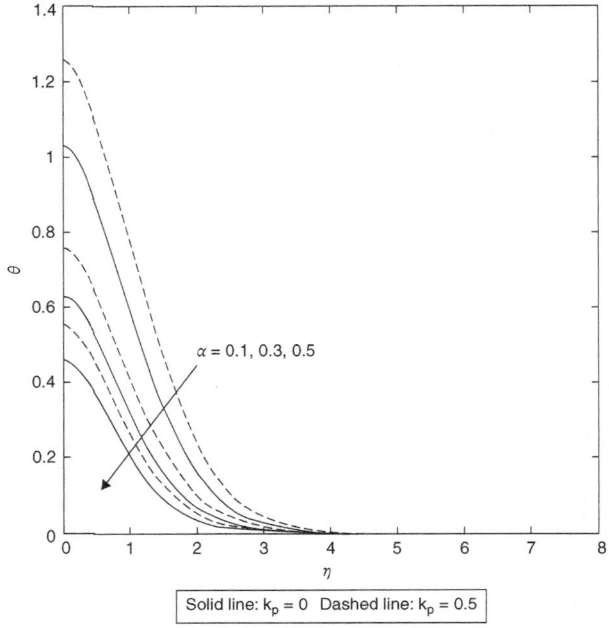

Figure 3.3 Effects of α and k_p on $\theta(\eta)$.

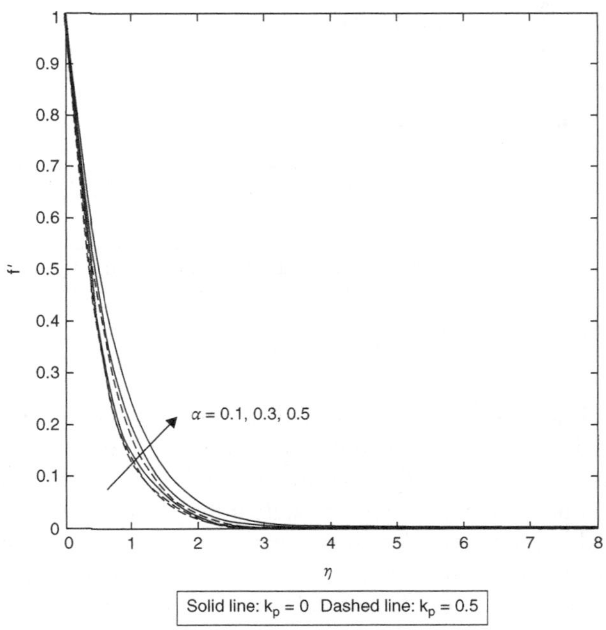

Figure 3.4 Effects of α and k_p on $f(\eta)$.

leading to a reduction in velocity (Figure 3.4)and enhancing the thermal (Figure 3.3) and concentration boundary layers(Figure 3.2). As shown in Figure 3.4, the fluid velocity rises with increasing values of α. Physically, a higher α reduces the fluid's viscosity, leading to an increase in velocity due to the diminished resistance within the liquid. Together, porosity and fluid characteristics control how quickly momentum, heat, and mass are transferred within the system, significantly affecting the overall distribution of velocity, temperature, and concentration across the flow field.

Figures 3.5–3.7 show the significant impact of magnetic and elastic parameters on concentration, temperature, and velocity profiles, respectively. An increase in the magnetic parameter, often represented by the magnetic field strength, tends to introduce a Lorentz force that acts as a resistive drag on the fluid motion. This reduces the velocity profile (Figure 3.7) and leads to the thickening of both the momentum and TBL. As the fluid slows down, the temperature within the flow generally rises due to enhanced Joule heating effects (Figure 3.6). On the other hand, higher magnetic fields can

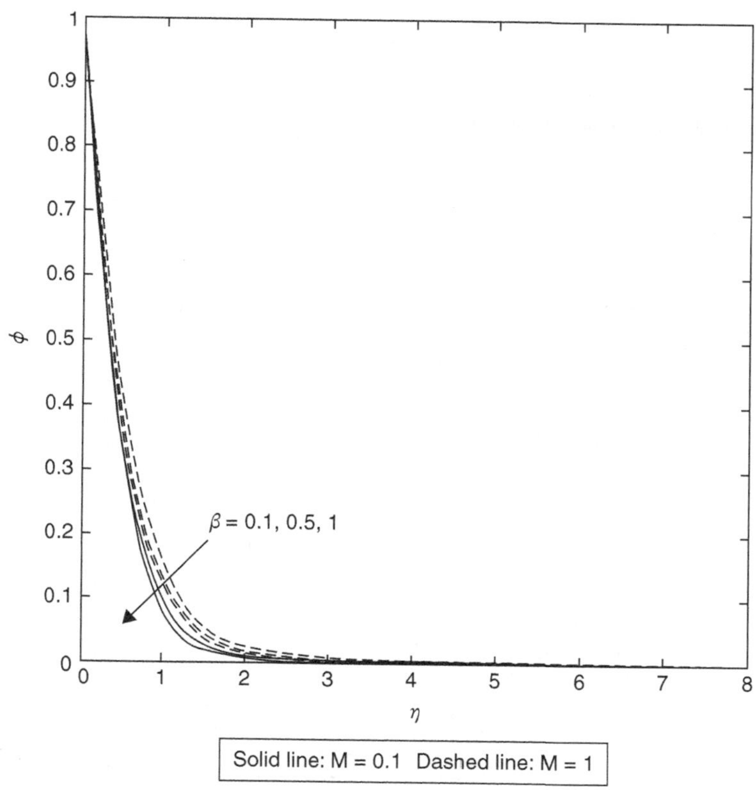

Figure 3.5 Effects of β and M on $\phi(\eta)$.

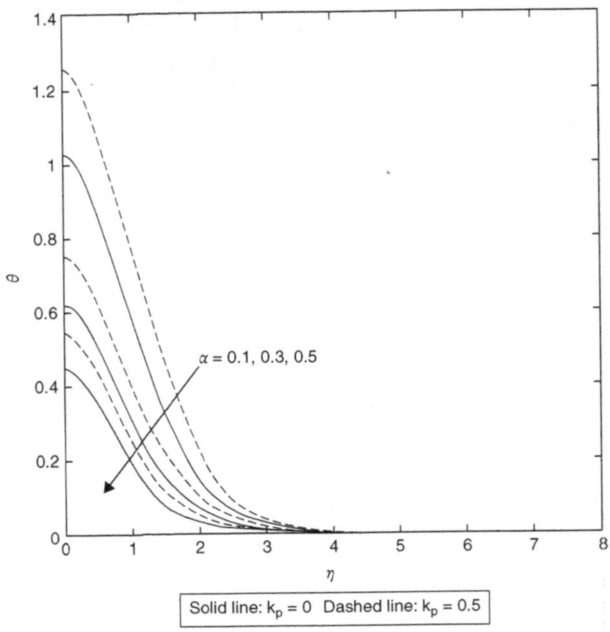

Figure 3.6 Effects of β and M on $\theta(\eta)$.

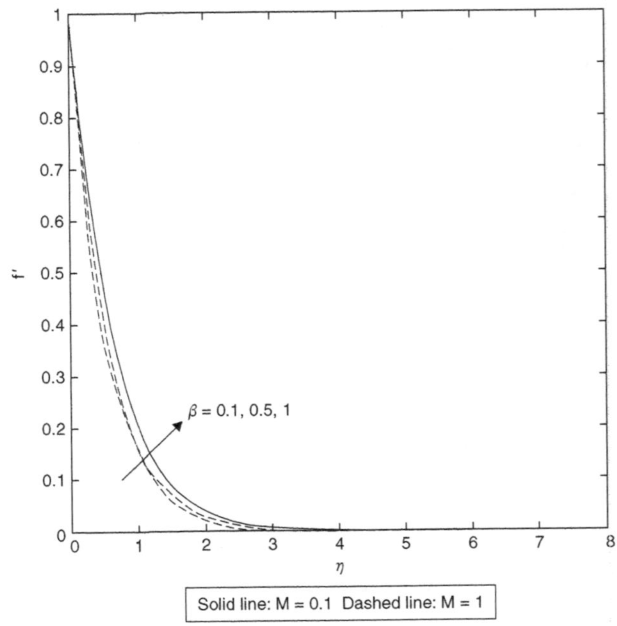

Figure 3.7 Effects of β and M on $f'(\eta)$.

suppress the concentration boundary layer, leading to lower species diffusion rates (Figure 3.5). Elastic parameter, which characterizes the fluid's ability to resist deformation, typically work to counteract changes in flow structure. Greater elasticity tends to stabilize the flow, reducing velocity fluctuations, while also influencing the thermal and concentration fields by modifying energy and mass transport rates. Overall, both magnetic and elastic effects play crucial roles in controlling the behavior of temperature, velocity, and concentration in complex fluid systems.

The effects of thermal buoyancy parameter (λ) and Biot number (γ) on concentration, temperature, and velocity profiles in fluid flows are displayed in Figures 3.8–3.10. An increase in λ enhances the buoyancy-driven forces within the flow, leading to an acceleration of fluid motion and higher velocity profiles, particularly near heated surfaces (Figure 3.10). This enhanced convection typically results in a reduction of temperature gradients, causing a thickening of the TBL(Figure 3.9). However, the stronger flow can also reduce the concentration boundary layer thickness, promoting sharper concentration gradients(Figure 3.8). The Biot number "measures the ratio of

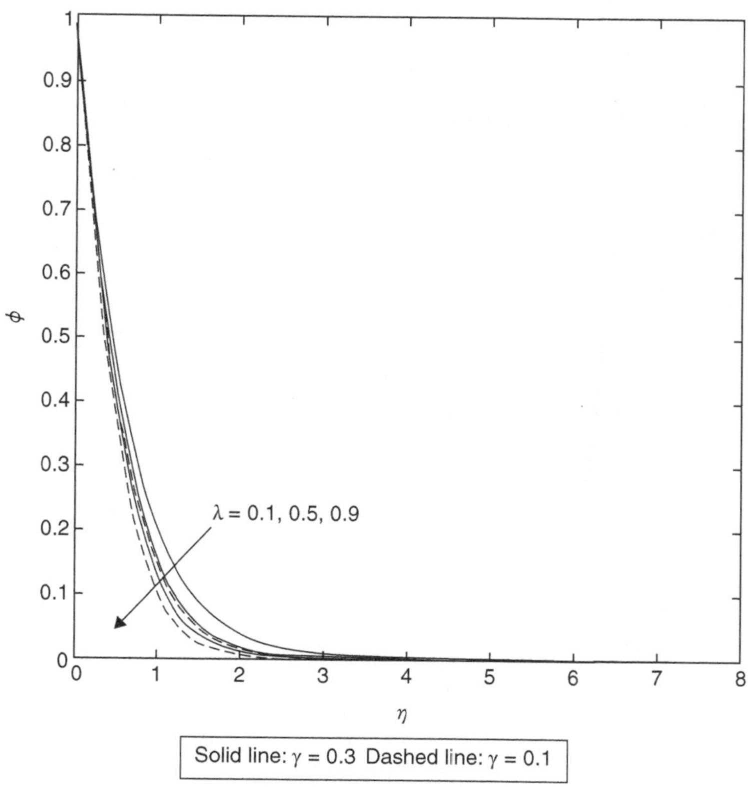

Figure 3.8 Effects of λ and γ on $\phi(\eta)$.

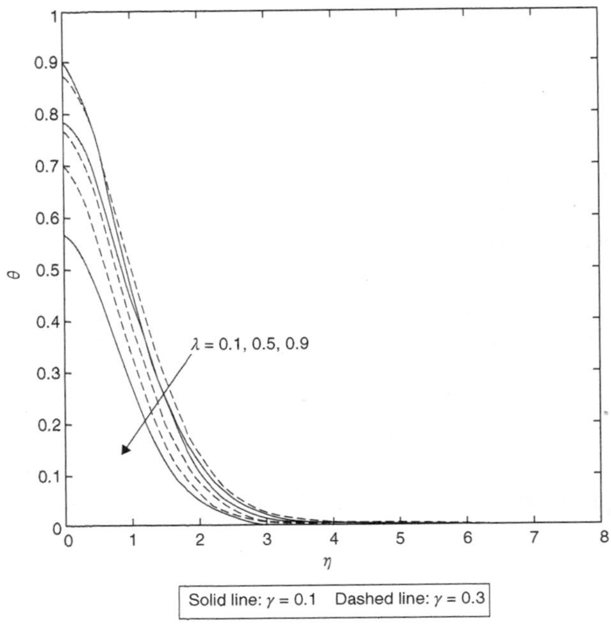

Figure 3.9 Effects of λ and γ on $\theta(\eta)$.

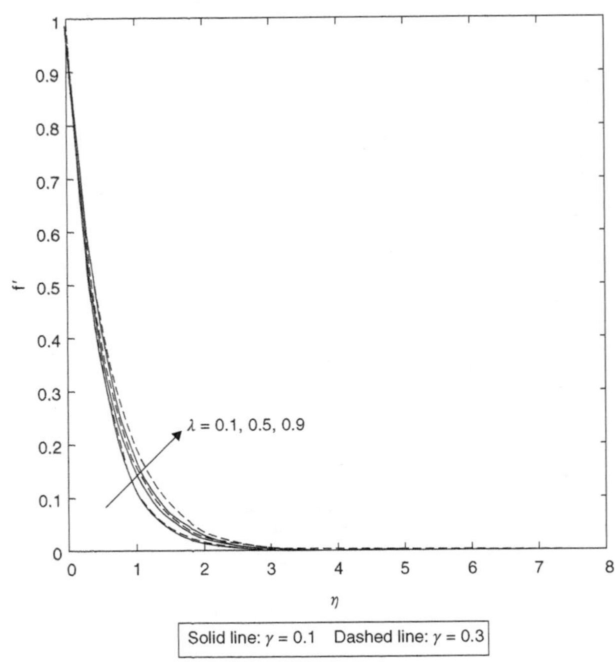

Figure 3.10 Effects of λ and γ on $f'(\eta)$.

internal thermal resistance to surface thermal resistance" plays a critical role. A higher Biot number indicates stronger convective heat exchange at the surface, which tends to elevate surface temperatures and increase the TBL thickness. Consequently, higher surface temperatures can influence concentration profiles through enhanced thermophoretic or diffusive effects. Overall, both the thermal buoyancy parameter and Biot number critically shape the transport characteristics of heat, mass, and momentum within the flow.

Figures 3.11 and 3.12 show the effects of Brownian motion (Nb) and thermophoresis (Nt) on concentration and temperature profiles, respectively.As the Brownian motion parameter increases, it enhances the random movement of nanoparticles, leading to rise the temperature of the fluid due to more vigorous thermal agitation. However, this also consequences in a decrease of the concentration distribution, as particles diffuse more rapidly from high to low concentration regions, causing the concentration boundary layer to decrease. In contrast, an increase in the thermophoresis

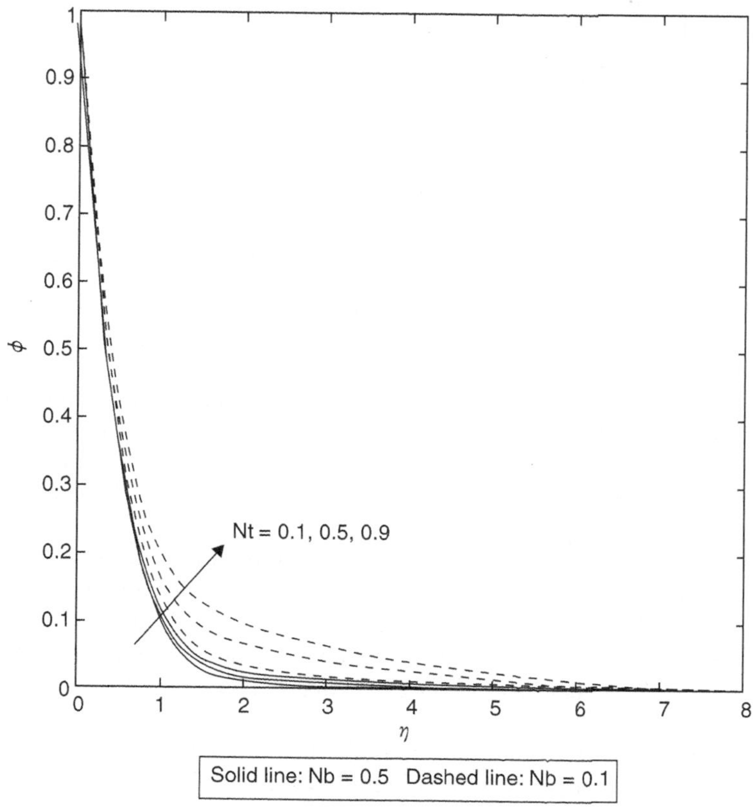

Figure 3.11 Effects of *Nt* and *Nb* on $\phi(\eta)$.

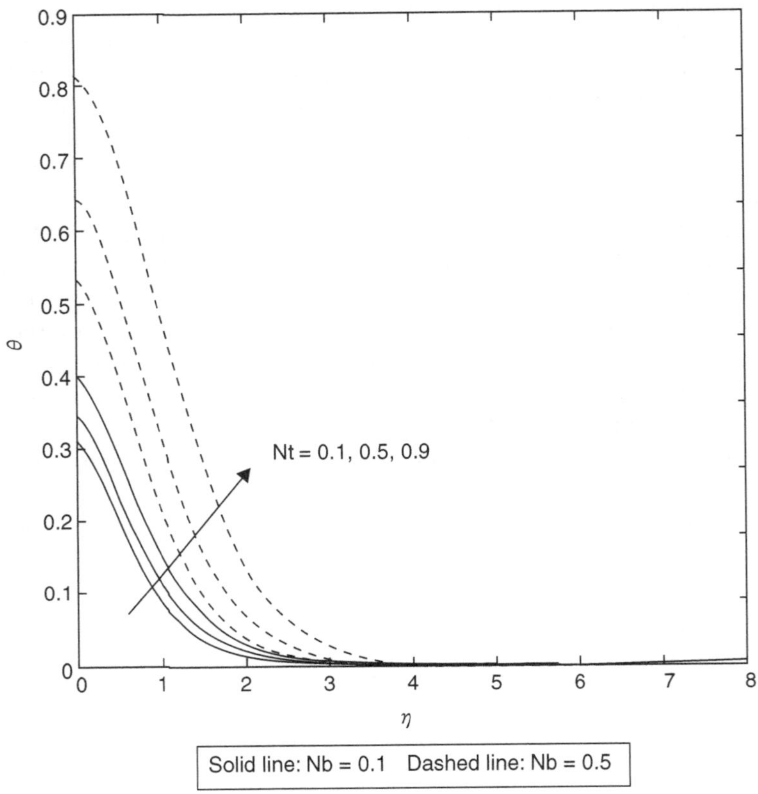

Figure 3.12 Effects of *Nt* and *Nb* on $\theta(\eta)$.

parameter causes both the temperature and concentration profiles to rise. Thermophoresis drives the particles from hotter regions to cooler ones, leading to an accumulation of particles and energy within the fluid, which thickens both the thermal and concentration boundary layers, thereby improving heat and mass transfer.

Figure 3.13 shows that a considerable rise in the radiation parameter (R) results in a greater thermal energy transmission, which in turn creates a thicker TBL. Thermal radiation plays a crucial role in affecting the temperature distribution within the flow domain. Due to significant effects of radiation, additional mode of heat transfer is contributed alongside conduction and convection. An increase in radiation typically enhances the overall energy transfer, leading to a rise in the fluid temperature. As a result, the TBL becomes thicker, and the temperature gradients within the flow become more gradual. In radiatively dominant systems, especially at high temperatures or in optically thick media, radiation can even become the

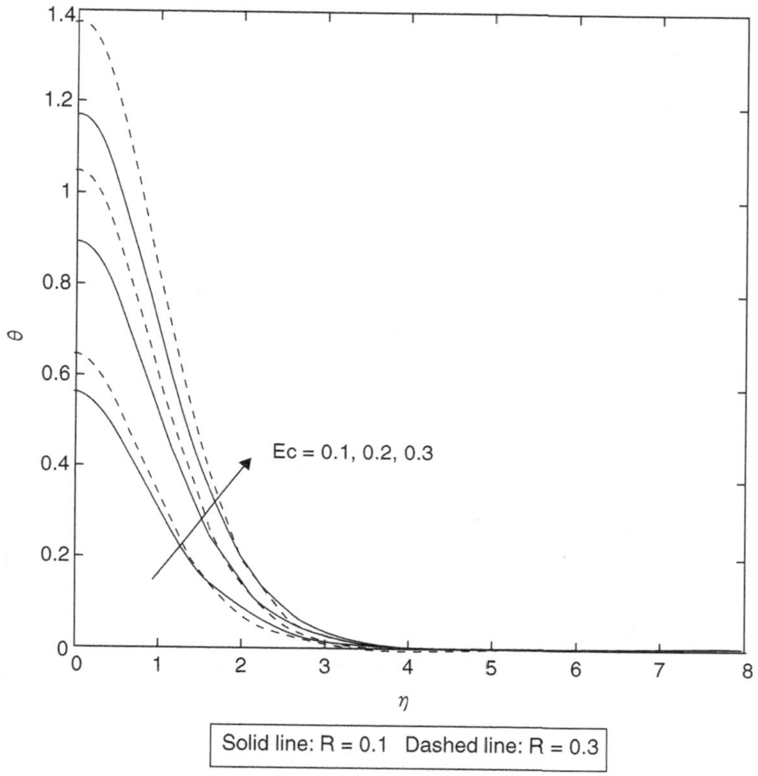

Figure 3.13 Effects of *Ec* and *R* on $\theta(\eta)$.

primary mechanism of heat transfer. Conversely, when radiation effects are weak, the temperature distribution is mainly controlled by conduction and convection processes. Overall, the presence of strong radiation tends to elevate temperatures and smooth out variations across the flow field. The Eckert number (Ec) significantly influences the temperature distribution in fluid flows. It represents the ratio of kinetic energy to thermal energy within the system. When the Eckert number increases, the effects of viscous dissipation become more pronounced, causing a renovation of mechanical energy into thermal energy. This results in additional heating of the fluid, foremost important to a rise in the overall temperature. Consequently, the TBLbecomes thicker, and the temperature gradients near the surface become less steep. In contrast, when the Eckert number is low, viscous dissipation is negligible, and the temperature distribution is primarily governed by external mechanisms such as conduction and convection. Therefore, a higher Eckert number tends to elevate the temperature across the flow field,

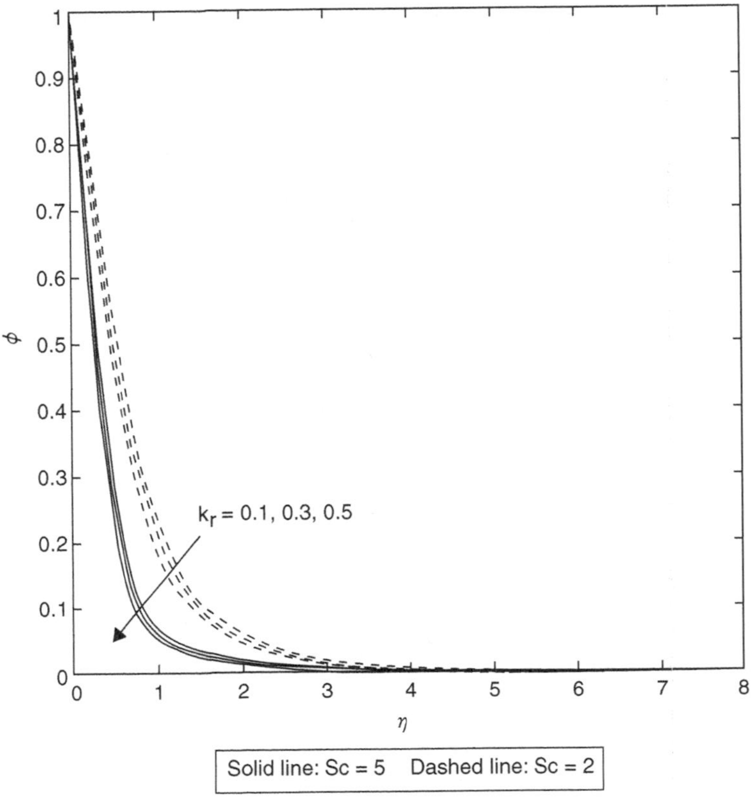

Figure 3.14 Effects of k_r and *Sc* on $\phi(\eta)$.

while a lower Eckert number maintains a temperature profile more closely aligned with external thermal conditions.

Figure 3.14 reveals that greater values of Sc (Schmidt number),that is, heavier species and chemical reaction parameter reduce the concentration profiles in same pattern. Physically, a rise in the chemical reaction parameters value produces disruptions in the fluid-species domain, which in turn induces molecular mobility and lowers concentration. Further, the k_r (chemical reaction parameter) shows a significant role in concentration distributions. An increase in k_r leads to a reduction in the concentration profile and a thinner boundary layer. Additionally, the Schmidt number reduces the concentration profile, as it represents the ratio of momentum diffusivity to mass diffusivity. With a rise in Schmidt number, mass diffusivity increases, consequently reduces the solutal boundary layer thickness.

Table 3.2 Values of $f''(0)$, $-\theta'(0)$ and $-\phi'(0)$ when $M = k_p = 1, Pr = 0.71, n = 2$ $Sc = 0.6$

α	β	λ	Nb	Nt	E_C	γ	k_r	$f''(0)$	$-\theta'(0)$	$-\phi'(0)$
0.1	0.1	0.1	0.1	0.1	0.1	0.1	0.1	-0.809671	0.093315	0.966063
0.3								-0.568545	0.099859	1.023789
0.5								-0.451017	0.102874	1.051753
	0.3							-0.449584	0.102895	1.052040
	0.5							-0.448189	0.102917	1.052321
	1.0							-0.444854	0.102968	1.052998
		0.3						-0.432493	0.103034	1.054064
		0.5						-0.420224	0.103098	1.055120
		1.0						-0.389938	0.103250	1.057718
			0.2					-0.382747	0.102342	1.085527
			0.3					-0.374647	0.101338	1.095509
			0.5					-0.355330	0.099006	1.105087
				0.2				-0.354459	0.098913	1.097858
				0.5				-0.351742	0.098624	1.076792
				0.9				-0.347859	0.098217	1.050225
					0.2			-0.366950	0.101039	1.029284
					0.3			-0.384687	0.103674	1.009820
					0.5			-0.413456	0.104713	0.980635
						0.2		-0.327032	0.190967	0.932763
						0.3		-0.246689	0.244508	0.911678
						0.5		-0.118833	0.339039	0.880934
							0.3	-0.414595	0.108160	1.145392
							0.5	-0.412800	0.107910	1.297682
							-0.1	-0.419115	0.108788	0.778311
							-0.3	-0.422152	0.109213	0.540814
							-0.5	-0.426185	0.109763	0.227303

The values of skin friction coefficient, Nusselt number, and Sherwood number are computed and shown in Table 3.2. It is seen that fluid parameter, elastic parameter and thermal buoyancy parameter increases the magnitude of $f''(0)$, $-\theta'(0)$,and $-\phi'(0)$ whereas Brownian motion have same effect on $-\theta'(0)$ but reverse effect is perceived in case of $-\phi'(0)$. Further, destructive and constructive chemical reaction have opposite effects on Sherwood number.

3.5 CONCLUSIONS AND FUTURE SCOPE

The study concludes that the combined effects of buoyancy force, Brownian motion, thermophoresis, Joule heating, and viscous dissipation play crucial roles in determining the behavior of Prandtl fluid flow. These factors significantly influence the temperature, concentration, and velocity profiles, ultimately affecting the overall efficiency of heat and mass transfer in fluid systems. Key findings include the following:

- Buoyancy force accelerates fluid flow, thickens the thermal boundary layer, and decreases concentration layers.
- Brownian motion raises fluid temperature but reduces concentration profiles due to increased particle diffusion.
- Thermophoresis enhances heat and mass transfer by thickening both thermal and concentration boundary layers.
- Joule heating and viscous dissipation increase temperature, contributing to thicker thermal boundary layers and more efficient heat transfer.
- The interaction of these factors determines the performance of fluid systems in practical applications such as cooling and heat exchange processes, highlighting their importance in optimizing fluid dynamics for engineering applications.

The present study opens several promising directions for future research. Further investigations can explore the influence of variable viscosity and thermal conductivity on the behavior of Prandtl fluids under the influence of buoyancy forces, Brownian motion, and thermophoretic effects. Extending the analysis to three-dimensional and unsteady flow configurations could provide deeper insights into practical engineering applications. Incorporating chemical reactions, nanoparticle aggregation, or hybrid NFs might reveal more complex heat and mass transfer mechanisms. The role of nonlinear Joule heating and non-Newtonian rheology under magnetic fields of varying strengths could also be examined for more realistic modeling. Moreover, experimental validation of the theoretical predictions, especially under extreme conditions of viscous dissipation and strong thermophoretic forces, would strengthen the applicability of the results. Future studies might also focus on optimizing heat transfer in industrial processes, biomedical engineering, or microfluidic systems where such combined effects are crucial.

REFERENCES

Asad, S. (2023). Nonlinear stretched flow a radiative MHD Prandtl fluid with entropy generation and mixed convection. *Frontiers in Physics*, 11, 1150457.

Biswal, M. M., Swain, K., Dash, G. C., & Ojha, K. L. (2023). Study of radiative magneto-non-Newtonian fluid flow over a nonlinearly elongating sheet with Soret and Dufour effects. *Numerical Heat Transfer, Part A: Applications*, 83(4), 331–342.

Choi, S. U. S. (1995). Enhancing thermal conductivity of fluids with nanoparticles: In *Developments and Applications of Non-Newtonian Flows. ASME*, 99–105.

Eid, M. R., Mabood, F., Mahny, K. L. (2021). On 3D Prandtl NF flow with higher-order chemical reaction. *Proceedings of the Institution of Mechanical Engineering Part C: Journal of Mechanical Engineering Science*, 235(19), 3962–3974.

Ganesh Kumar, K., Krishnamurthy, M. R., & Rudraswamy, N. G. (2019). Boundary layer flow and melting heat transfer of Prandtl fluid over a stretching surface by considering Joule heating effect. *Multidiscipline Modeling in Materials and Structures*, 15(2), 337–352.

Ghennai, A., & Bessaïh, R. (2021). Heat transfer enhancement in a cubical cavity filled with a hybrid NF. *Heat Transfer*, 50(2), 1658–1678.

Habib, D., Salmat, N., Hussain, S., Ali, B., & Abdal, S. (2021). Significance of Stephen blowing and Lorentz force on dynamics of Prandtl NF via Keller box approach. *International Journal of Communications in Heat and Mass Transfer*, 128, 105599.

Jena, S., Swain, K., Ibrahim, S. M., Sreenivasulu, P., & Lorenzini, G. (2024). Three-dimensional MHD rotating flow of radiative NF over a stretched sheet with homogeneous-heterogeneous chemical reactions. *Journal of Engineering Thermophysics*, 33 (2), 336–353.

Kavya, S., & Nagendramma, V. (2023). Opposing and assisting flow of a hybrid newtonian/non-newtonian NFPast a stretching cylinder. *Journal of Mines, Metals and Fuels*, 71(8), 1045–1057.

Keblinski, P., Phillpot, S. R., Choi, S. U. S., & Eastman, J. A. (2002). Mechanisms of heat flow in suspensions of nano-sized particles (NFs). *International Journal of Heat and Mass Transfer*, 45(4), 855–863.

Khan, I., Malik, M. Y., Hussain, A., & Salahuddin, T. (2017). Effect of homogenous-heterogeneous reactions on MHD Prandtl fluid flow over a stretching sheet. *Results in Physics*, 7, 4226–4231.

Khan, U., Zaib, A., Ishak, A., Sherif, El-Sayed M., Sarris, Ioannis E., Eldin, Sayed M., & Pop, I. (2023). Analysis of assisting and opposing flows of the Eyring-Powell fluid on the wall jet nanoparticles with significant impacts of irregular heat source/sink. *Case Studies in Thermal Engineering*, 49, 103209.

Kuik, G.A.M. van. (2004). The flow induced by Prandtl's self-similar vortex sheet spirals at infinite distance from the spiral kernel. *European Journal of Mechanics – B/Fluids*, 23(4), 607–616.

Nagaraja, K. V., Khan, U., Madhukesh, J. K. et al. (2023). Heat and mass transfer analysis of assisting and opposing radiative flow conveying ternary hybrid NF over an exponentially stretching surface. *Science Reports*, 13, 14795.

Raju, C. S. K., & Sandeep, N. (2016). Opposing and assisting flow characteristics of radiative Casson fluid due to cone in the presence of induced magnetic field.*InternationalJournal of Advances in Science and Technology*, 88, 43–62.

Ramanjini, V., Gopi Krishna, G., Ramanuja, M., Sree, H. K., & Mishra, S. R. (2023). Unsteady Prandtl NFic flow through stretching sheet due to mixed convection, first-order slip, and chemical reaction. *Journal of the Nigerian Society of Physical Sciences*, 5, 1140.

Santosh, C., & Parida, S. K. (2023). Heat transfer analysis of MHD viscous NF flow over a nonlinearly stretching sheet with heat source/sink: A numerical study. *Numerical Heat Transfer, Part A: Applications*, 1–20.

Shaw, S., Patra, A., Misra, A., & Nayak, M. K. (2022). Assisting/opposing/forced convection flow on entropy-optimized MHD NFs with variable viscosity: Interfacial layer and shape effects. *Heat Transfer*, 51(1), 578–603.

Siddappa, B., &Abel, S. (1985). Non-Newtonian flow past a stretching plate. *Zeitschrift für angewandte Mathematik und Physik*, 36, 890–892.

Swain, K., Mahanthesh, B., & Mebarek-Oudina, F. (2021). Heat transport and stagnation-point flow of magnetized nanoliquid with variable thermal conductivity, Brownian moment, and thermophoresis aspects. *Heat Transfer*, 50(1), 754–767.

Vajravelu, K., & Rollins, D. (1992). Heat transfer in an electrically conducting fluid over a stretching surface. *International Journal of Non-linear Mechanics*, 27(2), 265–277.

Wang, X. Q.,& Mujumdar, A. S.(2007). Heat transfer characteristics of NFs: a review. *International Journal of Thermal Sciences*, 46(1), 1–19.

Yaseen, M., Manoj Kumar, & Rawat, S. K. (2021). Assisting and opposing flow of a MHD hybrid NF flow past a permeable moving surface with heat source/sink and thermal radiation. *Partial Differential Equations in Applied Mathematics*, 4, 100168.

Zafar, S. S., Khan, U., Ali, F., Eldin, Sayed M., Mohammed Saeed, A., Zaib, A., & Galal, Ahmed M. (2023). Irreversibility analysis of radiative flow of Prandtl NF over a stretched surface in Darcy-Forchheimer medium with activation energy and chemical reaction.*Heliyon*, 9, e14877.

Chapter 4

Magnetic field effects on Casson–Maxwell nanofluid flow over a stretching surface in porous media

Seema Tinker, Om Prakash, and Piyush Kumar Gupta

Nomenclature

Symbol	Description
u, v	Velocity components (m s^{-1})
P_r	Prandtl number
D_u	Dufour Number
S_r	Soret number
C	Concentration (mol m^{-3})
B_0	Uniform magnetic field (Wb m^{-2})
K	Permeability of the medium (H m^{-1})
k_0	Fluid relaxation time (s)
K_p	Porosity parameter
T	Fluid temperature (K)
M	Magnetic field parameter
C_p	Specific heat (J Kg^{-1} K^{-1})
D_B	Brownian motion scale factor (m^2 s^{-1}) $(m^2\ s^{-1})$
D_T	Thermal diffusivity (m^2 s^{-1})
D_{CT}	Soret diffusivity (Kg m^{-1} s^{-1} K^{-1})
D_{TC}	Dufour diffusivity (m^2 s^{-1})
N_b	Brownian motion parameter
N_t	Thermophoresis parameter
L_1	Momentum slip length (m)
L_2	Thermal slip length (m)
L_3	Concentrations slip length (m)
Greek Symbols	
σ	Electrical conductivity (S m^{-1})
ρ	Density (Kg m^{-3})
γ	Maxwell fluid parameter
β	Casson fluid parameter
η	Dimensionless similarity variable
ϕ	Dimensionless concentration
θ	Dimensionless temperature
α	Thermal diffusivity (m^2 s^{-1})
υ	Kinematic viscosity (Kg m^{-1} s^{-1}) $(kg\ m^{-1}\ s^{-1})$
τ	Heat generation rate (W m^{-2}) $(W\,m^{-2})$
λ	Velocity slip parameter
δ	Thermal slip parameter
χ	Concentration slip parameter
Subscripts	
s	Surface conditions
f	Surrounding conditions

4.1 INTRODUCTION

Magnetohydrodynamics (MHD) and porous media, in conjunction with stretching sheets, form a fascinating area of study with significant implications in various fields of science and engineering. MHD flow provides

DOI: 10.1201/9781003642169-4

a variety of phenomena with applications in astrophysics, plasma physics, and engineering (Gregory et al., 2016). Porous media, characterized by their interconnected voids, play a crucial role in filtration, groundwater flow, and energy transmission. The study of fluid flow over a stretching sheet, often encountered in industrial processes like polymer extrusion and metal forming. Combining these concepts, the investigation of MHD flow over an extending surface opens up new avenues for research with potential applications in diverse areas such as geothermal energy extraction, microfluidic devices, and material processing. MHD flow in spongy media was deliberated by Geindreau and Auriault (2002), deriving a modified Darcy's law with a permeability tensor dependent on the magnetic field. Kocic et al. (2023) analyzed MHD micropolar fluid flow between plates in porous media, finding solutions for flow characteristics and examining magnetic field influence. Heat transmission rate for MHD Maxwell nanofluid flow was analyzed by Hayat et al. (2011) in a porous channel with radiation using HAM to solve the transformed equations. Mishra et al. (2013) examined the MHD stream of a viscoelastic fluid with fluctuating permeability and suction, finding that elasticity and magnetic forces slowed the flow, while a heat source reduced friction. Sharma et al. (2022) investigated MHD nanofluid flow over a stretching surface, finding that magnetic and suction forces slowed the flow, while radiation and heat sources increased temperature and heat transfer. Gupta et al. (2024) analyzed heat transfer near a vertical sheet, examining the effects of magnetic field, radiation, and nanoparticle concentration to optimize various applications. Sharma et al. (2020) examined MHD stream over a vertical surface, considering heat/mass transfer, radiation, chemical reactions, and Joule heating. Numerical methods showed that higher Biot numbers increased temperature and concentration.

The Casson model (Casson, 1959) describes the flow behavior of certain non-Newtonian fluids, like some paints, inks, and blood that exhibit a yield stress. (Khan et al. (2016) analyzed MHD Casson fluid flow in a permeable channel with a squeezing effect, revealing how various parameters influenced fluid velocity and skin friction. For an expanding sheet, Alhadhrami et al. (2021) inspected Casson fluid flow in spongy media, finding that magnetic and porosity effects enhance heat transfer while Soret and Dufour numbers influence mass and heat transfer rates, respectively. Dash et al. (1996) analyzed Casson fluid flow in porous tubes, modeling blocked arteries, and examined how permeability and yield stress affected flow characteristics. Jaliliet al. (2023) analyzed Casson fluid flow in spongy media under a magnetic field and radiation, finding significant impacts of the magnetic and electric fields, and Casson parameters on fluid properties. Khan et al. (2019) analyzed the MHD natural convection of a Casson nanofluid over a moving plate in a spongy medium with heat generation, chemical reactions, and radiation. It examined two magnetic field orientations and presented exact solutions, visualizing parameter effects on flow and heat/mass transfer. Rao et al. (2020) numerically investigated MHD flow and heat/mass transfer over a stretching surface, finding that a higher stretching ratio increased transfer rates, with non-Newtonian fluids performing better.

For Casson fluid, Kodi and Mopuri (2022) examined permeable media under a magnetic field, exploring the variation on flow characteristics due to chemical reaction. Increased magnetic field, inclination, and Casson parameter slowed the flow. Patel (2022) analyzed MHD Casson nanofluid flow in porous media, considering radiation, heat source, Dufour/Soret effects, and chemical reactions. The study found that the Soret number accelerates flow, while chemical reactions and porosity reduce skin friction. MHD Casson nanofluid flow in spongy media with radiation and heat source/sink was studied by Sharma, Mishra et al. (2020). Dufour and Soret effects, along with chemical reactions, significantly influenced the flow and heat/mass transfer characteristics. Sharma et al. (2024) numerically analyzed MHD Casson nanofluid flow in a spongy medium with radiation, heat source, Soret–Dufour effects, and chemical reactions, revealing how these factors influence fluid behavior and heat transfer. Das et al. (2020) numerically analyzed unsteady Casson fluid flow in porous media under a magnetic field, considering radiation, heat source, Dufour–Soret effects, and chemical reactions. In porous media, MHD flow of Casson nanofluid was numerically analyzed by Prakash (2025) over an elongating sheet, considering radiation, heat source, and chemical reactions. Magnetic field slowed the flow, while radiation and heat source increased temperature. Chemical reactions affected mass transfer.

Maxwell fluids are viscoelastic, exhibiting both viscous and elastic behavior. This time-dependent deformation, or stress relaxation, makes them useful in applications like lubricants, sealants, and inks where controlled flow and response are crucial. Their increasing use stems from their ability to function under extreme conditions, offers tunable viscosities, and provides performance advantages over purely viscous or elastic materials. Maxwell nanofluid flow was studied by Umavathi and Mohite (2016). Parameter effects on stability and heat/mass transport, including transient behavior, were investigated using linear/non-linear stability analyses and a modified Darcy–Maxwell model. Sowmiya and Kumar (2023) explored the Maxwell fluid flow with microorganisms in a spongy medium over an extending cylinder, considering activation energy, thermal radiation, and other factors. It analyzed the variation of fluid flow and heat/mass transfer. Muhammad et al. (2017) examined how a Maxwell nanofluid flows through a porous material when influenced by a magnetic field and a stretching surface. It considered heat transfer and nanoparticle movement, finding how these factors affect the thermal and mass transmission of fluid. Ramzan et al. (2016) analyzed the mixed convective flow of a Maxwell nanofluid flow, heat, and mass transfer rate.

Khan et al. (2018) examined how a magnetic field, porous medium, and heat source/sink affected UCMF flow over a shrinking sheet. Increased Deborah number and magnetic field strength reduced velocity but increased temperature. Porosity enhanced both. The heat source increased and the heat sink decreased fluid temperature. Maxwell nanofluid flow with a magnetic field, considering heat source, radiation, and chemical reactions was studied by Shah et al. (2022). Numerical solutions revealed that the fluid flow and heat/mass transfer were affected due to diverse parameters. In a porous medium, Olanrewaju et al. (2021) numerically analyzed unsteady Maxwell nanofluid flow under radiation and heat source. Bayones et al. (2024) analyzed the Maxwell nanofluid flow under a magnetic field, considering Soret effects, chemical reactions, and Joule heating. It examined how parameters like magnetic field strength and chemical reaction rates affected fluid flow and heat/mass transfer. Nuwairan et al. (2022) numerically analyzed the MHD Darcy flow of a Maxwell nanofluid in porous media, considering thermophoresis, nonlinear radiation, and heat source/sink. The results showed Deborah number's impact on flow and Soret/Dufour effects on mass transfer. Samuel (2022) investigated radiative Maxwell fluid flow with chemical reaction and Soret effect over a stretching sheet, considering temperature-dependent viscosity. Higher Frank–Kamenetskii and permeability parameters enhanced temperature, while the Soret number increased concentration.

Our research fills a gap in existing knowledge by studying Casson–Maxwell nanofluid flows over a stretching surface under a magnetic field in porous media. No one has looked at this combined situation before. We specifically examine how factors like thermophoresis, Brownian motion, the Dufour and Soret effects, and different slip conditions affect the fluid's speed, temperature, concentration, friction, and heat/mass transfer. This work helps us better understand these complex fluid interactions and predict how they behave in various scenarios.

4.2 DEVELOPMENT OF THE MODEL

This study investigates how a magnetic field and crosswise diffusion affect the flow of Casson–Maxwell fluid near a stretching sheet, considering the added influences of thermophoresis and Brownian motion. Our main goal is to understand the behavior of this fluid under these conditions, including the effects of slip parameters. To simplify the problem some assumptions are: 2D, electrically conductive, viscous, and incompressible flow. Based on these assumptions, we derive govern equations as per Babu et al. (2024) (see Figure 4.1 for a visual representation).

$$u_x + v_y = 0, \tag{4.1}$$

$$uu_x + vu_y = \upsilon\left(1+\frac{1}{\beta}\right)u_{yy} - \frac{\sigma B_0^2}{\rho}u - \frac{\upsilon}{K}u - k_0 \left(u^2 u_{xx} + v^2 u_{yy} + 2uvu_{xy}\right), \tag{4.2}$$

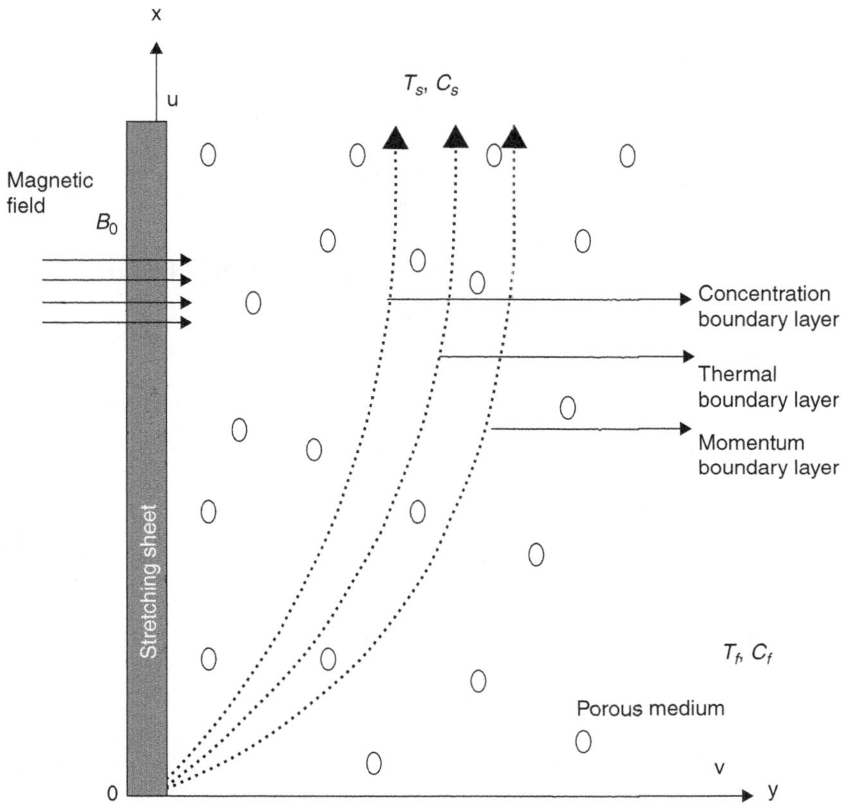

Figure 4.1 Physical model.

$$uT_x + vT_y = \alpha T_{yy} + \tau\left(D_B C_y T_y + \frac{D_T}{T_f} T_y^{\,2}\right) + D_{TC} C_{yy} + \frac{Q_0}{\rho C_p}\left(T - T_f\right), \tag{4.3}$$

$$uC_x + vC_y = D_B C_{yy} + \frac{D_T}{T_f} T_{yy} + D_{CT} T_{yy} - k_r\left(C - C_f\right), \tag{4.4}$$

Boundary conditions:

$$v = 0, L_1 \frac{\partial u}{\partial y} = u - ax, = L_2 \frac{\partial T}{\partial y} = T - T_s, L_3 \frac{\partial C}{\partial y} = C - C_s, \text{ at } y = 0, \tag{4.5}$$

$$u \to 0, T - T_f \to 0, C - C_f \to 0, \ as\, y \to \infty \tag{4.6}$$

$$\begin{aligned} u &= axf'(\eta), v = -\sqrt{a\upsilon} f(\eta), \ \eta = y\sqrt{\frac{a}{\upsilon}}, \frac{T}{T_f} - 1 \\ &= \left(\frac{T_s}{T_f} - 1\right)\theta(\eta), \frac{C}{C_f} - 1 = \left(\frac{C_s}{C_f} - 1\right)\phi(\eta), \end{aligned} \tag{4.7}$$

Using Eq. (4.7), the reduced form of Eqs. (4.2), (4.3) and (4.4) are

$$f'''\left(1 + \frac{1}{\beta}\right) + f''f - f'^2 - f'\left(M + K_p\right) - \gamma(f'''f^2 - 2f''ff) = 0, \tag{4.8}$$

$$\theta'' + P_r\left(f\theta' + N_t\theta'^2 + N_b\theta'\phi' + Q\theta\right) + D_u\phi'' = 0, \tag{4.9}$$

$$\phi'' + S_c f\phi' + S_r S_c \theta'' + \frac{N_t}{N_b}\theta'' - K_c S_c \phi = 0, \tag{4.10}$$

and reduced form of Eqs. (4.5) and (4.6) are

$$\begin{aligned} f(0) &= 0, f'(0) - 1 = \lambda f''(0), \theta(0) - 1 \\ &= \delta\theta'(0), \phi(0) - 1 = \chi\phi'(0), \end{aligned} \tag{4.11}$$

$$f'(l) \to 0, \theta(l) \to 0, \phi(l) \to 0, as\ l \to \infty, \tag{4.12}$$

Here,

$$M = \frac{\sigma B_0^2}{a\rho}, K_p = \frac{\upsilon}{aK}, \gamma = ak_0, P_r = \frac{\upsilon}{\alpha}, N_b = \frac{\tau D_B\left(C_s - C_f\right)}{\upsilon},$$

$$N_t = \frac{\tau D_T\left(T_s - T_f\right)}{T_f\upsilon}, D_u = \frac{D_{TC}\left(C_s - C_f\right)}{\alpha\left(T_s - T_f\right)},$$

$$Q = \frac{Q_0}{a\rho C_p}, S_c = \frac{\upsilon}{D_B}, \ S_r = \frac{D_{CT}\left(T_s - T_f\right)}{\left(C_s - C_f\right)\upsilon}, \ S_c = \frac{\upsilon}{D_B},$$

$$K_c = \frac{k_r}{a}, \ \lambda = L_1\sqrt{\frac{a}{\upsilon}}, \ \delta = L_2\sqrt{\frac{a}{\upsilon}}, \ \chi = L_3\sqrt{\frac{a}{\upsilon}},$$

4.3 METHODOLOGY

The fifth-order method through MATLAB solver bvp5c employs to solve equations (4.8)–(4.10) numerically.

$$f = \xi_1, f' = \xi_2, f'' = \xi_3, \theta = \xi_4, \theta' = \xi_5, \phi = \xi_6, \phi' = \xi_7, \tag{4.13}$$

Using Eq. (4.13), Eqs. (4.8), (4.9), and (4.10) become

$$f''' = \xi_3' = \frac{\xi_2^2 - \xi_1\xi_3 + M\xi_2 + K_p\xi_2 - 2\gamma\xi_1\xi_2\xi_3}{\left(1 + \frac{1}{\beta}\right) - \gamma y_1^2}, \tag{4.14}$$

$$\theta'' = \xi_5' = \frac{-P_r\xi_1\xi_5 - P_rN_b\xi_5\xi_7 - P_rN_t\xi_5^2 - P_rQ\xi_4 + D_uS_c\xi_1\xi_7 - D_uK_cS_c\xi_6}{1 - D_u\left(S_rS_c + \frac{N_t}{N_b}\right)}, \tag{4.15}$$

$$\phi' = \xi_7' = -\left(S_rS_c + \frac{N_t}{N_b}\right)\left(\frac{-P_r\xi_1\xi_5 - P_rN_b\xi_5\xi_7 - P_rN_t\xi_5^2 - P_rQ\xi_4 + D_uS_c\xi_1\xi_7 - D_uK_cS_c\xi_6}{1 - D_u\left(S_rS_c + \frac{N_t}{N_b}\right)}\right) - S_c\xi_1\xi_7 + K_cS_c\xi_6, \tag{4.16}$$

And Eqs. (4.11) and (4.12) are,

$$\xi_1(0) = 0,\ \xi_2(0) = 1 + \lambda\xi_3(0), \xi_4(0) = 1 + \delta\xi_5(0), \xi_6(0) = 1 + \chi\xi_7(0), \tag{4.17}$$

$$\xi_2(\infty) \to 0,\ \xi_4(\infty) \to 0,\ \xi_6(\infty) \to 0, \tag{4.18}$$

4.4 ENGINEERING QUANTITY

Local drag coefficient $(cf_x) = (1+\gamma)f''(0)$, Local Nusselt number $(Nu_x) = -\left(\sqrt{\mathrm{Re}_x}\right)\theta'(0)$,

Local Sherwood number $(Sh_x) = -\left(\sqrt{\mathrm{Re}_x}\right)\phi'(0)$, and local Reynolds number $(\mathrm{Re}_x) = \frac{ax^2}{\upsilon}$,

4.5 RESULTS AND DISCUSSION

The outcomes of the numerical simulations are delineated through graphical representations. These visualizations provide a detailed insight into the complex interactions among the studied parameters. To conduct a thorough parametric analysis, a systematic computational investigation was performed. This involved the controlled variation of key engineering parameters across a defined range, enabling a comprehensive assessment of their respective influences on the fluid's behavior and take the values: $Q=-1, \gamma=S_c=\lambda=\delta=\chi=0.1, M=N_t=D_u=S_r=0.2, N_b=1,$ $K_p=\beta=\mathrm{P}_r=K_c=0.5.$

The fluid velocity profile is clearly illustrated in Figures 4.2–4.5. Firstly, as depicted in Figure 4.2, the velocity profile declines with magnetic field strength because Lorentz force resists the fluid's motion. The velocity diminishes with the Maxwell fluid parameter as per Figure 4.3 because the

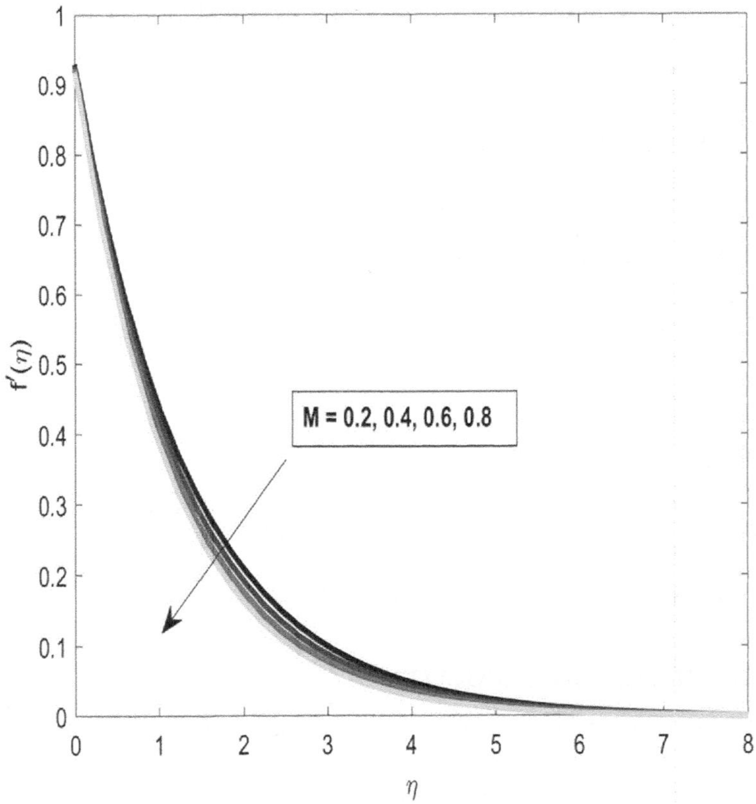

Figure 4.2 $f'(\eta)$ vs M.

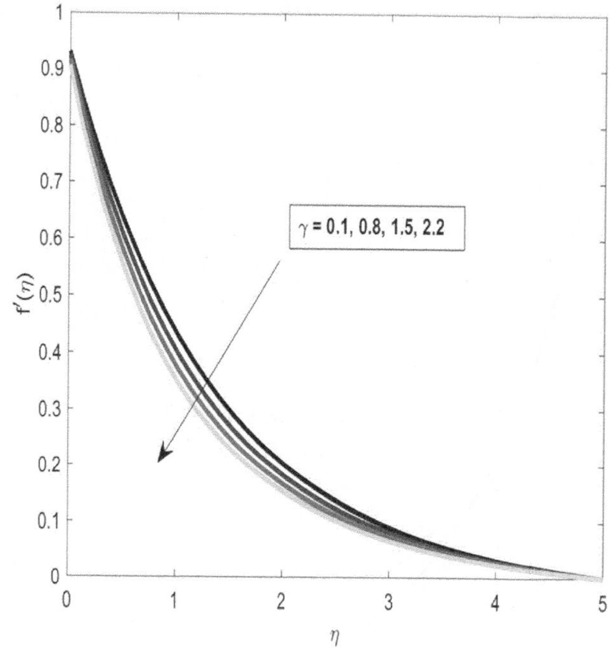

Figure 4.3 $f'(\eta)$ vs γ.

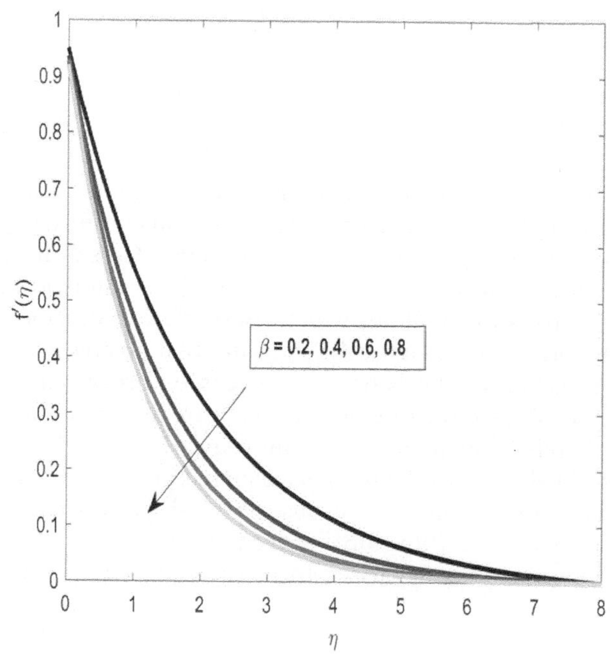

Figure 4.4 $f'(\eta)$ vs β.

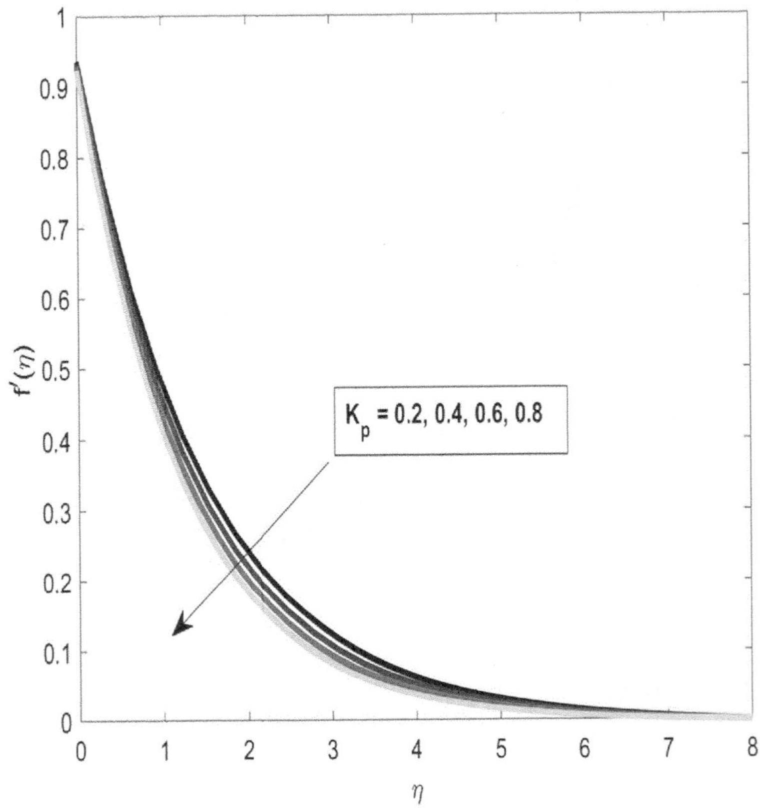

Figure 4.5 $f'(\eta)$ vs K_p.

higher Maxwell parameter corresponds to an extended relaxation time, which in turn hinders the efficient transfer of momentum within the fluid, and velocity reduces. Figure 4.4 reveals that higher Casson fluid parameter reduces the velocity profile because an increased Casson parameter implies a higher yield stress of the fluid, which resists flow and velocity decreases within the boundary layer. Figure 4.5 indicates that greater porosity leads to a reduction in velocity. This is because increased porosity introduces more resistance to the fluid flow, forcing it to navigate a more complex and tortuous path, thereby slowing its overall movement.

The thermal behavior of the fluid, as depicted in Figures 4.6–4.9, reveals distinct trends influenced by various physical parameters. Figure 4.6 illustrates an inverse relationship between the Prandtl number and temperature. This phenomenon arises from the reduced thermal diffusivity

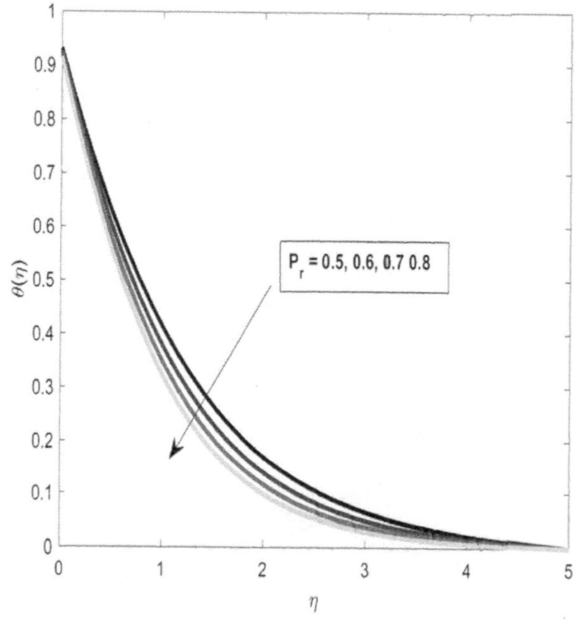

Figure 4.6 $\theta(\eta)$ vs P_r.

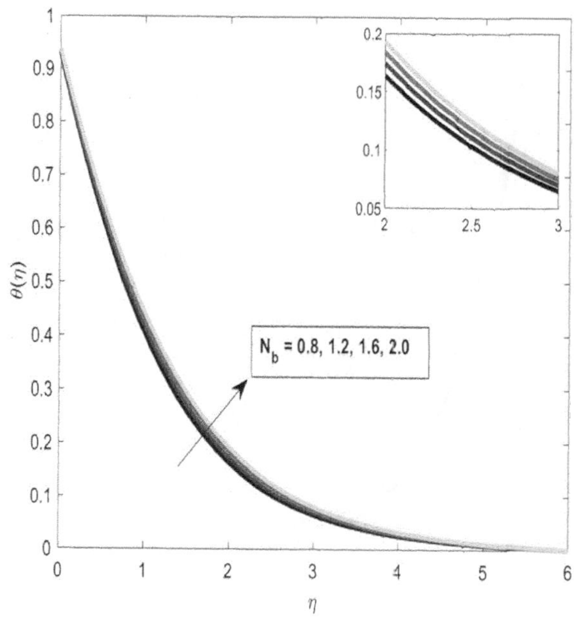

Figure 4.7 $\theta(\eta)$ vs N_b.

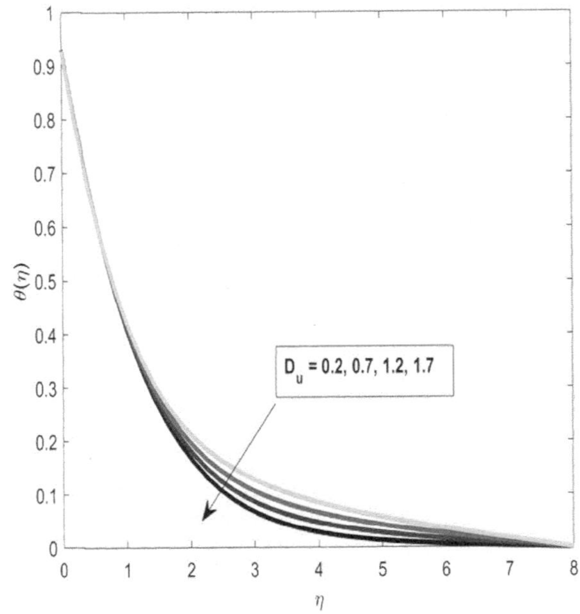

Figure 4.8 $\theta(\eta)$ vs D_u.

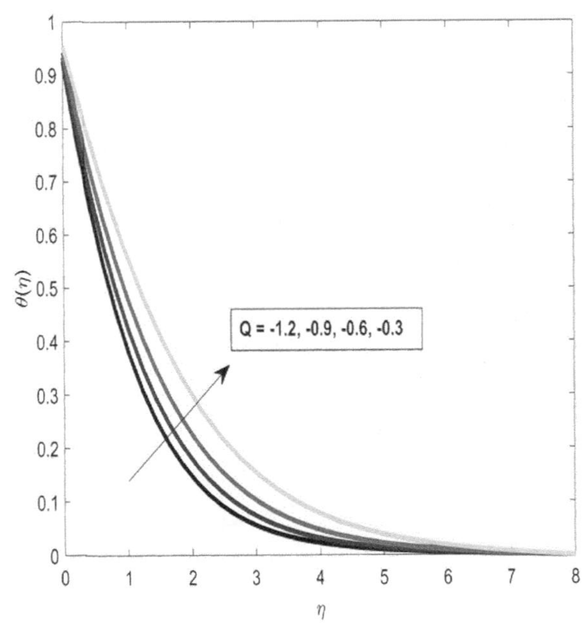

Figure 4.9 $\theta(\eta)$ vs Q.

associated with higher Prandtl numbers, which impedes heat transfer and results in steeper temperature gradients. The temperature is enhanced with the Brownian motion number shown in Figure 4.7. Elevated Brownian motion intensifies particle kinetic energy, leading to more frequent collisions and a subsequent temperature rise. In contrast, Figure 4.8 highlights a decrease in temperature with increasing Dufour number. This reduction is attributed to the enhanced energy diffusion driven by concentration gradients, which effectively extract thermal energy from the system. Figure 4.9 shows that the temperature profile varies with the heat source parameter. An increase in this parameter introduces additional energy into the fluid and its temperature increases.

The concentration profiles, depicted in Figures 4.10–4.14, reveal distinct trends. Figure 4.10 shows that concentration diminished with Brownian motion number. This phenomenon arises because heightened Brownian motion facilitates particle diffusion, promoting a more homogeneous distribution and consequently reducing localized concentrations. Figure 4.11 demonstrates that the concentration profile is amplified with an increase in the thermophoresis number. This enhancement is attributed to the thermophoretic force, which propels nanoparticles toward cooler regions, thereby accumulating them within the boundary layer due to prevailing

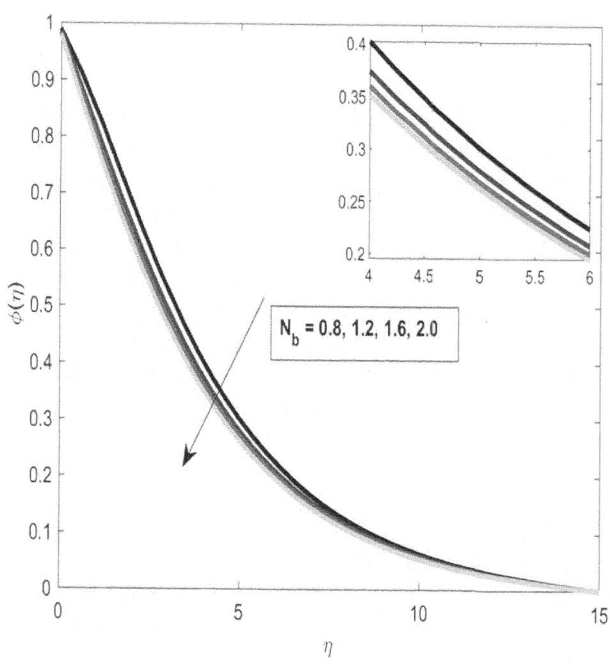

Figure 4.10 $\phi(\eta)$ vs N_b.

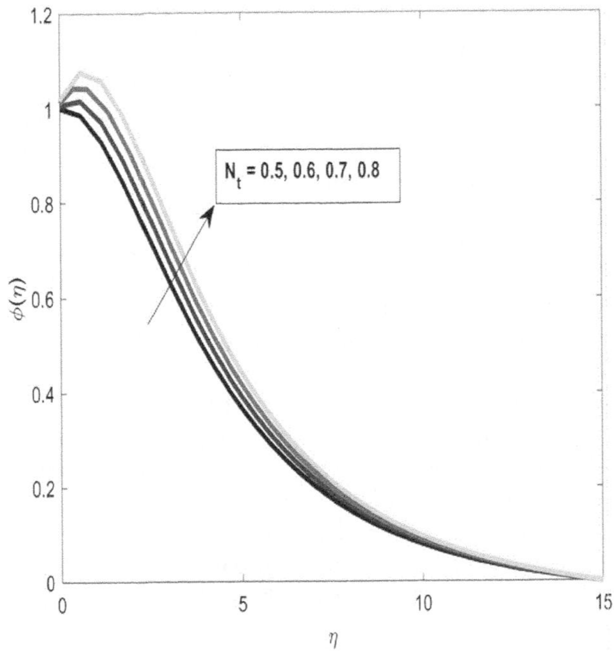

Figure 4.11 $\phi(\eta)$ vs N_t.

temperature gradients. Figure 4.12 indicates the inverse relation between concentration and chemical reaction. This is due to the enhanced consumption of the reacting species, leading to a reduction in the overall substance concentration within the system. Figure 4.13 highlights a direct correlation between the concentration profile and the Soret number. As the Soret number increases, thermal diffusion intensifies, driving more solute particles and subsequently elevating the fluid's concentration. Figure 4.14 illustrates that due to an increase in the Schmidt number, concentration decreases. This inverse relationship stems from the fact that higher Schmidt numbers correspond to lower mass diffusivity, resulting in slower concentration dispersion and thus diminished concentration profiles.

4.6 CONCLUSION

This study uniquely addressed the complex interplay of Casson–Maxwell nanofluid flow over a stretching surface under a magnetic field, Brownian motion, Dufour and Soret effects, and slip conditions; we established a comprehensive understanding of the fluid's flow characteristics. The findings are given below:

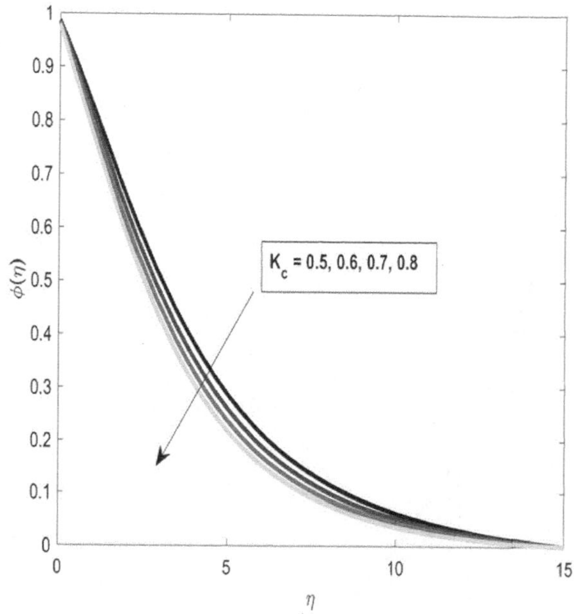

Figure 4.12 $\phi(\eta)$ vs $K_c\,K_c$.

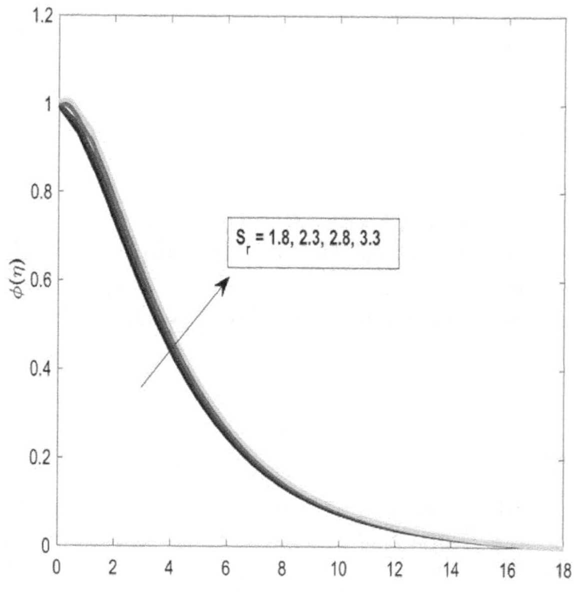

Figure 4.13 $\phi(\eta)$ vs S_r.

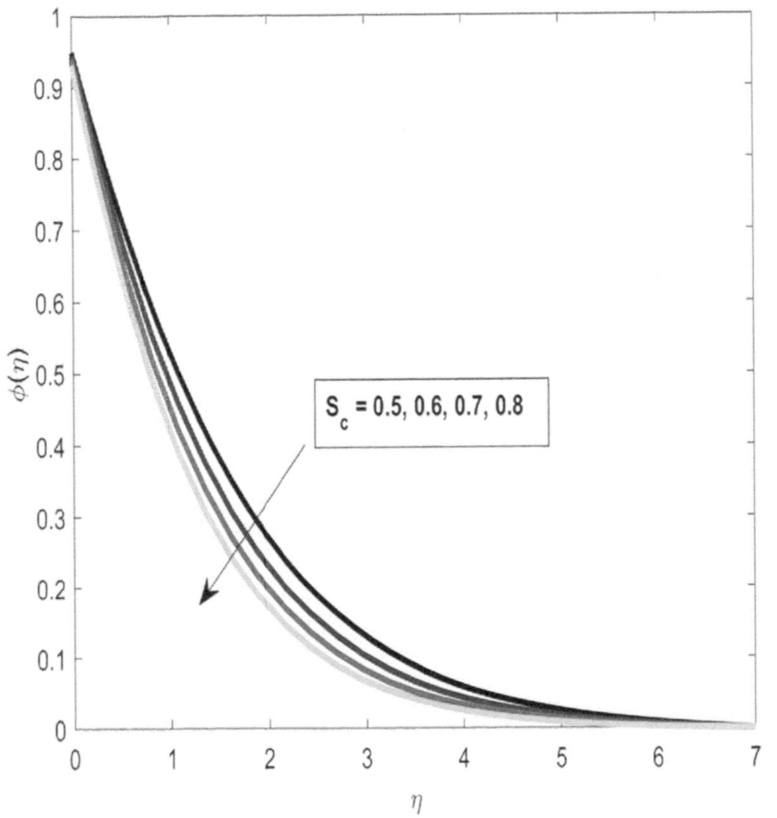

Figure 4.14 $\phi(\eta)$ vs S_c.

- The velocity decreased with M, γ, β and K_p.
- The temperature profile diminished with P_r, D_u, while it amplified with N_b, Q.
- The concentration profile declined with N_b, K_c, S_c, but improved with N_t, S_r.

REFERENCES

Alhadhrami, A., Vishalakshi, C.S., Prasanna, B.M., Sreenivasa, B.R., Alzahrani, H.A., Gowda, R.P. and Kumar, R.N., 2021. Numerical simulation of local thermal non-equilibrium effects on the flow and heat transfer of non-Newtonian Casson fluid in a porous media. *Case Studies in Thermal Engineering*, 28, 101483.

Al Nuwairan, M., Hafeez, A., Khalid, A., Souayeh, B., Alfadhli, N. and Alnaghmosh, A., 2022. Flow of Maxwell fluid with heat transfer through porous medium with thermophoresis particle deposition and Soret–Dufour effects: Numerical solution. *Coatings*, 12(10), 1567.

Babu, B.H., Swarnalathamma, B.V. and Krishna, M.V., 2024. Diffusion-thermo and thermo-diffusion impacts on MHD natural convection flow of Maxwell nanofluids over a stretching sheet. *Case Studies in Thermal Engineering*, 61, 105016.

Bayones, F.S., Abd-Alla, A.M. and Thabet, E.N., 2024. Magnetized dissipative Soret effect on nonlinear radiative Maxwell nanofluid flow with porosity, chemical reaction and Joule heating. *Waves in Random and Complex Media*, 34(6), 6270–6288.

Casson, N., 1959. Flow equation for pigment-oil suspensions of the printing ink-type. *Rheology of Disperse Systems*, 84–104.

Das, M., Mahanta, G. and Shaw, S., 2020. Heat and mass transfer effect on an unsteady MHD radiative chemically reactive Casson fluid over a stretching sheet in porous medium. *Heat Transfer*, 49(8), 4350–4369.

Dash, R.K., Mehta, K.N. and Jayaraman, G., 1996. Casson fluid flow in a pipe filled with a homogeneous porous medium. *International Journal of Engineering Science*, 34(10), 1145–1156.

Geindreau, C. and Auriault, J.L., 2002. Magnetohydrodynamic flows in porous media. *Journal of Fluid Mechanics*, 466, 343–363.

Gregory, T.S., Cheng, R., Tang, G., Mao, L. and Tse, Z.T.H., 2016. The magnetohydrodynamic effect and its associated material designs for biomedical applications: a state-of-the-art review. *Advanced Functional Materials*, 26(22), 3942–3952.

Gupta, P.K., Prakash, O. and Tinker, S., 2024. Unsteady Free Convection of Nanofluid Past an Infinite Vertical Sheet Under the Effects of Inclined Magnetic Field and Cogley Radiation. Recent Patents on Engineering.

Hayat, T., Sajjad, R., Abbas, Z., Sajid, M. and Hendi, A.A., 2011. Radiation effects on MHD flow of Maxwell fluid in a channel with porous medium. *International Journal of Heat and Mass Transfer*, 54(4), 854–862.

Jalili, P., Azar, A.A., Jalili, B. and Ganji, D.D., 2023. Study of nonlinear radiative heat transfer with magnetic field for non-Newtonian Casson fluid flow in a porous medium. *Results in Physics*, 48, 106371.

Khan, D., Khan, A., Khan, I., Ali, F., Karim, F.U. and Tlili, I., 2019. Effects of relative magnetic field, chemical reaction, heat generation and Newtonian heating on convection flow of Casson fluid over a moving vertical plate embedded in a porous medium. *Scientific Reports*, 9(1), 400.

Khan, H., Qayyum, M., Khan, O. and Ali, M., 2016. Unsteady squeezing flow of Casson fluid with magnetohydrodynamic effect and passing through porous medium. *Mathematical Problems in Engineering*, 2016(1), 4293721.

Khan, Z., Rasheed, H.U., Alkanhal, T.A., Ullah, M., Khan, I. and Tlili, I., 2018. Effect of magnetic field and heat source on upper-convected-maxwell fluid in a porous channel. *Open Physics*, 16(1), 917–928.

Kocić, M., Stamenković, Ž.,Petrović, J. and Bogdanović-Jovanović, J., 2023. MHD micropolar fluid flow in porous media. *Advances in Mechanical Engineering*, 15(6), 16878132231178436.

Kodi, R. and Mopuri, O., 2022. Unsteady MHD oscillatory Casson fluid flow past an inclined vertical porous plate in the presence of chemical reaction with heat absorption and Soret effects. *Heat Transfer*, 51(1), 733–752.

Mishra, S.R., Dash, G.C. and Acharya, M., 2013. Mass and heat transfer effect on MHD flow of a visco-elastic fluid through porous medium with oscillatory suction and heat source. *International Journal of Heat and Mass Transfer*, 57(2), 433–438.

Muhammad, T., Alsaedi, A., Shehzad, S.A. and Hayat, T., 2017.A revised model for Darcy–Forchheimer flow of Maxwell nanofluid subject to convective boundary condition. *Chinese Journal of Physics*, 55(3), 963–976.

Olanrewaju, A.M., Salawu, S.O., Olanrewaju, P.O. and Amoo, S.A., 2021. Unsteady radiative magnetohydromagnetic flow and entropy generation of maxwell nanofluid in a porous medium with Arrhenius chemical kinetic. *Cogent Engineering*, 8(1), 1942639.

Patel, H.R., 2022. Soret and heat generation effects on unsteady MHD Casson fluid flow in porous medium. *Waves in Random and Complex Media*, 1–24.

Prakash, O., 2025. Dissipative stagnation point flow of the Casson nanofluid over a stretching sheet under the influence of porous media, binary chemical reaction and variable surface heat flux condition. *Pramana*, 99(1), 31.

Ramzan, M., Bilal, M., Chung, J.D. and Farooq, U., 2016.Mixed convective flow of Maxwell nanofluid past a porous vertical stretched surface—An optimal solution. *Results in Physics*, 6, 1072–1079.

Rao, P.S., Prakash, O., Mishra, S.R. and Sharma, R.P., 2020.Similarity solution of three-dimensional MHD radiative Casson nanofluid motion over a stretching surface with chemical and diffusion-thermo effects. *Heat Transfer*, 49(4), 1842–1862.

Samuel, D.J., 2022. Numerical investigations of thermal radiation and activation energy imparts on chemically reactive Maxwell fluid flow over an exothermal stretching sheet in a porous medium. *International Journal of Applied and Computational Mathematics*, 8(3), 148.

Shah, S., Rafiq, N., Abdullah, F.A., Atif, S.M. and Abbas, M., 2022.Slip and radiative effects on MHD Maxwell nanofluid with non-Fourier and non-Fick laws in a porous medium. *Case Studies in Thermal Engineering*, 30, 101779.

Sharma, K., Kumar, L., Singh, A. and Kumar Joshi, V., 2024. Dynamics of MHD Casson fluid in non-Darcy porous medium: the impact of thermal radiation, Dufour–Soret, and chemical reaction. *Modern Physics Letters B*, 38(16), 2341007.

Sharma, R.P., Mishra, S.R. and Tinker, S., 2020. Influence of Soret and Dufour effect on MHD flow over an exponential stretching sheet: A numerical study. *Indian Journal of Pure & Applied Physics*, 58(7), 558–568.

Sharma, R.P., Prakash, O., Rashidi, I., Mishra, S.R., Rao, P.S. and Karimi, F., 2022. Nonlinear thermal radiation and heat source effects on unsteady electrical MHD motion of nanofluid past a stretching surface with binary chemical reaction. *The European Physical Journal Plus*, 137(3), 297.

Sharma, R.P., Tinker, S., Gireesha, B.J. and Nagaraja, B., 2020. Effect of convective heat and mass conditions in magnetohydrodynamic boundary layer flow with joule heating and thermal radiation, *International Journal of Applied Mechanics and Engineering*, 25(3), 103–116.

Sowmiya, C. and Kumar, B.R., 2023. MHD Maxwell nanofluid flow over a stretching cylinder in porous media with microorganisms and activation energy. *Journal of Magnetism and Magnetic Materials*, 582, 171032.

Umavathi, J.C. and Mohite, M.B., 2016. Convective transport in a porous medium layer saturated with a Maxwell nanofluid. *Journal of King Saud University-Engineering Sciences*, 28(1), 56–68.

Chapter 5

Hybrid nanofluid heat transfer

Williamson–Maxwell interactions

Seema Tinker, Om Prakash, and Piyush Kumar Gupta

Nomenclature

Symbol	Description	Symbol	Description
k	Thermal conductivity (W m^{-1} K^{-1}	We	Williamson term
k_p	Permeability of medium (H m^{-1})	F	Radiation term
Γ	Williamson term	M	Magnetic parameter
τ	Heat generation rate (W m^{-2})	Kp	Porosity parameter
σ	Electrical Conductivity (S m^{-1})	Ec	Eckert number
β^*	Thermal extension coefficient (K^{-1})	Δ	Mixed convection number
B_0	Magnetic field (Wb m^{-2})	B	Deborah term
D_T	Thermal diffusivity (m^2 s^{-1})	Nt	Thermophoresis number
λ	Relaxation time (s)	Q	Heat source
ρ	Density (Kg m^{-3})	Pr	Prandtl number
C_p	Specific heat(J Kg^{-1} K^{-1})		

5.1 INTRODUCTION

Magnetohydrodynamic (MHD) flow within nanofluids has provided a crucial domain of inquiry, driven by its diverse applications across engineering. These applications span from the augmentation of heat transfer and the manipulation of microfluidic systems to advancements in energy generation and biomedical technologies. Consequently, this dynamic research area has attracted substantial academic attention, leading to a wealth of investigations that seek to elucidate the underlying mechanisms governing MHD nanofluid flow and its potential for practical implementation. For a hybrid nanofluid, Alnahdi and Gul (2024) studied MHD flow over an extended surface with convective boundary conditions. Ragulkumar et al. (2023) examined dissipative, free convective, MHD flow over a vertical cone, considering radiation, chemical reaction, and mass flux. Sheikholeslami et al. (2021) investigated MHD-free convection and radiation in a CuO–water nanofluid-filled porous cavity, which were simulated using CVFEM, validating heat transfer results across varying parameters.

 DOI: 10.1201/9781003642169-5

Haddad et al. (2012) investigated CuO–water nanofluid natural convection (Rayleigh-Bénard problem) using a non-homogeneous model, finding enhanced heat transfer with Brownian motion and thermophoresis, relevant to MHD applications. Dey and Borah (2023) numerically investigated heat transfer on MHD flow over an escalating sheet with a second-grade fluid model, considering absorption, chemical reaction, Brownian motion, and thermophoresis. Alipour et al. (2023) numerically investigated and optimized MHD heat-fluid transfer in a ternary nanofluid-filled circular cavity, revealing nanoparticle shape's significant impact on Nusselt number. Ullah et al. (2021) explored nanofluids for MHD heat transfer enhancement, particularly with a stretching plate, considering Joule heating and viscous dissipation. The investigation of the flow of Williamson fluid, particularly considering the combined effects of mixed convection, thermal radiation, and magnetic fields, has emerged as a significant area of research.

Williamson (1948) established the fundamental characteristics of this fluid model. Mixed convection, encompassing both natural and forced convection, plays a crucial role in many practical scenarios, such as heat exchangers and solar collectors. The interplay between these factors—Williamson fluid behavior, nanofluid properties, mixed convection mechanisms, magnetic field influence, and radiative heat transfer—makes this a rich area of investigation with numerous potential applications. Unsteady free convective MHD flow of Williamson fluid over an extending surface in porous media was explored by Shateyi and Muzara (2020), revealing flow characteristics vary with various parameters. Shawky et al. (2019) examined 2D hydromagnetic Williamson nanofluid flow over a stretching sheet in porous media, considering heat/mass transfer, radiation, and other effects, solved numerically. For lid-driven cavity, Janjanam et al. (2021) numerically investigated heat transfer with mixed convection flow of Aluina–water nanofluid, analyzing the impact of Reynolds and Grashof numbers.

Faizan et al. (2024) analyzed the flow of Williamson fluid with radiation, chemical reactions, and microorganisms under magnetic mixed convection, finding enhanced heat transfer with radiation, thermophoresis, and Brownian motion. Kothandapani and Prakash (2015) theoretically investigated Williamson nanofluid peristaltic transport in a tapered asymmetric channel with thermal radiation. Analytical solutions and numerical computations explored the influence of various parameters, including the Weissenberg number and Grashof number, on flow characteristics and trapping. Hamid and Khan (2018) investigated time-dependent Williamson nanofluid flow over a radially stretching surface, considering a variable magnetic field, mixed convection, and zero nanoparticle mass flux, analyzing heat transfer and fluid flow. Faisal et al. (2021) investigated MHD Williamson nanofluid flow over an unsteady stretching surface in porous media, analyzing heat and mass transfer under varying thermal conditions and parameters, and finding enhanced transfer with certain parameter increases.

Sharma et al. (2022) analyzed MHD Williamson nanofluid flow past a stretching sheet, incorporating heat transfer, chemical reaction, Brownian motion, thermophoresis, radiation, and magnetic field effects, finding increased temperature with cross-diffusion and dissipation. MHD flow of Williamson nanofluidin porous media was investigated by Dawar et al. (2021), considering heat/mass transfer with prescribed temperature/flux, revealing the influence of unsteadiness, expansion ratio, and Williamson parameter. Jamshed et al. (2021) investigated Ag-Cu/EO flow over a slippery, porous widening surface, considering nanoparticle shape and radiation, finding enhanced heat transfer compared to conventional nanofluids. Sharma et al. (2023) analyzed the MHD Williamson fluid stream over a widening sheet, considering nonlinear thermal radiation, Joule heating, and chemical reactions, finding that radiation enhanced heat and mass transfer.

The investigation of Maxwell nanofluid heat and mass transfer, considering the coupled effects of mixed convection, magnetic fields, and thermal radiation, constitutes a significant area of research due to its relevance in diverse engineering applications. Maxwell fluid, a viscoelastic fluid model, captures the time-dependent behavior of certain non-Newtonian fluids, making it suitable for describing the flow of polymer melts and other complex fluids (Maxwell, 2023). In porous media, Khan et al. (2024) investigated MHD flow of Maxwell hybrid nanofluid, analyzing heat transfer with varying parameters and demonstrating enhanced Nusselt numbers compared to conventional nanofluids. Ahmed et al. (2019) inspected MHD mixed convection heat transference of Maxwell nanofluids over a stretching sheet in a porous medium, considering radiation, magnetic field, and heat source/sink, finding decreased velocity with Darcy, Deborah numbers, and magnetic field, and increased Nusselt number with radiation. Gupta et al. (2024) analyzed the unsteady natural convection of Cu–H_2O nanofluid over a vertical sheet, examining magnetic inclination, Cogley radiation, and heat source effects on flow properties. For Maxwell nanofluid flow past a widening sheet, heat transfer rates were investigated by Ramesh et al. (2015) with chemical reaction, radiation, and magnetic field. The combined effects of electric and magnetic fields on heat and mass transfer for the unsteady flow of a Maxwell fluid were examined by Khan et al. (2018).

For a partially open wavy porous cavity, Prakash et al. (2023) investigated MHD-free convection flow using FDM. Prasannakumara (2021) studied Maxwell nanofluid flow on an elongating sheet with a magnetic dipole using the Cattaneo–Christov model and found that heat transfer increased with thermal relaxation and volume fraction. Biswas et al. (2022) examined MHD Maxwell nanofluid flow over a stretched sheet, considering Brownian motion, thermal radiation, and chemical reactions, revealing increased fluid velocity and thermal boundary layer thickness with radiation. To analyze heat and mass transfer, Khan, Khan et al. (2024) investigated Maxell–Williamson–Casson nanofluid flow, using the BVP4C method and validating results. Sharma et al. (2024) analyzed heat transfer in the MHD stagnation point flow of Al_2O_3-Cu/H_2O above a stretching/shrinking sheet, considering slips, non-uniform heat source/sink, and radiation.

Our research, along with a review of existing studies, investigates the combined effects of Williamson and Maxwell nanofluids, which has not been studied by anyone so far. This is a significant oversight, as understanding how these two distinct non-Newtonian fluid behaviors interact can lead to substantial improvements in various industrial processes. This research examines the combined heat transfer effects of Williamson and Maxwell nanofluids, extending beyond individual fluid behaviors. This integrated approach, incorporating Cogley radiation, an angled magnetic field, and internal heat generation, offers a more realistic model for complex fluid systems. The findings provide critical insights for enhancing thermal management in applications like electronic cooling, chemical processing, and pipeline transport, potentially leading to significant industrial advancements.

5.2 PROBLEM FORMULATION

This study explores how magnetic fields and temperature affect heat transport in the combined stream of Williamson–Maxwell nanofluids, flowing over an extending sheet embedded in a porous material. Imagine a sheet with tiny holes being stretched; this is our porous sheet. We're looking at how the flow of these special fluids changes as the sheet stretches, influenced by both heat and a magnetic field B_0 applied at an angle α. u, v are the velocities along perpendicular axes. Temperature T_s, T_f areas are as per Figure 5.1. The governing PDEs for the proposed 2D laminar steady flow are given by Khan et al. (2024).

$$u_x + v_y = 0, \tag{5.1}$$

$$\begin{aligned} uu_x + vu_y &= \upsilon u_{yy} + \sqrt{2}\,\upsilon\Gamma u_y u_{yy} - \frac{u\sigma B_0^2 \sin^2\alpha}{\rho} - \frac{\upsilon u}{k_p} \\ &\quad + g\beta^*\left(T - T_f\right) - \lambda\left(u^2 u_{xx} + v^2 u_{yy} + 2uvu_{xy}\right), \end{aligned} \tag{5.2}$$

$$\begin{aligned} uT_x + vT_y &= \frac{k}{\rho C_p} T_{yy} - \frac{1}{\rho C_p}\left(q_r\right)_y + \tau\frac{D_T}{T_\infty} T_y^{\,2} \\ &\quad + \frac{Q_0(T - T_f)}{\rho C_p} + \frac{\mu}{\rho C_p}\left(u_y^2 + \frac{\Gamma}{\sqrt{2}} u_y^3\right), \end{aligned} \tag{5.3}$$

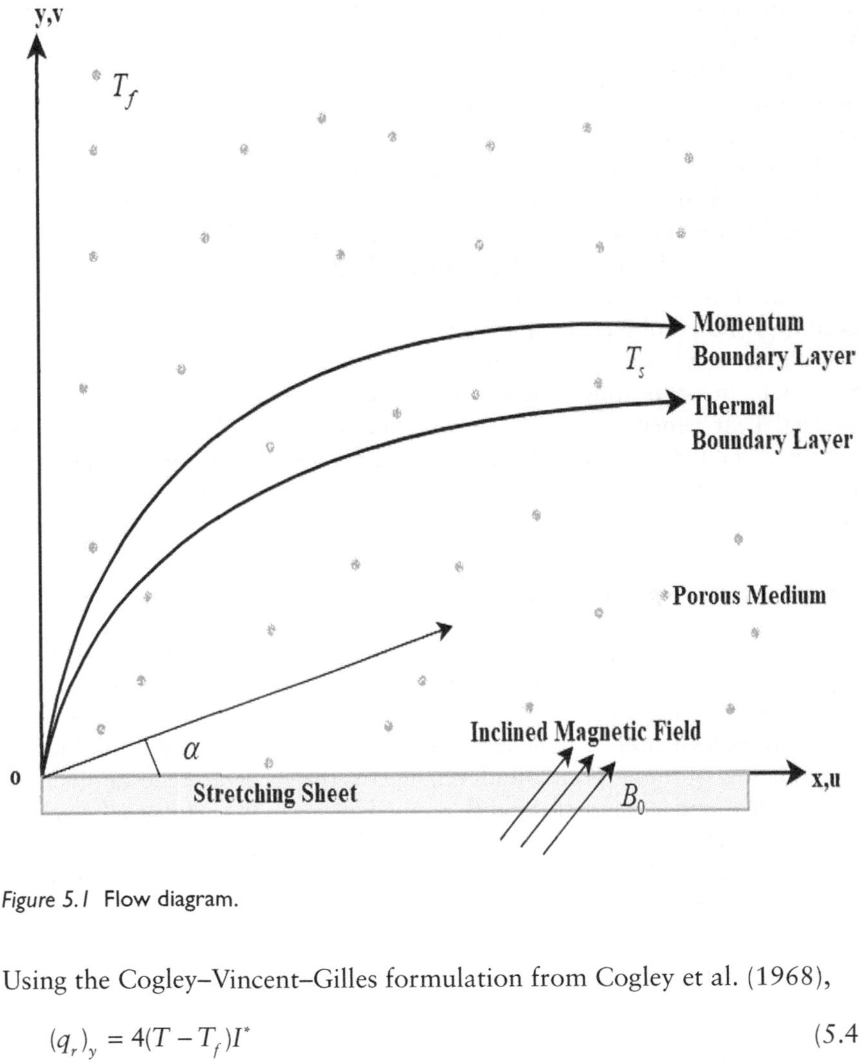

Figure 5.1 Flow diagram.

Using the Cogley–Vincent–Gilles formulation from Cogley et al. (1968),

$$(q_r)_y = 4(T - T_f)I^* \tag{5.4}$$

Boundary conditions:

$$at\, y = 0 \begin{cases} u = ax + \lambda_1 u_y \left(1 + \dfrac{\Gamma}{\sqrt{2}} u_y\right), \\ v = 0, T = T_s, \end{cases} \text{ and as } y \to \infty \begin{cases} u \to 0 \\ T \to T_f, \end{cases} \tag{5.5}$$

Similarity transformations:

$$u = axf'(\eta), v = -\sqrt{a\upsilon}f(\eta), \frac{T}{T_f} = 1 + \left(\frac{T_s}{T_f} - 1\right)\theta(\eta), \eta = \sqrt{\frac{a}{\upsilon}}y, \tag{5.6}$$

Using Eq. (5.6), Eqs. (5.2) and (5.3) can be written as

$$\begin{aligned} &f''' - f'^2 + Wef'''f'' + ff'' - B(f^2f''' - 2ff'f'') \\ &\quad - (M + Kp)f' + \Delta\theta = 0, \end{aligned} \tag{5.7}$$

$$\frac{1}{\Pr}\theta'' + f\theta' + Nt\theta'^2 + (Q - F)\theta + Ec\left(f''^2 + \frac{We}{2}f''^3\right) = 0, \tag{5.8}$$

And Eq. (5.5) becomes

$$\begin{aligned} &f(0) = 0,\ \theta(0) = 1, && f'(\infty) \to 0, \\ &f'(0) = 1 + \lambda f''(0)(Wef''(0) + 1),\ \ and && \theta(\infty) \to 0, \end{aligned} \tag{5.9}$$

$$\begin{aligned} &We = \Gamma x\sqrt{\frac{2a^3}{\upsilon}}, M = \frac{\sigma B_0^2 \sin^2\alpha}{a\rho}, Kp = \frac{\upsilon}{ka}, \Delta = \frac{g\beta^*\left(T_s - T_f\right)}{a^2x}, B = \lambda a, \\ &Nt = \frac{D_T\left(T_s - T_f\right)}{T_\infty \upsilon}, Q = \frac{Q_0}{\rho C_p a}, F = \frac{4I^*}{\rho C_p a}, Ec = \frac{a^2x^2}{C_p\left(T_s - T_f\right)}, \Pr = \frac{\mu C_p}{k}, \end{aligned}$$

5.3 PHYSICAL ENTITIES

Here is the local Reynolds number:

$$\mathrm{Re}_x = \frac{ax^2}{\upsilon}$$

Nusselt number and drag coefficient are given by

$$Nu_x = \frac{xq_w}{k(T_w - T_\infty)},\ and\ Cf_x = \frac{\tau_w}{\rho a^2x^2},$$

and

$$Nu_x \mathrm{Re}_x^{-0.5} = -\theta'(0), \text{ and } Cf_x \mathrm{Re}_x^{0.5} = f''(0). \tag{5.10}$$

5.4 METHODOLOGY

To solve the modified ODEs (Eqs. (5.7) and (5.8)),

$$f = \xi_1, f' = \xi_2, f'' = \xi_3, \theta = \xi_4, \theta' = \xi_5, \tag{5.11}$$

using Eq. (11) convert them. RK fourth-order method was applied using bvp4c MATLAB program to obtain the solution. Appropriate initial conditions are applied to the system before the numerical solution process begins.

Using Eq. (5.11), Eqs. (5.7) and (5.8) become

$$\begin{aligned} f''' = \xi_3{}' &= \frac{\xi_2^2 - \xi_1\xi_3 + M\xi_2 + Kp\xi_2 - \Delta\xi_4 - 2B\xi_1\xi_2\xi_3}{1 - By_1^2 + Wey_3}, \\ \theta'' = \xi_5{}' &= -\Pr\left[\xi_1\xi_5 + Nt\xi_5^2 - F\xi_4 + Q\xi_4 + Ec\left(\xi_3^2 + \frac{We}{2}\xi_3^3\right)\right], \end{aligned} \tag{5.12}$$

and Eq. (5.9) is changed to

$$\begin{aligned} &\xi_1(i) = 0,\ \xi_2(i) = 1 + \lambda\left[\xi_3(i) + \frac{We}{2}\xi_3^2(i)\right], \xi_4(i) = 1,\ at\ i = 0 \\ &\xi_2(j) \to i,\ \xi_4(j) \to i, for\ i = 0\ and\ j \to \infty \end{aligned} \tag{5.13}$$

5.5 RESULTS AND DISCUSSION

This study helps us understand how the fluid's properties change. We're looking at how things like velocity and temperature change, as shown in the graphs 5.2 – 5.9. We'll discuss these changes in detail, focusing on specific values of some dimensionless numbers:

$$We = 0.01,$$

$$Nt = Ec = B = \lambda = 0.1, M = \Delta = 0.2, \Pr = 0.3, Kp = F = 0.5, Q = -1.5.$$

Figures 5.2–5.5 illustrate the velocity profiles. Figure 5.2 illustrates the inverse relation between the magnetic field and velocity. This is due to the Lorentz force generated by the magnetic field acting on the moving conductive fluid, which opposes the fluid's motion. Figure 5.3 demonstrates the inverse relationship between velocity and the porosity parameter. Higher porosity implies more obstacles and complex flow paths, resulting in increased resistance and thus lower fluid velocity. Figure 5.4 reveals that the velocity increases with the mixed convection. A higher mixed convection

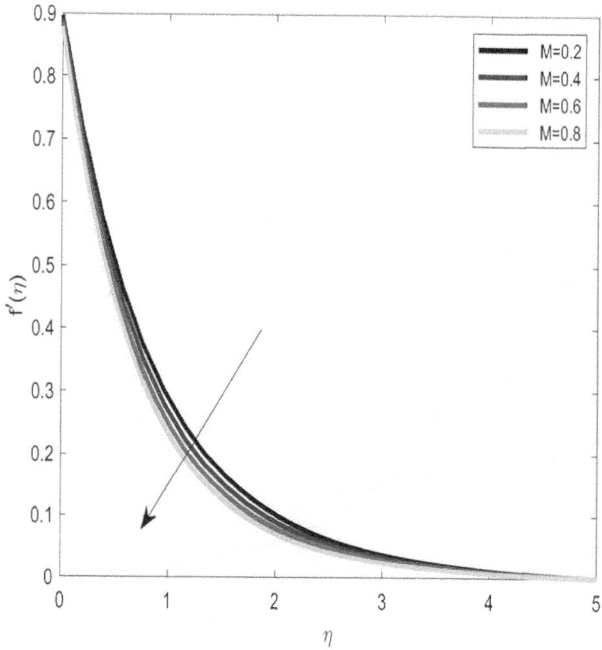

Figure 5.2 $f'(\eta)$ vs M.

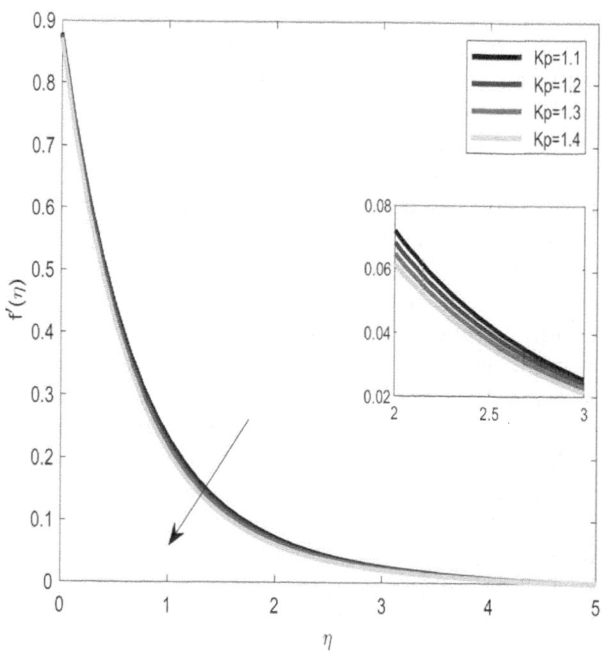

Figure 5.3 $f'(\eta)$ vs *Kp*.

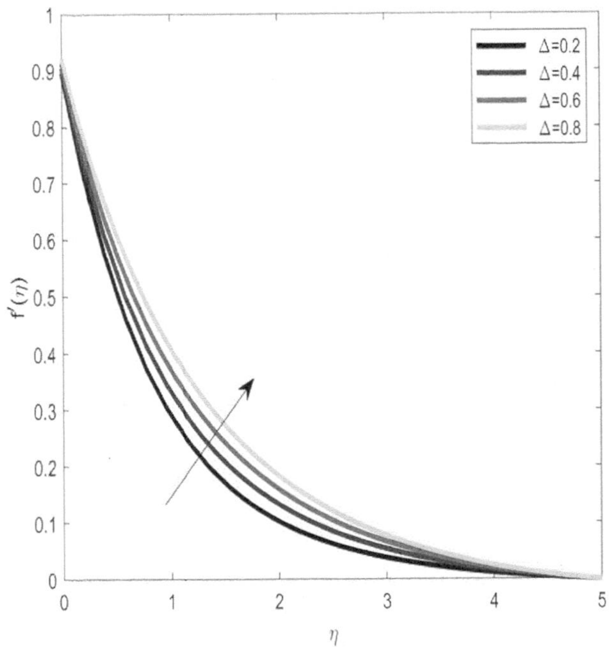

Figure 5.4 $f'(\eta)$ vs Δ.

number indicates stronger buoyancy forces, which enhance fluid motion and increase velocity. Figure 5.5 shows that the velocity profile decreases with Deborah's number. This is because a higher Deborah number signifies more elastic fluid behavior, and increased elasticity leads to greater resistance to flow, thereby reducing velocity.

Figures 5.6–5.9 illustrate temperature variations. Figure 5.6 shows that as the heat source strengthens, it transfers more thermal energy. This increased energy input raises the kinetic energy of particles, resulting in higher temperatures. Figure 5.7 demonstrates that temperature decreases with increasing Prandtl number. A higher Prandtl number signifies slower heat diffusion, leading to less effective heat distribution and consequently lower temperatures. Figure 5.8 depicts the temperature profile's dependence on the thermophoresis parameter. A higher thermophoresis parameter indicates stronger particle movement toward colder regions. This particle accumulation carries energy, thus increasing the temperature. Figure 5.9 shows that the temperature profile decreases with the radiation parameter. A higher radiation parameter means more radiative heat transfer to surroundings, leading to a net decrease in temperature. Table 5.1 shows the influence of nanofluid physical properties.

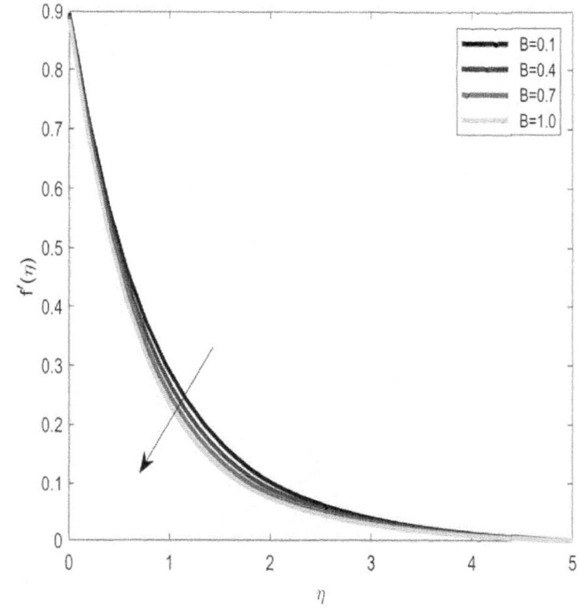

Figure 5.5 $f'(\eta)$ vs *B*.

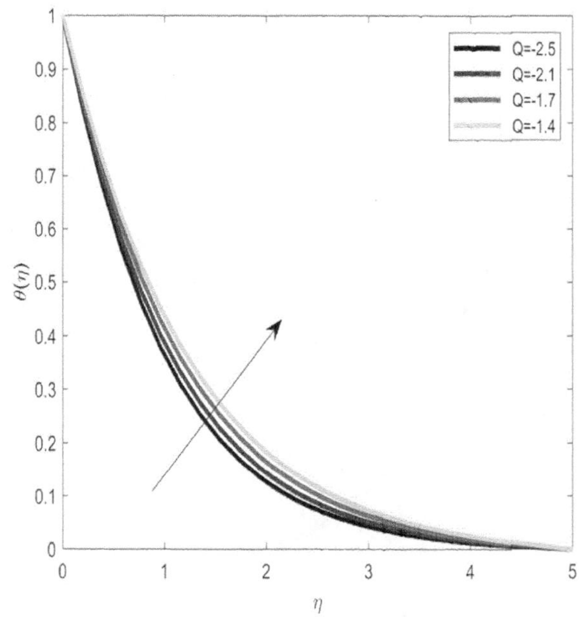

Figure 5.6 $\theta(\eta)$ vs *Q*.

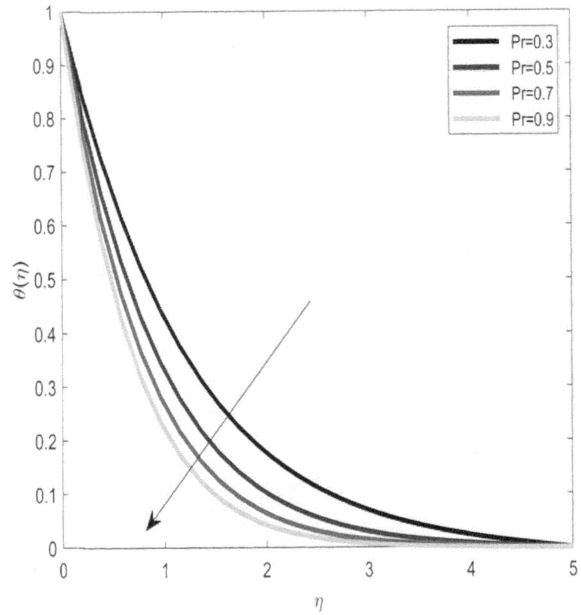

Figure 5.7 $\theta(\eta)$ vs Pr.

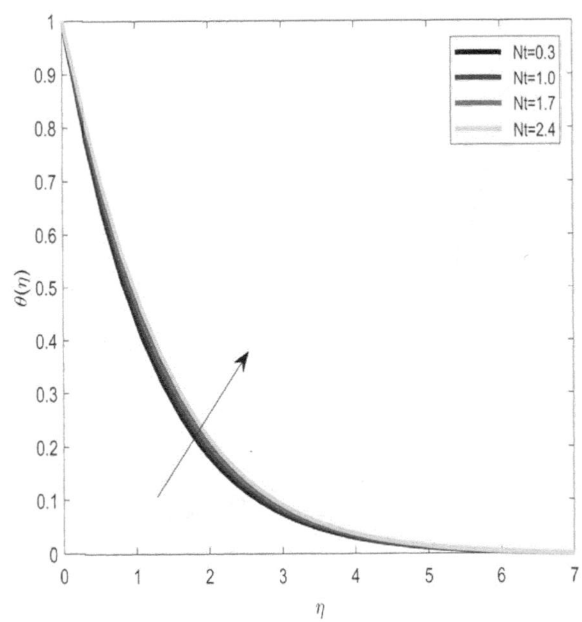

Figure 5.8 $\theta(\eta)$ vs Nt.

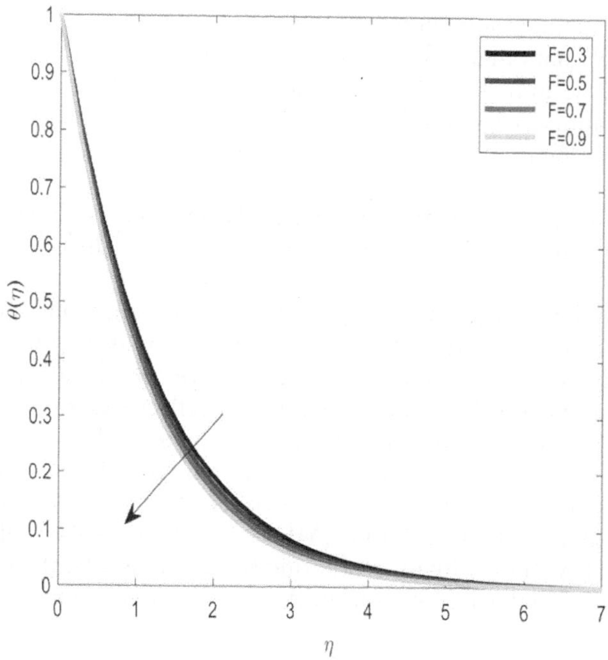

Figure 5.9 $\theta(\eta)$ vs F.

Table 5.1 Values of $f''(0)$ and $-\theta'(0)$ with various parameters

M	Δ	Pr	Kp	F	Q	Nt	B	We	$f''(0)$	$-\theta'(0)$
0.2	0.2	0.3	0.5	0.5	-1.5	0.1	0.1	0.01	-1.0526	0.8070
0.3									-1.0835	0.8058
	0.3								-1.0116	0.8088
		0.4							-1.0582	0.9298
			0.6						-1.0835	0.8058
				0.6					-1.0534	0.8251
					-1.0				-1.0480	0.7097
						0.2			-1.0523	0.7986
							0.2		-1.0651	0.8063
								0.02	-1.0564	0.8070

5.6 CONCLUSION

This study examines how thermophoresis affects heat transfer in Williamson and Maxwell nanofluids flowing over an enlarging sheet. Specifically, it investigates the influence of Cogley radiation and a heat source within a porous medium on the fluid flow. The following conclusions are drawn from this investigation:

- Velocity profile enhances with Δ and diminishes with M, Kp, B.
- Temperature profile enhances with Nt, Q and reduces with $\Pr, F$.

This study lays the groundwork for further investigations. Future work can be extended by incorporating chemical reactions as it would be particularly relevant for applications involving reactive flows. Furthermore, analyzing a non-uniform heat source would the reflect real-time situation, as the heat input is rarely uniform in practical situations. These enhancements would not only make the model more comprehensive but also broaden its applicability to real-world scenarios such as industrial cooling, chemical processing, magnetohydrodynamic (MHD) generators, solar collectors, and nanomaterial synthesis. By incorporating these extensions, the research can provide more valuable insights for the design of various engineering systems.

REFERENCES

Ahmed, S.E., Mohamed, R.A., AbdElraheem, M.A. and Soliman, M.S., 2019. Magnetohydrodynamic Maxwell nanofluids flow over a stretching surface through a porous medium: effects of non-linear thermal radiation, convective boundary conditions and heat generation/absorption. *International Journal of Aerospace and Mechanical Engineering*, 13(6), 436–443.

Alipour, N., Jafari, B. and Hosseinzadeh, K., 2023. Thermal analysis and optimization approach for ternary nanofluid flow in a novel porous cavity by considering nanoparticle shape factor. *Heliyon*, 9(12), e22257.

Alnahdi, A.S. and Gul, T., 2024. MHD hybrid nanofluid flow over a stretched surface with convective boundary conditions: Applications of heat transfer. *Modern Physics Letters B*, 38(08), 2450046.

Biswas, R., Hossain, M.S., Islam, R., Ahmmed, S.F., Mishra, S.R. and Afikuzzaman, M., 2022.Computational treatment of MHD Maxwell nanofluid flow across a stretching sheet considering higher-order chemical reaction and thermal radiation. *Journal of Computational Mathematics and Data Science*, 4, 100048.

Cogley, A.C., Vincent, W.G. and Gilles, S.E., 1968. Differential approximation for radiative transfer in a nongrey gas near equilibrium. *AIAA Journal*, 6(3), 551–553.

Dawar, A., Shah, Ź. and Islam, S., 2021. Mathematical modeling and study of MHD flow of Williamson nanofluid over a nonlinear stretching plate with activation energy. *Heat Transfer*, 50(3), 2558–2570.

Dey, D. and Borah, R., 2023.MHD nanofluid flow with energy transfer over a porous stretching surface by using a second-grade fluid model. *Numerical Heat Transfer, Part A: Applications*, 1–15.

Faisal, M., Ahmad, I. and Javed, T., 2021. Numerical assessments of prescribed heat sources on unsteady 3D flow of Williamson nanoliquid through porous media. *Special Topics & Reviews in Porous Media: An International Journal*, 12(2), 71–92.

Faizan, M., Zafar, S.S., Ali, F., Lone, S.A. and Saeed, A., 2024.Modeling of magneto bioconvection Williamson nanoliquid through modified heat flux: the significance of entropy. *Partial Differential Equations in Applied Mathematics*, 100746.

Gupta, P.K., Prakash, O. and Tinker, S., 2024. Unsteady Free Convection of Nanofluid Past an Infinite Vertical Sheet Under the Effects of Inclined Magnetic Field and Cogley Radiation. Recent Patents on Engineering.

Haddad, Z., Abu-Nada, E., Oztop, H.F. and Mataoui, A., 2012. Natural convection in nanofluids: are the thermophoresis and Brownian motion effects significant in nanofluid heat transfer enhancement?. *International Journal of Thermal Sciences*, 57, 152–162.

Hamid, A. and Khan, M., 2018. Unsteady mixed convective flow of Williamson nanofluid with heat transfer in the presence of variable thermal conductivity and magnetic field. *Journal of Molecular Liquids*, 260, 436–446.

Jamshed, W., Devi, S.U. and Nisar, K.S., 2021. Single phase based study of Ag-Cu/EO Williamson hybrid nanofluid flow over a stretching surface with shape factor. *PhysicaScripta*, 96(6), 065202.

Janjanam, N., Nimmagadda, R., Asirvatham, L.G., Harish, R. and Wongwises, S., 2021. Conjugate heat transfer performance of stepped lid-driven cavity with Al_2O_3/water nanofluid under forced and mixed convection. *SN Applied Sciences*, 3(6), 605.

Khan, H., Haneef, M., Shah, Z., Islam, S., Khan, W. and Muhammad, S., 2018. The combined magneto hydrodynamic and electric field effect on an unsteady Maxwell nanofluid flow over a stretching surface under the influence of variable heat and thermal radiation. *Applied Sciences*, 8(2), 160.

Khan, K.A., Vivas-Cortez, M., Ishfaq, K., Javed, M.F., Raza, N., Nisar, K.S. and Abdel-Aty, A.H., 2024. Exploring the numerical simulation of Maxwell nanofluid flow over a stretching sheet with the influence of chemical reactions and thermal radiation. *Results in Physics*, 60, 107635.

Khan, Z.H., Khan, W.A., Ibrahim, S.M., Swain, K. and Huang, Z., 2024. Impact of multiple slips and thermal radiation on heat and mass transfer in MHD Maxwell hybrid nanofluid flow over porous stretching sheet. *Case Studies in Thermal Engineering*, 61, 104906.

Kothandapani, M. and Prakash, J., 2015. Effects of thermal radiation parameter and magnetic field on the peristaltic motion of Williamson nanofluids in a tapered asymmetric channel. *International Journal of Heat and Mass Transfer*, 81, 234–245.

Maxwell, J.C., 2003. On the dynamical theory of gases. In *The Kinetic Theory of Gases: An Anthology of Classic Papers with Historical Commentary*, 197–261. London: Imperial College Press.

Prakash, O., Barman, P., Rao, P.S. and Sharma, R.P., 2023. MHD free convection in a partially open wavy porous cavity filled with nanofluid. *Numerical Heat Transfer, Part A: Applications*, 84(5), 449–463.

Prasannakumara, B.C., 2021. Numerical simulation of heat transport in Maxwell nanofluid flow over a stretching sheet considering magnetic dipole effect. *Partial Differential Equations in Applied Mathematics*, 4, 100064.

Ragulkumar, E., Palani, G., Sambath, P. and Chamkha, A.J., 2023. Dissipative MHD free convective nanofluid flow past a vertical cone under radiative chemical reaction with mass flux. *Scientific Reports*, 13(1), 2878.

Ramesh, G.K., Gireesha, B.J., Hayat, T. and Alsaedi, A., 2015. MHD flow of Maxwell fluid over a stretching sheet in the presence of nanoparticles, thermal radiation and chemical reaction: a numerical study. *Journal of Nanofluids*, 4(1), 100–106.

Sharma, R.P., Mishra, S.R., Tinker, S. and Kulshrestha, B.K., 2022. Effect of nonlinear thermal radiation and binary chemical reaction on the Williamson nanofluid flow past a linearly stretching sheet. *International Journal of Applied and Computational Mathematics*, 8(4), 71.

Sharma, R.P., Mishra, S.R., Tinker, S. and Kulshrestha, B.K., 2023. Exploration of radiative and dissipative heat on Williamson nanofluid flow in conjunction with convective boundary condition. *Journal of Nanofluids*, 12(1), 223–230.

Sharma, R.P., Prakash, O. and Shafi, S., 2024. MHD hybrid nanofluid and slip flow towards an exponentially stretching/shrinking sheet with nonlinear thermal radiation and nonuniform heat source/sink. *Modern Physics Letters B*, 38(23), 2450195.

Shateyi, S. and Muzara, H., 2020. On the numerical analysis of unsteady MHD boundary layer flow of Williamson fluid over a stretching sheet and heat and mass transfers. *Computation*, 8(2), 55.

Shawky, H.M., Eldabe, N.T., Kamel, K.A. and Abd-Aziz, E.A., 2019. MHD flow with heat and mass transfer of Williamson nanofluid over stretching sheet through porous medium. *Microsystem Technologies*, 25, 1155–1169.

Sheikholeslami, M., Ijaz Khan, M., Chu, Y.M., Kadry, S. and Khan, W.A., 2021. CVFEM based numerical investigation and mathematical modeling of surface dependent magnetized copper-oxide nanofluid flow using new model of porous space. *Numerical Methods for Partial Differential Equations*, 37(2), 1481–1494.

Ullah, H., Hayat, T., Ahmad, S., Alhodaly, M.S. and Momani, S., 2021. Numerical simulation of MHD hybrid nanofluid flow by a stretchable surface. *Chinese Journal of Physics*, 71, 597–609.

Williamson, R. V. (1948). The flow of pseudoplastic materials. *Industrial & Engineering Chemistry*, 40(6), 1027–1031.

Chapter 6

Enhanced thermal performance of magnetized Maxwell hybrid nanofluid transportation over a radiative stretching sheet with exponential heat source

Subhajit Panda, Folarin Oluwaseun, Titilayo M. Agbaje, S. Baag, and P. K. Pattnaik

6.1 INTRODUCTION

A Maxwell hybridized fluid is a viscoelastic, non-Newtonian fluid that follows Maxwell's rheological model. It is created by dispersing multiple nanoparticles into a base fluid, enhancing thermal conduction and heat dissipation effectiveness. Due to its enhanced thermal and mechanical properties, it is widely applied in cooling technologies, biomedical systems, and energy transport applications. Jeelani and Abbas [1] evaluated the thermal and species transport in a hybrid nanofluid circulating within a porous domain using a non-Newtonian Maxwell liquid. Magneto fluidic and suction effects are involved, along with solar radiation, to enhance heat transfer. The researchers employ the finite difference method combined utilizing the shooting method to handle the coupled partial differential equations (PDEs). Abbas and Nawaz [2] examined the implication of magnetic intensity, thermal energy exchange, and non-Newtonian attributes concerning the motion and thermal dispersion of Maxwell hybrid nanofluids (MHNF). The researchers employ the shooting technique to solve the governing equations, ensuring precise computational outcomes. Pattnaik et al. [3] examined the free convection of a viscoelastic nanofluid with electrical conductivity over an expanding configuration structure. Heat transfer is influenced by Brownian motion and thermophoresis, which arise from cross-diffusion and thermal radiation effects. They are solved by governing equations using the perturbation method alongside a numerical technique to ensure precise computation in their research.

A stretching sheet in a hybrid nanofluid analysis represents a surface that extends continuously, affecting convective flow and heat exchange dynamics. The incorporation of hybrid nanoparticles boosts thermal conductance, enhancing energy transport efficiency. This phenomenon is crucial in industries like polymer extrusion, fiber spinning, and coating processes,

DOI: 10.1201/9781003642169-6

where managing fluid dynamics and heat dissipation is essential. Nazari et al. [4] examined the effect of mass suction on radiative flow influenced by an inclined buoyancy force over a stretchable or shrinkable sheet. The analysis considers water-based Cu–Al_2O_3 hybrid nanofluids. A numerical method is employed by the researchers to solve the governing ordinary differential equations. Jamrus et al. [5] explored the stagnation zone motion of a ternary hybrid nanofluid (THNF) along a sheet undergoing exponential stretching or shrinking. The researchers transformed the governing continuity, momentum, and energy equations into ordinary differential equations to facilitate their analysis by using a numerical method. Baag et al. [6] investigated the magnetically influenced flow of a viscous fluid over an expanding sheet, accounting for thermal buoyancy, a varying heat emission/sink, and chemical species.

Thermal radiation in hybrid nanofluids involves heat transfer through electromagnetic waves, playing a key role in high-temperature systems. The inclusion of hybrid nanoparticles enhances radiative heat transfer, boosting thermal performance. This process is vital in applications like solar energy, aerospace technology, and electronic cooling, where efficient heat regulation is required. Saleem et al. [7] examined entropy formation in a stratified cross-hybrid nanomaterial flow around a stretchable cylinder, influenced by gyrotactic microorganisms. The hybrid nanofluid was created by dispersing titanium dioxide and silver nanoparticles into a water-based cross fluid. The momentum equation was formulated by incorporating Darcy–Forchheimer effects and surface permeability. Galal et al. [8] reported the contribution of radiative heat and slip constraints pertaining to the electromagnetic flow of a tetra-hybrid nanoparticle suspension via an inclined revolving disk, considering local thermal nonequilibrium limitations. Tinker et al. [9] assessed energy transfer in the evolving flow of a hybrid nanofluid over a stretching or shrinking sheet, considering the effects of thermal radiation and a heat source or sink. A numerical approach was used to solve the governing equations by the researchers in their research.

A heat source or sink in hybrid nanofluids refers to the process of adding or removing thermal energy, affecting temperature distribution and fluid flow. The inclusion of hybrid nanoparticles enhances thermal conductivity, optimizing heat absorption or dissipation. This mechanism is essential in cooling systems, energy storage, and industrial heat exchangers for effective thermal regulation. Alharabi et al. [10] explored the influence of various parameters on transverse and axial velocities, drag force, temperature, and heat transfer. In the case of opposing flow, two distinct solutions emerged, whereas assisting flow resulted in a single solution. Additionally, the Lorentz force contributed to alterations in flow behavior and heat transfer properties. Ramasekhar et al. [11] studied the temperature profile improves with an amplification in the magnetism and thermal radiation factors. Higher Casson fluid and radiation parameters also enhance skin friction and the Nusselt number patterns.

6.1.1 Motivation for the study

The study reported in the literature lacks in several aspects, which leads to investigate the thermal behavior of a magnetized hybrid nanofluid to improve heat transfer efficiency. Moreover, the impact of radiative exponentially expanding surface on convective flow and energy transfer is explored to enhance the thermal performance. Also, the slip velocity along with convective heating conditions is also useful in enhancing flow characteristic. The finding contributes to energy-efficient lubrication and high-performance cooling mechanisms.

6.2 MATHEMATICAL MODEL AND FORMULATION

The motion of a kerosene oil-based hybrid nanofluid (Go+Ag) is analyzed under steady-state and incompressible conditions across a longitudinally extendable plate, with the fluid flowing through a finely porous medium. The sheet's wall stretches symmetrically from the origin along its sides at a specific speed $U_w = cx$. The framework of Maxwell fluid describes the emulsion as a non-Newtonian fluid. The components of fluid velocity are u, v and temperature is T The schematic of the flow is sketched in Figure 6.1.

Based on the above analysis, the equations regulating flow and heat transfer are ([12]):

$$\partial_x(u) + \partial_y(v) = 0, \tag{6.1}$$

$$\begin{aligned} u\partial_x(u) + v\partial_y(u) = \Bigg\{ & \frac{\mu_{hnf}}{\rho_{hnf}} \partial_{yy}(u) - \frac{\sigma_{hnf}}{\rho_{hnf}} B_0^2 u \\ & - \left[2uv\partial_{xy}(u) + u^2\partial_{xx}(u) + v^2\partial_{yy}(u) \right] \lambda, \end{aligned} \tag{6.2}$$

$$\begin{aligned} u\partial_x(T) + v\partial_y(T) = \Bigg\{ & \left(\frac{k_{hnf}}{(\rho c_p)_{hnf}} + \frac{1}{(\rho c_p)_{hnf}} \frac{16\sigma^* T_\infty^*}{3k^*} \right) \partial_{yy}(T) \\ & + \frac{Q_0 e^{-n\eta}}{(\rho c_p)_{hnf}} (T - T_\infty) \end{aligned} \tag{6.3}$$

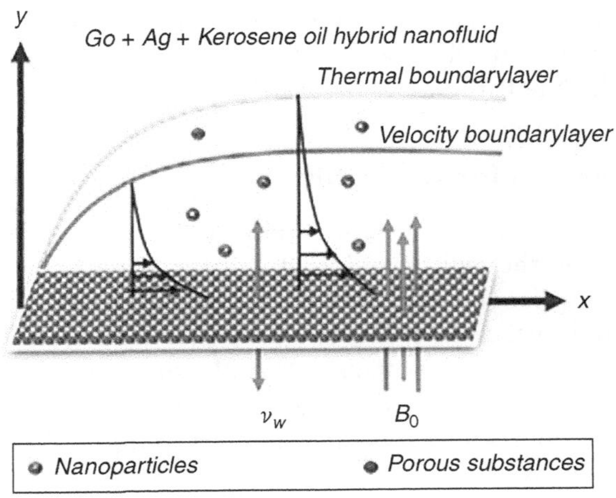

Figure 6.1 Flow geometry.

The corresponding boundary condition are:

$$\begin{aligned}&as\ y=0,\ \begin{cases}\left[u,v,-k_{hnf}\partial_y(T)\right]=\left[U_w+L\partial_y(u)=\right.\\ \left.cx+L\partial_y(u),v_w,h_f(T-T_\infty)\right],\end{cases}\\ &as\ y\to\infty,\ [u,T]\to[0,T_\infty]\end{aligned} \tag{6.4}$$

v_w corresponds to a consistent velocity along the wall, where negative v_w specifies injection and positive v_w indicates suction.

Implementing the similarity approach,

$$\begin{aligned}\eta&=\sqrt{\frac{c}{v_f}}y,u=cxf'(\eta),-k_{hnf}\partial_y(T)\\ &=h_f(T-T_\infty),v=-\sqrt{cv_f}f(\eta).\end{aligned} \tag{6.5}$$

Eqs. (6.2–6.4) can take the form,

$$(\mu_m+\beta\rho_m f^2)f'''-\rho_m(f'^2-ff'')-(M\sigma_m)f'-2\beta\rho_m f f' f''=0, \tag{6.6}$$

$$(k_m+Rd)\theta''+\Pr\left((\rho c_p)_m(f\theta')+He^{-n\eta}\theta\right)=0, \tag{6.7}$$

$$at\ \eta = 0, \quad \begin{cases} f(\eta) = f_w, f'(\eta) = 1 + Pf''(\eta), \\ k_m \theta'(\eta) = B_i\left(1 - \theta(\eta)\right), \end{cases} \quad (6.8)$$
$$as\ \eta \to \infty,\ f'(\infty) \to 0, \theta(\infty) \to 0.$$

where

$$\mu_m = \frac{\mu_{hnf}}{\mu_f}, \rho_m = \frac{\rho_{hnf}}{\rho_f}, \sigma_m = \frac{\sigma_{hnf}}{\sigma_f}, k_m = \frac{k_{hnf}}{k_f}, (\rho c_p)_m = \frac{(\rho c_p)_{hnf}}{(\rho c_p)_f}.$$

Skin friction coefficient expressions and local Nusselt number are defined as:

$$C_f = \frac{\tau_w}{\rho_f U_w^2}, Nu = \frac{x q_w}{k_f\left(T_w - T_\infty\right)}, \quad (6.9)$$

where

$$\tau_w = \mu_{hnf}\left[\partial_y (u)\right]_{y=0}, q_w = k_{hnf}\left[\partial_y (T)\right]_{y=0}$$

Employing similarity approach,

$$\begin{cases} C_f = C_{fx} \mathrm{Re}_x^{0.5} = \mu_m f''(0) \\ Nu = Nu_x \mathrm{Re}_x^{-0.5} = -\left(k_m + Rd\right)\theta'(0) \end{cases} \quad 6.10)$$

6.3 RESULTS AND DISCUSSION

This study investigates the numerical simulation of magnetized Maxwell hybrid nanofluid transportation over a radiative stretching sheet with exponential heat source. Table 6.1 describes the list of non-dimensional parameters. Table 6.2 portrays the thermophysical models for hybrid and nanofluid. Table 6.3 shows the comparison of different values of Pr with a previous study. Figure 6.2 depicts the velocity characteristics of the fluid for the pure fluid, nanofluid and hybrid nanofluid. Physically, the suction/injection parameter represents the effect of mass flux at the surface of the stretchable sheet. When $f_w > 0$, suction occurs, drawing fluid toward the surface. This action removes momentum from the boundary layer, resulting in thinner velocity profiles and a sharper decline in velocity. In contrast, when $f_w < 0$, injection occurs where additional fluid is introduced into the boundary layer. The underlying physics behind these observations can be explained through momentum transfer mechanisms.

Table 6.1 List of non-dimensional parameters

Parameters	*Symbols*	*Expression*
Magnetic parameter	M	$\frac{B_0^2\sigma_f}{c\rho_f}$
Darcy number	Da	$\frac{\upsilon_f}{ck_p}$
Exponential heat source/sink parameter	H	$\frac{Q_0}{c(\rho c_p)_f}$
Prandtl number	Pr	$\frac{\upsilon_f(\rho c_p)_f}{k_f}$
Velocity slip parameter	P	$L\sqrt{c/\upsilon_f}$
Radiation parameter	Rd	$\frac{16\sigma^* T_\infty^*}{3k_f k^*}$
Suction/injection	f_w	$-\frac{v_w}{\sqrt{c\upsilon_f}}$
Thermal Biot number	B_i	$\frac{h}{k_f}\sqrt{\frac{\upsilon_f}{c}}$
Maxwell Deborah number	β	λc

Table 6.2 Thermophysical models

Attributes	*Hybrid nanofluid*	*Nanofluid*
Viscosity (μ)	$\mu_{hnf} = 0.904e^{14.8\phi_1}\mu_{nf}$	$\mu_{nf} = 0.904e^{14.8\phi_2}\mu_f$
Fluid density (ρ)	$\rho_{hnf} = \left[(1-\phi_2) + \phi_2\frac{\rho_{s_2}}{\rho_{nf}}\right]\rho_{nf}$	$\rho_{nf} = \left[(1-\phi_1) + \phi_1\frac{\rho_{s_1}}{\rho_f}\right]\rho_f$
Electrical conductivity (σ)	$\sigma_{hnf} = \left[\frac{\sigma_{s_2} + 2\sigma_{nf} + 2\phi_2(\sigma_{s_2} - \sigma_{nf})}{\sigma_{s_2} + 2\sigma_{nf} - \phi_2(\sigma_{s_2} - \sigma_{nf})}\right]\sigma_{nf}$	$\sigma_{nf} = \left[\frac{\sigma_{s_1} + 2\sigma_f + 2\phi_1(\sigma_{s_1} - \sigma_f)}{\sigma_{s_1} + 2\sigma_f - \phi_1(\sigma_{s_1} - \sigma_f)}\right]\sigma_f$
Thermal conductivity (k)	$k_{hnf} = \left[\frac{2k_{nf} + k_{s_2} + 2\phi_2(k_{s_2} - k_{nf})}{2k_{nf} + k_{s_2} - \phi_2(k_{s_2} - k_{nf})}\right]k_{nf}$	$k_{nf} = \left[\frac{2k_f + k_{s_2} + 2\phi_2(k_{s_2} - k_f)}{2k_f + k_{s_2} - \phi_2(k_{s_2} - k_f)}\right]k_f$
Thermal capacity (ρc_p)	$(\rho c_p)_{hnf} = \left[(1-\phi_2) + \phi_2\frac{(\rho c_p)_{s_2}}{(\rho c_p)_{nf}}\right](\rho c_p)_{nf}$	$(\rho c_p)_{nf} = \left[(1-\phi_1) + \phi_1\frac{(\rho c_p)_{s_1}}{(\rho c_p)_f}\right](\rho c_p)_f$

Table 6.3 Thermophysical characteristics

Thermophysical properties	*Density* (ρ) (kg/m^3)	*Thermal conductivity* (k) $(W/m \cdot K^{-1})$	*Heat capacity* (c_p) (J/K)	*Electrical conductivity (S/m)*
Go	1800	5000	717	6.3×10^{7}
Ag	10500	429	235	63×10^{-6}
Kerosene oil	783	0.145	2090	21×10^{-6}

Table 6.4 Comparison of different values of Pr with a previous study

Pr	*Ahmad et al. [12]*	*Current study*
2	0.9111	0.91136568756
6.13	1.7595	1.75956778678
20	3.3538	3.3539696779

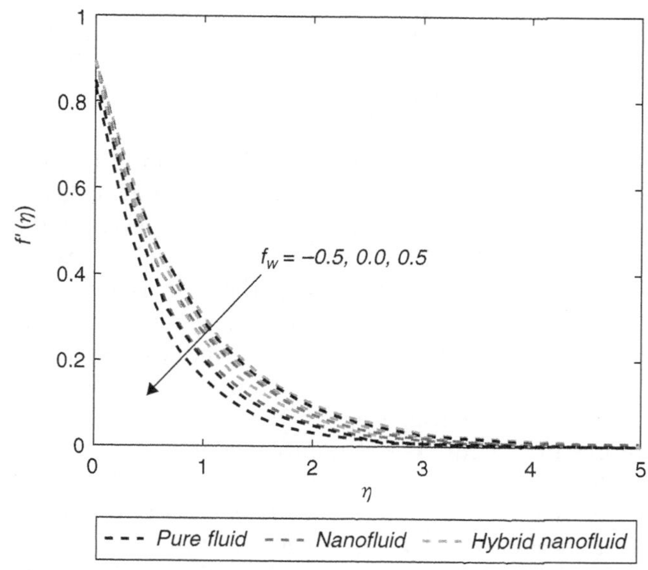

Figure 6.2 Deviation in $f'(\eta)$ for different f_w.

The presence of nanoparticles enhances the velocity, with the hybrid nanofluid exhibiting the highest velocity compared to the nanofluid and pure fluid. Figure 6.3 presents the implication of the magnetic term on the fluid momentum pattern. Physically, when a magnetic field is applied, Lorentz force is produced within the fluid, which is a form of frictional force and opposes the conducting fluid motion over the stretching sheet.

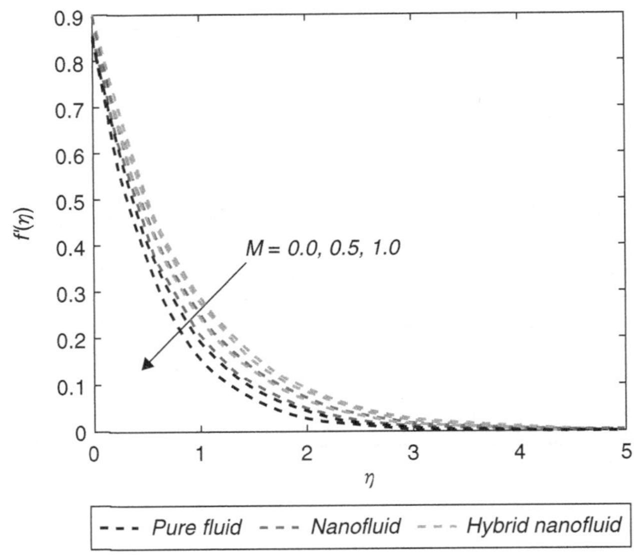

Figure 6.3 Deviation in $f'(\eta)$ for different M.

This force opposes the flow, leading to a suppression of velocity. As the magnetic parameter increases from $M = 0$ (no magnetic field) to $M \neq 2$ (strong magnetic field), the fluid slows down. This is because the increased magnetic parameter enhances the Lorentz force, which extracts kinetic energy from the fluid, thereby decreasing the flow velocity. However, in nanofluids and hybrid nanofluids, nanoparticles enhance electrical conductivity, reducing the resistive effect and, in some cases, increasing velocity by improving momentum diffusion. In contrast, pure fluids, with lower conductivity, experience stronger magnetic resistance, resulting in greater velocity suppression. Figure 6.4 demonstrates how the Maxwell number β influences the velocity distribution. Physically, the Maxwell fluid model considers both viscous effects and elastic deformation, which explains why velocity increases as β increases. According to this model, a longer relaxation time is associated with a greater β value, which indicates that the fluid keeps more of its initial motion before losing energy. The velocity across the boundary layer increases because of the conversion of this stored elastic energy into kinetic energy. In comparison among the three fluids, pure fluid has the lowest velocity, followed by nanofluid and hybrid nanofluid, which exhibits the biggest velocity enhancement as β increases. Figure 6.5 displays temperature profile in response to variations in the thermal radiation parameter, Rd. The fluid energy pattern rises in tandem with Rd. This is because thermal radiation lowers the rate of heat dissipation, more energy stays within the fluid, raising its temperature. Although its impact

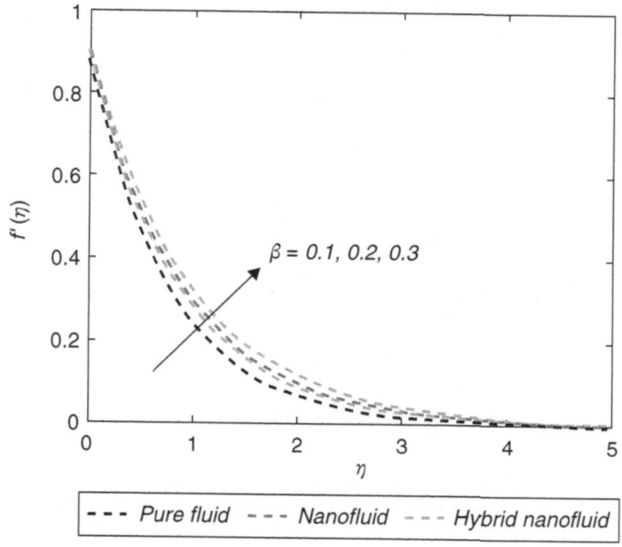

Figure 6.4 Deviation in $f'(\eta)$ for different β.

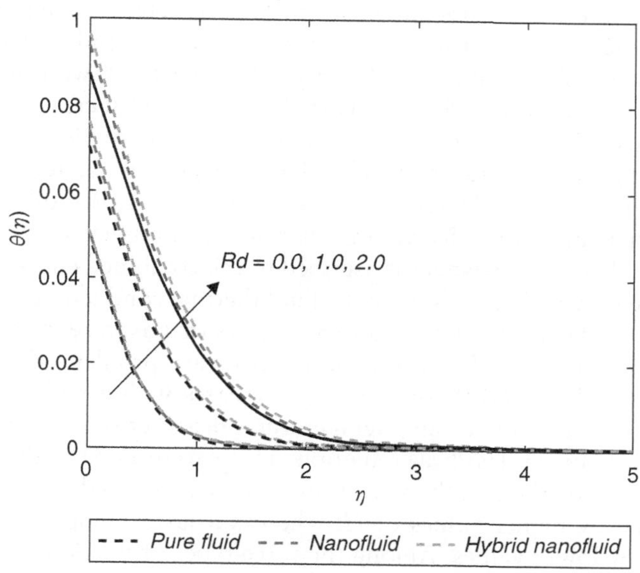

Figure 6.5 Deviation in $\theta(\eta)$ for different Rd.

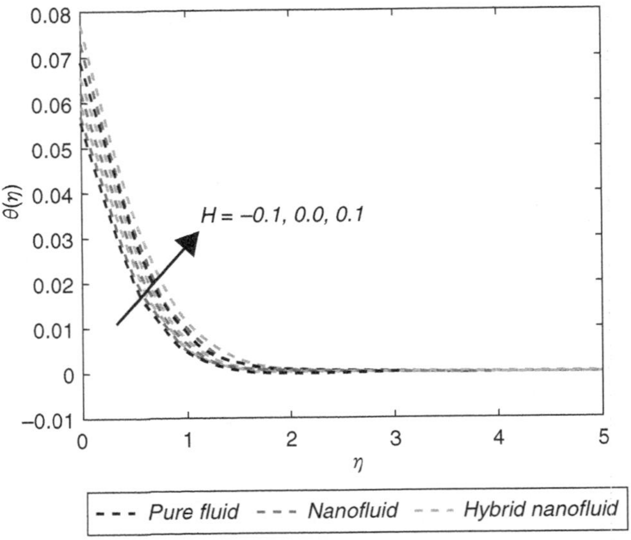

Figure 6.6 Deviation in $\theta(\eta)$ for different H.

varies depending on the fluid's composition, the effect of Rd is visible in all three of the fluid types shown in the figure. There is slightly noticeable difference in the thermal behavior of the various fluids when the temperature profiles are compared. The pure fluid has the lowest temperature distribution because it lacks nanoparticles that could increase its thermal conductivity. The nanofluid with single graphene oxide nanoparticles shows a higher temperature profile than the pure fluid, indicating that the nanoparticles have better heat transfer. The hybrid nanofluid exhibits the highest temperature distribution because it combines graphene oxide with silver. This improvement is ascribed to the combined effects of several nanoparticles, which greatly raise the fluid thermal conductivity and enable more effective heat retention. Figure 6.6 shows that as the exponential heat source/sink parameter H varies, the temperature profiles exhibit noticeable changes. The fluid thermal energy increases with a higher H, raising the temperature. Conversely, temperature distributions decrease as H decreases, indicating less heat retention. This variation has a direct impact on the boundary layer's thickness; an extended thermal boundary layer is produced by higher values of H, whereas a faster temperature drop is produced by lower values. Among the three types of fluids studied, hybrid nanofluids consistently display the highest temperature profiles, followed by nanofluids and then pure fluids. This pattern demonstrates how hybrid nanofluids, which take advantage of the combined thermal conductivity

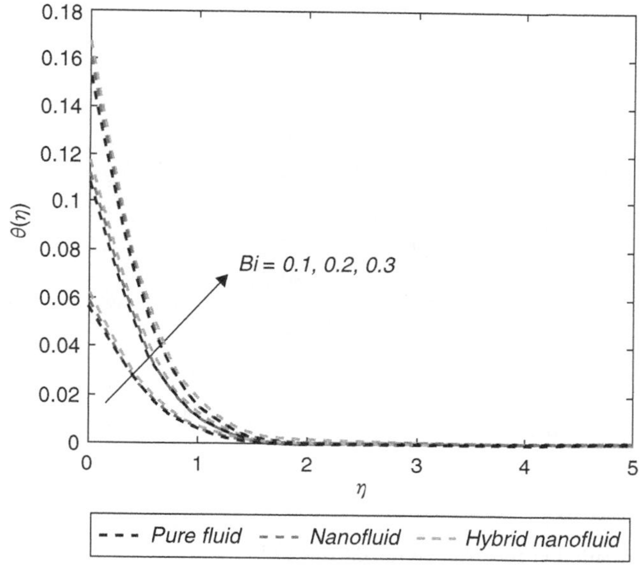

Figure 6.7 Deviation in $\theta(\eta)$ for different *Bi*.

effects of the graphene oxide and silver nanoparticles, have superior heat transfer capabilities. Figure 6.7 showcases that as the thermal Biot number, *Bi*, amplifies, the temperature of the fluids decreases more rapidly. The energy field is maximum near the sheet surface and gradually declines toward the free stream. The Biot number simply represents the proportion of thermal conduction within the fluid to convective heat transfer at the fluid surface. Higher *Bi* values speed up the fluid's rate of heat removal and indicate stronger convective heat transfer at the fluid surface. This suggests better cooling effects and explains why temperature profiles sharply decline as *Bi* increases. Lower temperatures across the fluid domain result from more heat dissipating quickly due to the surface's higher convective efficiency. The estimation of various factors on the shear stress presenting skin friction and the heat transfer rate in the form of the Nusselt number is depicted graphically. Figure 6.8 displays the diversified activity of the factors like magnetization because of the inclusion of applied magnetic field, the suction which appears for the fluid motion across a porous boundary and the non-Newtonian Maxwell parameter. The illustrated figure highlights that the magnetic resistance inducing the Lorentz force contributes to the enhancement of the velocity gradient, which gradually enhances the skin friction coefficient. Further, the increasing suction, that is, for the implication of the external forces and the elevating non-Newtonian attribute of the fluid enhances the skin friction at the surface. The rate of heat flow

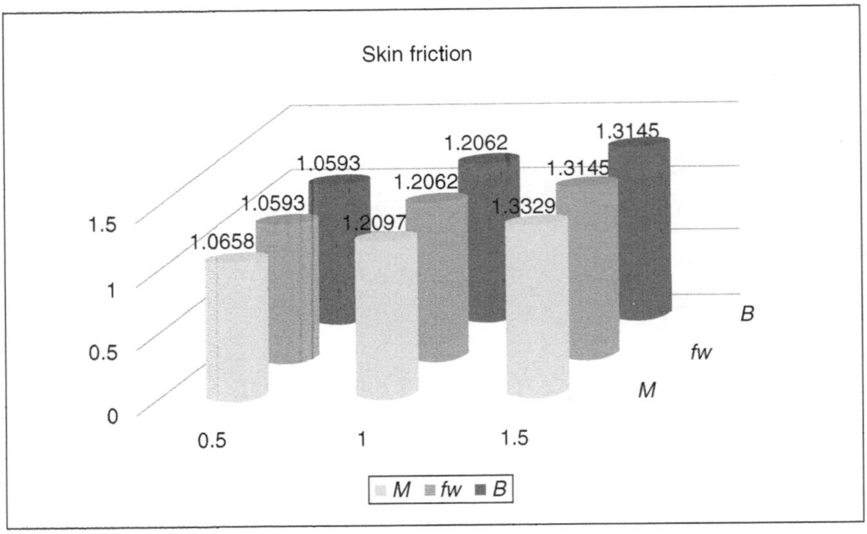

Figure 6.8 Deviation in skin friction for different M, f_w, β.

is also affected by the factors like thermal radiation, heat source, and the thermal Biot number, which is depicted in Figure 6.9. The response of the system to thermal radiation is because of the insertion of the nonlinear heat flux generally favors in augmenting the rate of heat flow, which upsurges the numerical values of the Nusselt number at the surface. But the imposition of amplifying heat source opposes the transport rate, leading to a significant deceleration in the Nusselt number. The thermal Biot number is obtained for the assumption of energy flux boundary condition and the rate of change in the temperature favors in augmenting heat dissipation with amplifying Biot number and this induces a considerable elevation in the pattern of Nusselt number for the increasing Biot number.

6.4 CONCLUDING REMARKS

A numerical investigation on the combined impact of grapheme oxide and silver nanocomposites in the host liquid kerosene is reported in this chapter. The flow of non-Newtonian Maxwell nanofluid with the interaction of thermal radiation, exponential space-based heat source along with convective condition refines the flow patterns. The mathematical model presented here is transformed into ordinary and the model is handled numerically. The results of the contributing factors are displayed using graphs and the important outcomes are reported as the coinciding numerical values in this particular scenario verifies the obtained results and exhibits

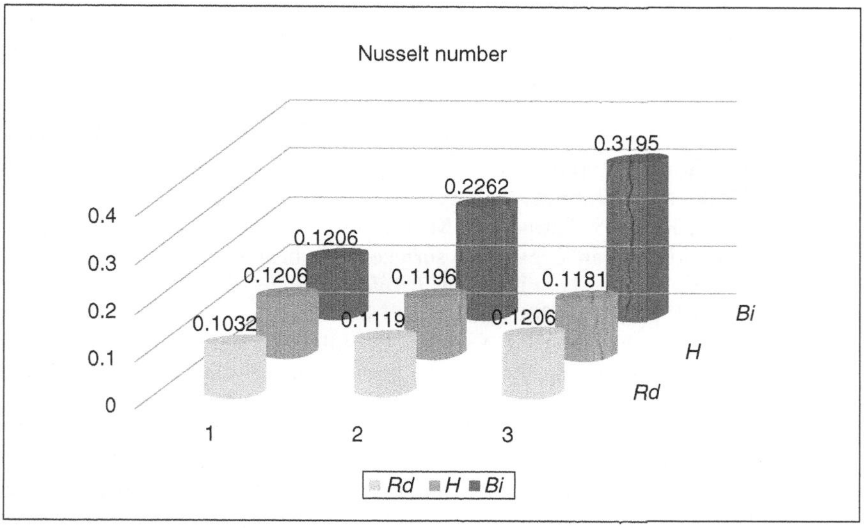

Figure 6.9 Deviation in Nusselt number for different *Rd*, *H*, *Bi*.

the convergence parameters for the selected methodology. The inclusion of one or more nanoparticles enhances the density as well as the thermal conductivity, which dominates the density and the fluid velocity escalates notably. The elevating suction favors controlling the fluid velocity and the increasing magnetization also reduces the confining layer thickness. The non-Newtonian parameter overshoots the fluid momentum and its patterns are asymptotic throughout the domain. The radiative heat for the inclusion of nonlinear heat flux enriches the fluid temperature and also favors the enhancing the heat transfer rate.

REFERENCES

[1] Jeelani MB, Abbas A. Energy transport in MHD Maxwell hybrid nanofluid flow over inclined stretching porous sheet with effects of chemical reaction, solar radiation and porous medium. *Case Stud Therm Eng* 2025;68:105915. https://doi.org/10.1016/J.CSITE.2025.105915.

[2] Abbas T, Nawaz R. Numerical simulation of steady incompressible flow of Maxwell hybrid nanofluid (MgO–Ag/H_2O) with thermal and flow behavior analysis. *Nano-Struct Nano-Objects* 2025;41:101460. https://doi.org/10.1016/J.NANOSO.2025.101460.

[3] Pattnaik P, Mishra SR, Panda S, Syed S, Muduli K. Hybrid methodology for the computational behaviour of thermal radiation and chemical reaction on viscoelastic nanofluid flow. *Math Probl Eng* 2022;2022. https://doi.org/10.1155/2022/2227811.

[4] Nazari NL, Ishak A, Khan A, Zaib A, Hussain SD, Elattar S. The implication of thermal radiation in a mixed convection hybrid nanofluid flow past an inclined stretching/shrinking sheet with mass suction. *J Radiat Res Appl Sci* 2025;18:101420. https://doi.org/10.1016/J.JRRAS.2025.101420.
[5] Jamrus FN, Waini I, Khan U, Ishak A. Stagnation slip flow of ternary hybrid nanofluid over an exponentially shrinking/Stretching sheet with joule heating, MHD, and thermal radiation effects. *Chinese J Phys* 2025;94:518–39. https://doi.org/10.1016/J.CJPH.2025.01.042.
[6] Baag S, Panda S, Pattnaik P, Mishra SR. Free convection of conducting nanofluid past an expanding surface with heat source with convective heating boundary condition. *Int J Ambient Energy* 2022;44:1–22. https://doi.org/10.1080/01430750.2022.2156607.
[7] Saleem M, Khouqeer GA, Haq F, Sallah M. Entropy modeling in bioconvective stratified magnetohydrodynamic cross hybrid nanofluid flow with thermal radiation. *J Radiat Res Appl Sci* 2025;18:101415. https://doi.org/10.1016/J.JRRAS.2025.101415.
[8] Galal AM, Khan I, Abbas M, Faqihi AA, Sajjad MS. Thermal radiation and local thermal non-equilibrium effects on MHD chemical reactive flow of tetra hybrid nanofluid with velocity slip conditions. *J Radiat Res Appl Sci* 2025;18:101405. https://doi.org/10.1016/J.JRRAS.2025.101405.
[9] Tinker S, Mishra SR, Pattnaik PK, Sharma RP. Simulation of time-dependent radiative heat motion over a stretching/shrinking sheet of hybrid Nanofluid: Stability analysis for dual solutions. *Proc Inst Mech Eng Part N J Nanomater Nanoeng Nanosyst* 2022;236:19–30. https://doi.org/10.1177/23977914211069021.
[10] Alharbi LF, Ishak A, Khan U, Waini I, Ahmed SE. Analysis of irregular heat source/sink of a stagnation-point flow toward a vertical plate induced by spherical hybrid nanofluids. *Case Stud Therm Eng* 2024;64:105551. https://doi.org/10.1016/J.CSITE.2024.105551.
[11] Ramasekhar G, Mebarek-Oudina F, Suneetha S, Vaidya H, Selvi PD. Computational simulation of Casson hybrid nanofluid flow with Rosseland approximation and uneven heat source/sink. *Int J Thermofluids* 2024;24:100893. https://doi.org/10.1016/J.IJFT.2024.100893.
[12] F. Ahmad, S. Abdal, H. Ayed, S. Hussain, S. Salim, A.O. Almatroud, The improved thermal efficiency of Maxwell hybrid nanofluid comprising of graphene oxide plus silver/kerosene oil over stretching sheet, *Case StudThermal Eng* 27 (2021) 101257.

Chapter 7

Ree–Eyring hybrid nanofluid flow bounded by double-revolving disks with Darcy–Forchheimer medium and variable thermal conductivity

Heat transfer analysis

Sk Enamul and Surender Ontela

Nomenclature

r, Θ, z	Cylindrical coordinates
u,v,w	Velocity components (m/s)
p	Pressure ($kg/m/s^2$)
C_p	Specific heat (J/kg/K)
f',g,f	Dimensionless radial, tangential and axial velocities
k	Thermal conductivity (W/m/K)
T	Fluid's temperature (K)
Pr	Prandtl number
Re	Reynolds number
M	Magnetic field component
We	Weissenberg number
K	Darcy permeability
F_1	Non-uniform inertia coefficient for porous media
Fr	Inertia coefficient parameter
C_b	Coefficient of drag

Greek Symbol

ρ	Density (kg/m^3)
σ	Electrical conductivity (A/V/m)
μ	Dynamic viscosity (kg/m/s)
ν	Kinematic viscosity (m^2/s)
ϵ	Variable thermal conductivity parameter
θ	Dimensionless temperature
η	Similarity variable
ϕ_1	The volume fraction of TiO_2
ϕ_2	The volume fraction of $CoFe_2O_4$

DOI: 10.1201/9781003642169-7

7.1 INTRODUCTION

Non-Newtonian (N-N) fluid is a topic that sparks vigorous debate in scientific communities due to its significant relevance in everyday life. The Ree–Eyring (RE) fluid is a notable example of a N-N fluid. The model of the RE fluid was established to clarify the N-N behavior exhibited by certain fluids. The model is particularly useful for describing shear-thinning fluids, which exhibit a drop in viscosity as the shear rate rises. The RE has applications in many domains, such as polymer processing, studying biological fluids, industrial manufacturing processes, and approaches for better oil recovery. Based on this, the recent study focuses on the flow behavior of the RE fluid. The RE fluid model is based on the study conducted by Eyring (1936 and Ree and Eyring (1955). Their research focused on understanding fluids' intricate rheological properties that display viscous and elastic attributes under varying flow circumstances. Ijaz et al. (2018) studied the RE model in elastohydrodynamic theory rather than the Newtonian model to align with high-speed and high-viscosity lubricant film data accurately. In their study, Wang et al. (1991) derived a formula to calculate the reduction factor for the thickness of thermal and RE films. The research conducted by Zahir et al. (2024) focuses on analyzing the magnetohydrodynamic (MHD) behavior of the RE fluid, considering the impact of the Hall effect, inside the confines of two parallel plates. In addition, Rajashekhar et al. (2023) analyzed the diverse characteristics of the RE fluid. Li et al. (2024) highlight the attributes of the flow of RE fluid through the incorporation of ion slip and Hall effects. Vergne et al. (1990) examined the flow of RE fluid in a steady flow state. Shawky (2009) conducted a thorough examination of the control of various physical characteristicson the flow of RE fluid. The work conducted by Musawi and Abdulhadi (2023) investigates the impact of MHD on the flow of RE fluid across a revolving frame. Additionally, Khan et al. (2020) investigate the RE fluid flow across a rolling disk.

Hybrid nanofluid (HNF) is an advanced kind of fluid that has a base fluid with nanoparticles (NP) suspended in it. These nanoparticles are made up of many types of nanomaterials. These nanoparticles may consist of various nanomaterials, such as metal, metal oxide, carbon-based compounds, or other types of nanomaterials. The primary goal of developing HNFs is to enhance the thermal properties of conventional nanofluids, which usually consist of a single kind of nanoparticle. In order to rectify this deficiency, NP, these are particles that vary in size from 1 to 100 nm are added to a standard fluid to enhance its thermal conductivity (TC). The name "nanofluid" (NF) was first used by Choi (1998), who noticed its remarkable properties, including higher heat conductivity, improved rheological features, and greater stability compared to fluids containing micronized particles. Hamilton and Crosser (1962) performed a thermal study to investigate the influence of particle shape, volume, and conductivity on the thermal of a heterogeneous material. The research undertaken by Jagadesh et al. (2024) focuses on analyzing the behavior of a nanofluid with radiative RE properties in a divergent channel that is inclined. The analysis considers several factors, such as energy production, activation energy, viscous dissipation, and Ohmic dissipation. Loganathan et al. (2022) study RE nanofluid dynamics on a convection-heated Riga plate under heat source and radiation from the heat. Additionally, Jamshed et al., (2021) analyzed the thermal efficiency of the HNF flow of the RE model using the Keller box method. Engine oil, sometimes called the "lifeblood" of vehicles, performs the crucial functions of lubricating and cleaning the engine, which are necessary for the proper operation of automobiles. Nanoparticles of titanium dioxide (TiO_2) are known for their exceptional friction-reducing and durable qualities. Their promise lies in improving the friction-reducing capabilities of the oil and offering more robust protection for engine parts. Cobalt ferrite ($CoFe_2O_4$) nanoparticles possess magnetic properties that make them suitable for the production of magnetorheological fluids. In their study, Rashad et al. (2023) focus on the HNF Eyring–Powell (E-P) flow through a perpendicular plate. Ogunseye et al. (2020) examined the HNF flow of the E-P model in their investigation. The RE nanofluid flow bounded by stretchable spinning disks was carried out by Al-Mdallal et al. (2021). The work carried out by Ramzan et al. (2022) examines the transmission of heat of an HNF flowing across a single porous disk, with a focus on the shape factor (SF) of the NP. Khashi'ie et al. (2021) examined the nanoparticle's shape in the HNF that flowed in the direction of the Riga plate. Devi and Devi (2016) investigated the transmission of heat of the HNF flowing across a stretched sheet by exploiting the SF of the nanoparticles. Recent investigations have extensively explored the domain of HNF. The referenced sources have cited the studies mentioned (Enamul & Ontela, 2024; Gowda et al., 2021; Gul et al., 2021; Hosseinzadeh et al., 2021; Raju et al., 2021; Shoaib et al., 2021; SK & Ontela, 2024).

A porous medium is defined as a material characterized by the presence of numerous small openings or voids within its structure. The pores are usually filled with a fluid. This medium facilitates the movement and interaction of these fluids. Having a comprehensive understanding of porous media plays a crucial role in several scientific and engineering disciplines, such as hydrology, petroleum engineering, environmental engineering, and materials science. The Darcy–Forchheimer model is used to elucidate the movement of fluids through porous materials. This model is crucial for precisely forecasting the flow characteristics in many applications that require porous media, such as petroleum engineering, chemical engineering, biomedical engineering, and renewable engineering. PurnaChandar Rao et al. (2021) investigated the flow of RE fluid across an inclined Darcy–Forchheimer porous plate. It is observed that the Weissenberg and Forchheimer values reduce surface friction drag. Shafiq et al. (2023)

focus on the flow of an RE fluid with Darcy–Forchheimer over a flexible, extended surface, considering the convective boundaries. The irreversibility of RE nanofluid flow was the subject of a study that was done by Zafar et al. (2024). They also considered the Darcy–Forchheimer porous medium toward a stretched surface. A study on the thermal properties was conducted by Alzahrani and Khan (2022) focused on the RE fluid flow. They have considered the presence of a nonlinear heat source/sink, which was both contained by a double-spinning disk. Mishra et al. (2024) undertook an investigation into the phenomenon of NF flow over a porous surface. Seth et al. (2018) studied the flow over in a revolving frame, considering a notable porous effect of Darcy–Forchheimer.

Heat transfer (HT) analysis is a crucial component of engineering that guarantees the effective and secure functioning of diverse systems. Engineers may enhance performance, safety, and energy efficiency in several sectors, such as electronics and aircraft, by comprehending and implementing the concepts of conduction, convection, and radiation. Based on the previous work, our study specifically examined the heat transport analysis of a RE HNF. Ali et al. (2024) experimented to clarify the mass and HT characteristics of a RE HNF in a stretched flow condition. Almusawi and Abdulhadi (2021) analyze the heat transmission of the RE fluid in a revolving frame and investigate the effects of MHD. PurnaChandar Rao et al. (2022) carried out an inquiry into the HT characteristics of RE NF flow over an angled plate, considering the effects of quadratic thermal radiation. The energy equation incorporates the non-Fourier heat flow model. Baithalu and Mishra (2023a, 2023b) performed research to improve the reliability of HT analysis by using the response surface approach. In their study, Ontela et al. (2023) optimized the HT rate. It is found that the HT is highly responsive to the Biot number. Hamid et al. (2018) investigate the turbulent flow of Williamson nanofluid, including both convection and HT. The research considers the impact of different thermal conductivity levels in conjunction with the effects of a magnetic field (MF).

Based on the literature survey, it can be inferred that there are several articles available that investigate the flow of HNFs between double parallel rotating disks under different geometries and over a single spinning disk. However, the range of discussion gets narrower when we particularly focus on the movement of Ree–Eyring HNFs between double-rolling disks, with an investigation of the transference of heat analysis in a porous medium. Engine oil is crucial for preserving the internal combustion engine's maximum output and efficiency. In addition to its main purpose of lubricating, it also plays a role in dissipating heat, preventing corrosion, conditioning seals, preserving engine cleanliness, and extending the overall lifespan of the engine. The research investigates the movement of an N-N Ree–Eyring HNF (composed of TiO_2 + $CoFe_2O_4$/EO) that is contained between two rolling disks. The study considers the presence of a Darcy–Forchheimer medium, which affects the flow and also considers the varying TC of the medium and shape factor. The research introduces new findings and highlights areas that need further investigation. The novelty of the present approach lies in the utilization of Ree–Eyring hybrid nanofluids, which synergistically combine two distinct nanoparticles to potentially enhance TC and augment HT performance beyond the capabilities of conventional single-component nanofluids. The research utilizes the Eyring fluid model to represent N-N behaviors correctly. It examines a double-revolving disk configuration that is applicable to industrial applications. Additionally, the study uses the Darcy–Forchheimer model to consider both viscous and inertial effects in the porous medium. In addition, it takes into account varying thermal conductivity in order to model thermal behaviors that are more realistic. The combined impacts of these factors contribute to a thorough comprehension of the system, setting this research apart from previous studies.

7.2 MATHEMATICAL FORMULATION

The formula for the stress tensor of the RE HNF is presented as follows (Shawky, 2009):

$$\overline{S}_{ij} = \left[\mu_{hnf} + \frac{1}{\beta\lambda} sinh^{-1}\left(\frac{\lambda}{c_1}\right)\right] A_1 \tag{7.1}$$

where $\lambda = \sqrt{\frac{1}{2} tr\left(A_1^2\right)}$

Since $\sinh^{-1}\left(\frac{\lambda}{c_1}\right) \approx \frac{\lambda}{c_1}$ for $\frac{\lambda}{c_1} \leq 1$, we get $\overline{S}_{ij} = \left[\mu_{hnf} + \frac{1}{\beta\lambda} * \left(\frac{\lambda}{c_1}\right)\right] A_1$.

$$\text{Therefore, } \overline{S}_{ij} = \left[\mu_{hnf} + \frac{1}{\beta c_1}\right] A_1 \tag{7.2}$$

Here, μ_{hnf} is the dynamic viscosity of the HNF, β, and c_1 are material constants, A_1 denotes the first Rivlin–Erickson tensor.

The present work is based on the flow properties of an R–E hybrid nanofluid confined between two Darcy–Forchheimer porous disks. A continuous, symmetrical, and incompressible motion characterizes the planned condition of the flow. The bottom disk is located at a vertical location of $z=0$ whereas the top disk is positioned at a vertical coordinate of $z = d$, as shown in Figure 7.1. The bottom disk maintains an angular velocity of Ω_1 around the z-axis, whereas the top disk rotates at a velocity of Ω_2. Furthermore, S_2

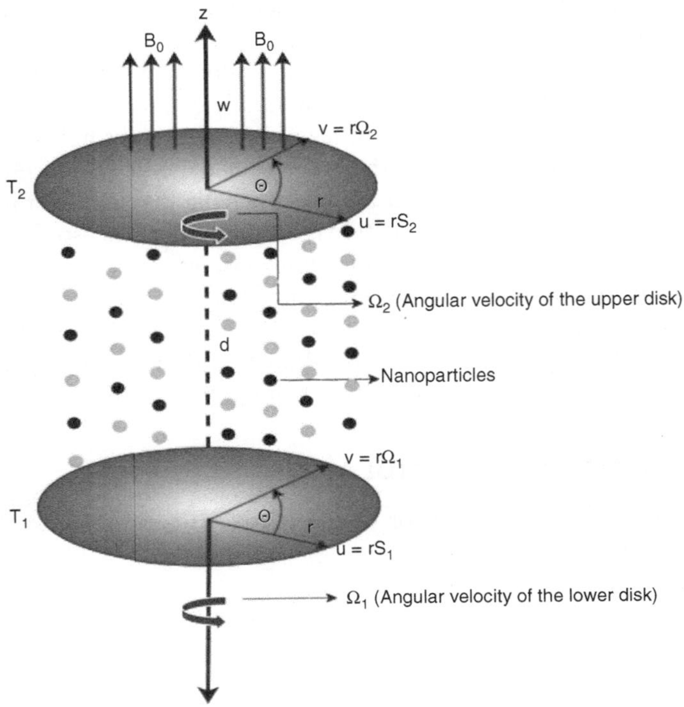

Figure 7.1 Schematic diagram.

and S_1 are the stretching rates of the upper and bottom disks, respectively. In cylindrical dimensions (r,Θ,z), the velocity vector is given by $(u,v,w\ )$.

The equations that determine the system are given in the literature (Khan et al., 2020; Al-Mdallal et al., 2021; Seth et al., 2018):

Mass conservation equation:

$$\frac{1}{r}\frac{\partial}{\partial r}(ru)+\frac{\partial w}{\partial z}=0 \tag{7.3}$$

Radial-momentum equation:

$$\begin{aligned}\rho_{hnf}\left(u\frac{\partial u}{\partial r}+w\frac{\partial u}{\partial z}-\frac{v^2}{r}\right)&=-\frac{\partial p}{\partial r}+\frac{\partial \overline{S}_{rr}}{\partial r}+\frac{\partial \overline{S}_{zr}}{\partial z}\\&+\frac{1}{r}\left(\overline{S}_{rr}-\overline{S}_{\Theta\Theta}\right)-\sigma_{hnf}B_0^2u-\frac{\mu_{hnf}}{\mathrm{K}}u-\rho_{hnf}F_1u^2\end{aligned} \tag{7.4}$$

Angular-momentum equation:

$$\rho_{hnf}\left(u\frac{\partial v}{\partial r}+w\frac{\partial v}{\partial z}+\frac{uv}{r}\right)=\frac{\partial \overline{S}_{r\Theta}}{\partial r}+\frac{\partial \overline{S}_{z\Theta}}{\partial z}+\frac{2\overline{S}_{r\Theta}}{r}$$
$$-\sigma_{hnf}B_0^2 v-\frac{\mu_{hnf}}{\mathrm{K}}v-\rho_{hnf}F_1 v^2 \qquad (7.5)$$

Axial-momentum equation:

$$\rho_{hnf}\left(u\frac{\partial w}{\partial r}+w\frac{\partial w}{\partial z}\right)=-\frac{\partial p}{\partial z}+\frac{\partial \overline{S}_{rz}}{\partial r}+\frac{\partial \overline{S}_{zz}}{\partial z}+\frac{\overline{S}_{rz}}{r} \qquad (7.6)$$

Energy equation:

$$\left(\rho C_p\right)_{hnf}\left(u\frac{\partial T}{\partial r}+w\frac{\partial T}{\partial z}\right)=\frac{\partial}{\partial r}\left(k_{hnf}(T)\frac{\partial T}{\partial r}\right)$$
$$+\frac{k_{hnf}(T)}{r}\frac{\partial T}{\partial r}+\frac{\partial}{\partial z}\left(k_{hnf}(T)\frac{\partial T}{\partial z}\right)-\Delta.q_r+I \qquad (7.7)$$

in the presence of boundary conditions (BC)

$$u=rS_1, v=r\Omega_1, w=0, T=T_1 \ \ at\ z=0$$

$$u=rS_2, v=r\Omega_2, w=0, T=T_2 \ \ at\ z=d \qquad (7.8)$$

The boundary conditions under consideration depict a scenario involving two spinning and radially expanding disks operating at distinct temperatures. This configuration serves to model practical systems, including revolving disk reactors, turbomachinery heat exchangers, and microelectronic cooling devices, where the regulation of flow and heat transfer between the spinning surfaces is of paramount importance.

In the equation (7.7) q_r is given by $q_r=-\frac{16\sigma^* T_2^3}{3k^*}\frac{\partial T}{\partial z}$ and the variable thermal conductivity $k_{hnf}(T)$ is defined by (Hamid et al., 2018)

$$k_{hnf}(T)=k_{hnf}\left(1+\epsilon\frac{T-T_2}{T_1-T_2}\right)$$

And I is defined by

$$I = \left[\mu_{hnf} + \frac{1}{\beta c_1}\right]\left[2\left(\frac{\partial u}{\partial r}\right)^2 + 2\left(\frac{u}{r}\right)^2 + 2\left(\frac{\partial w}{\partial z}\right)^2 + \left(\frac{\partial v}{\partial r} - \frac{v}{r}\right)^2 + \left(\frac{\partial u}{\partial z} + \frac{\partial w}{\partial r}\right)^2 + \left(\frac{\partial v}{\partial z}\right)^2\right]$$

The above boundary conditions pertain to a fluid system arranged in a cylindrical coordinate system and bounded by two planes, $z = 0$ and $z = d$. At $z = 0$, the radial velocity u is equal to rS_1, and the tangential velocity v is equal to $r\Omega_1$. Both velocities are directly proportional to the radius r. There is no axial flow ($w = 0$), and the temperature T is equal to T_1. At $z = d$, the radial velocity u is directly proportional to r and is equal to rS_2. Similarly, the tangential velocity v is also proportional to r and is equal to ($r\Omega_2$). There is no axial flow ($w = 0$), and the temperature (T) is constant and equal to (T_2). This indicates the presence of rotational movement around the z-axis with varying speeds in terms of both distances from the axis and angle. Additionally, there may be temperature differences, which means that the system is suitable for conducting an analysis of fluid dynamics, including both thermal and rotational effects. The stress tensor's components are provided in:

$$\bar{S}_{rr} = 2\left[\mu_{hnf} + \frac{1}{\beta c_1}\right]\frac{\partial u}{\partial r}$$

$$\bar{S}_{\Theta\Theta} = 2\left[\mu_{hnf} + \frac{1}{\beta c_1}\right]\frac{u}{r}$$

$$\bar{S}_{zz} = 2\left[\mu_{hnf} + \frac{1}{\beta c_1}\right]\frac{\partial w}{\partial z}$$

$$\bar{S}_{r\Theta} = \bar{S}_{\Theta r} = \left[\mu_{hnf} + \frac{1}{\beta c_1}\right]\left(\frac{\partial v}{\partial r} - \frac{v}{r}\right)$$

$$\bar{S}_{rz} = \bar{S}_{zr} = \left[\mu_{hnf} + \frac{1}{\beta c_1}\right]\left(\frac{\partial u}{\partial z} + \frac{\partial w}{\partial r}\right)$$

$$\bar{S}_{z\Theta} = \bar{S}_{\Theta z} = \left[\mu_{hnf} + \frac{1}{\beta c_1}\right]\frac{\partial v}{\partial z}$$

7.3 THERMODYNAMIC CHARACTERISTICS

Based on Khashi'ie et al. (2021) and Raju et al. (2021), the thermodynamic characteristics of the HNF may be represented in the following manner:

$$\rho_{hnf} = \phi_2 \rho_{s_2} + \left[(1-\phi_1)\rho_f + \phi_1 \rho_{s_1} \right](1-\phi_2)$$

$$\frac{\mu_{hnf}}{\mu_f} = \frac{1}{(1-\phi_1)^{2.5}(1-\phi_2)^{2.5}}$$

$$\sigma_{nf} = \frac{\sigma_{s_1} + 2\phi_1\left(\sigma_{s_1} - \sigma_f\right) + 2\sigma_f}{\sigma_{s_1} + \phi_1\left(\sigma_f - \sigma_{s_1}\right) + 2\sigma_f} \times \sigma_f$$

$$\sigma_{hnf} = \frac{\sigma_{s_2} + 2\phi_2\left(\sigma_{s_2} - \sigma_{nf}\right) + 2\sigma_{nf}}{\sigma_{s_2} + \phi_2\left(\sigma_{nf} - \sigma_{s_2}\right) + 2\sigma_{nf}} \times \sigma_{nf}$$

$$k_{nf} = \frac{k_{s_1} - (1-s)\phi_1\left(k_{s_1} - k_f\right) - (1-s)k_f}{k_{s_1} + \phi_1\left(k_f - k_{s_1}\right) - (1-s)k_f} \times k_f$$

$$k_{hnf} = \frac{k_{s_2} - (1-s)\phi_2\left(k_{s_2} - k_{nf}\right) - (1-s)k_{nf}}{k_{s_2} + \phi_2\left(k_{nf} - k_{s_2}\right) - (1-s)k_{nf}} \times k_{nf}$$

$$\left(\rho C_p\right)_{hnf} = \left[\left(\rho C_p\right)_f (1-\phi_1) + \phi_1 \left(\rho C_p\right)_{s_1} \right](1-\phi_2) + \phi_2 \left(\rho C_p\right)_{s_2}$$

s_1 represents the first nanomaterial, TiO_2, whereas s_2 represents the subsequent nanomaterial, $CoFe_2O_4$. In this scenario, the symbol 'f' represents the base fluid, namely EO. The abbreviation 'hnf' refers to a hybrid nanofluid mixed with a basic fluid and two different NP. On the other hand, 'nf' denotes a nanofluid that contains just one kind of nanomaterial. The study delineates the shape factor (SF) of the nanoparticle (NP) as s, defined as 3 divided by ψ, with ψ denoting the sphericity of the NP. The volume percentage of the NP is adjusted to 0%, 10%, and 20%. The model discussed before seems to be linked to the Hamilton–Crosser (H-C) model. The H-C model, frequently employed to forecast the effective characteristics of composite materials, exhibits structural parallels with the model in concern.

Table 7.1 The thermophysical characteristics of engine oil (EO), titanium dioxide (TiO_2) (s_2), and cobalt ferrite ($CoFe_2O_4$) (s_2)

	Engine oil (EO) (base fluid)	*TiO_2*	*$CoFe_2O_4$*
ρ (kg/m^3)	863	4250	4907
C_p (J/kg K)	2048	686.2	700
k (W/mK)	1404×10^{-4}	8.9538	3.7
σ (s/m)	5.5×10^{-6}	2.38×10^{6}	5.51×10^{9}

(Khashi'ie et al., 2021; Raju et al., 2021)

Table 7.2 The influence of the shape factor on the nanoparticles

Shape factor			
s	3.0 Spherical	3.7 Brick	8.6 Blade
		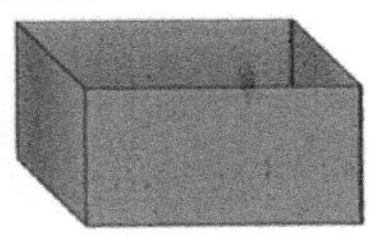	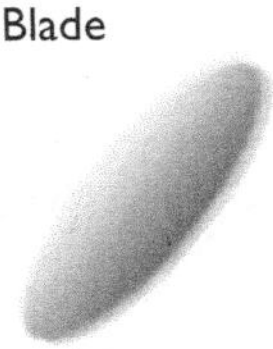

(Khashi'ie et al., 2021)

Tables 7.1 and 7.2 represent the thermophysical characteristics and the different shapes of the nanoparticles, respectively.

7.4 SIMILARITY CONVERSION

The presented dimensional equations are converted into non-dimensional equations with the Von Kármán similarity method (Al-Mdallal et al., 2021):

$$u = r\Omega_1 f'(\eta),\ v = r\Omega_1 g(\eta),\ w = -2d\Omega_1 f(\eta),$$
$$p = \rho_f \Omega_1 \nu_f \left[P(\eta) + \frac{1}{2}\frac{r^2}{d^2}\epsilon_1 \right]$$

$$\theta(\eta) = \frac{T - T_2}{T_1 - T_2},\ \eta = \frac{z}{d} \tag{7.9}$$

Upon applying a transformation (7.9) to equations (7.3–7.7), it has been seen that the equation of the continuity is fully satisfied, and the remaining equations become:

$$\begin{aligned}(a_5+We)f^{(iv)}+2(Re\,a_1)(ff'''+gg')-a_2M^2f''\\ -(a_5\lambda Re)f'-2(a_1FrRe)f'f''=0\end{aligned} \tag{7.10}$$

$$\begin{aligned}(a_5+We)g''+2(Re\,a_1)(fg'-f'g)-a_2M^2g\\ -(a_5Re\,\lambda)g-(a_1Fr\,Re)g^2=0\end{aligned} \tag{7.11}$$

$$\begin{aligned}\left(\frac{1}{Re\,Pr}\right)\left[a_4\{(1+\epsilon\,\theta)\theta''+\epsilon\,\theta'^2\}+Rd\,\theta''\right]+2(a_3)f\theta'\\ +\left(\frac{Ec}{Re\,A^2}\right)(a_5+We)\left[12f'^2+A^2(f''^2+g'^2)\right]=0\end{aligned} \tag{7.12}$$

Therefore, the BC specified by equation (7.8) are as follows:

$$\begin{aligned}&f'(0)=A_1, g(0)=1, f(0)=0, \theta(0)=1 \quad at \quad \eta=0\\ &f'(1)=A_2, g(1)=\Omega, f(1)=0, \theta(1)=0 \quad at \quad \eta=1\end{aligned} \tag{7.13}$$

where

$$a_1=\frac{\rho_{hnf}}{\rho_f},\quad a_2=\frac{\sigma_{hnf}}{\sigma_f},\quad a_3=\frac{(\rho C_p)_{hnf}}{(\rho C_p)_f},\quad a_4=\frac{k_{hnf}}{k_f}$$

$$a_5=\frac{\mu_{hnf}}{\mu_f},\quad M=\sqrt{\frac{\sigma_f B_0^2 d^2}{\mu_f}},\quad Re=\frac{\Omega d^2}{\nu_f},\quad \lambda=\frac{\nu_f}{\Omega_1 K},$$

$$\mathrm{We}=\frac{1}{\mu_f\beta c_1},\quad F_1=\frac{C_b}{r\sqrt{K}},\quad \mathrm{Rd}=\frac{16\sigma^* T_2{}^3}{3k^* k_f},\quad \mathrm{Pr}=\frac{\mu_f (C_p)_f}{k_f}$$

$$\mathrm{Fr}=\frac{C_b}{\sqrt{K}},\quad A=\frac{r}{d},\quad A_1=\frac{S_1}{\Omega_1},\quad A_2=\frac{S_2}{\Omega_1}$$

$$\Omega=\frac{\Omega_2}{\Omega_1}$$

7.5 PHYSICAL QUANTITIES

Skin friction refers to the resistive force produced by a fluid, which moves across a surface. The Nusselt number quantifies the effectiveness of convective HT within the framework of fluid dynamics interacting with a surface. This section presents the computation of the Nusselt number (Nu) and the skin friction coefficient (C_f) for both disks. The mathematical representation of the C_f in the radial direction at the lower disk is provided by (Al-Mdallal et al., 2021):

$$C_{f1} = \frac{\bar{S}_{zr}}{\rho_f \left(r\Omega_1\right)^2}\bigg|_{z=0} = \frac{1}{\rho_f \left(r\Omega_1\right)^2} \frac{r\Omega_1}{d}\left(\mu_{hnf} + \frac{1}{\beta c_1}\right) f''(0)$$

$$= \frac{\mu_f}{d\rho_f \left(r\Omega_1\right)^2}\left[\frac{\mu_{hnf}}{\mu_f} + \frac{1}{\mu_f \beta c_1}\right] f''(0)$$

$$= \frac{\nu_f}{dr\Omega_1}(\mathrm{a}_5 + \mathrm{We}) f''(0)$$

$$= \frac{1}{Re_r}(\mathrm{a}_5 + \mathrm{We}) f''(0) \tag{7.14}$$

Similarly, the skin friction for the bottom disk in the tangential direction,

$$C_{g^1} = \frac{1}{Re_r}(a_5 + \mathrm{We}) g'(0) \tag{7.15}$$

For both the lower and top disks, the Nusselt number represents(Al-Mdallal et al., 2021)

$$\mathrm{Nu}_1 = \frac{q_w|_{z=0}}{k_f\left(T_1 - T_2\right)} \quad \text{and } \mathrm{Nu}_2 = \frac{q_w|_{z=d}}{k_f\left(T_1 - T_2\right)}$$

where q_w is given by $q_w = -k_{hnf}(T)\frac{\partial T}{\partial z} - \frac{16\sigma^* T_2^3}{3k^*}\frac{\partial T}{\partial z}$

$$\text{Therefore, } \mathrm{Nu}_1 = -\left[a_4\left(1 + \epsilon\,\theta(0)\right) + Rd\right]\theta'(0) \tag{7.16}$$

Similarly, for the top disk

$$\mathrm{Nu}_2 = -\left[a_4\left(1+\epsilon\theta(1)\right)+Rd\right]\theta'(1) \tag{7.17}$$

7.6 METHODOLOGY

Utilizing the HAM, our objective is to address the nonlinear equations (7.10–7.12) alongside the specified BC (7.13) in order to obtain a set of semi-analytical solutions. Following the HAM technique, we choose the first guesses. The functions $f_0(\eta), g_0(\eta)$ and $\theta_0(\eta)$ correspond to the linear operators $\mathcal{L}_f$, $\mathcal{L}_g$ and $\mathcal{L}_\theta$ is defined as:

$$f_0(\eta) = A_1\eta-\left(2A_1+A_2\right)\eta^2+\left(A_1+A_2\right)\eta^3 \tag{7.18}$$

$$g_0(\eta) = (\Omega-1)+1 \tag{7.19}$$

$$\theta_0(\eta) = 1-\eta \tag{7.20}$$

Utilizing auxiliary linear operators:

$$\mathcal{L}_f(\eta) = f^{(iv)} \tag{7.21}$$

$$\mathcal{L}_g(\eta) = g'' \tag{7.22}$$

$$\mathcal{L}_{\cdot}(\eta) = \theta'' \tag{7.23}$$

Characterized by the following features:

$$\mathcal{L}_f\left(d_1+d_2\eta+d_3\eta^2+d_4\eta^3\right) = 0 \tag{7.24}$$

$$\mathcal{L}_g\left(d_5+d_6\eta\right) = 0 \tag{7.25}$$

$$\mathcal{L}_\theta\left(d_7+d_8\eta\right) = 0 \tag{7.26}$$

The values of the arbitrary variables $d_i\ (i = 1,2,,\ldots,8)$ has to be determined in accordance with the specified boundary constraint. Let us consider q as a parameter constrained within the interval (0, 1), while $\hbar_f$, $\hbar_g$, $\hbar_\theta$ represent constants that can assume any value. The subsequent step involves articulating the deformation equations of the zeroth order, which are delineated as follows:

$$\mathcal{L}_f\left[f(\eta,q)-f_0(\eta)\right](1-q)=q\hbar_f\mathcal{N}_g\left[F(\eta;q)\right] \tag{7.27}$$

$$\mathcal{L}_{\mathcal{G}}\left[G(\eta,q)-G_0(\eta)\right](1-q)=q\hbar_g\mathcal{N}_g\left[G(\eta;q)\right] \tag{7.28}$$

$$\mathcal{L}_\theta\left[\theta(\eta,q)-\theta_0(\eta)\right](1-q)=q\hbar_\theta\mathcal{N}_\theta\left[\theta(\eta;q)\right] \tag{7.29}$$

And

$$f(0;q)=0, g(0;q)=1, f'(0;q)=A_1, \theta(0;q)=1,$$

$$f(1;q)=0, g(1;q)=\Omega, f'(1;q)=A_2, \theta(1;q)=0 \tag{7.30}$$

The operators with non-linearity $\mathcal{N}_{\mathcal{F}}, \mathcal{N}_{\mathcal{G}}, \mathcal{N}_\theta$ are defined as follows:

$$\begin{aligned}\mathcal{N}_{\mathcal{F}} =&(a_5+We)f^{(iv)}+2(Re\,a_1)(ff'''+gg')-a_2M^2f''\\&-(a_5\lambda Re)f'-2(a_1FrRe)f'f''\end{aligned} \tag{7.31}$$

$$\begin{aligned}\mathcal{N}_{\mathcal{G}} =&(a_5+We)g''+2(Re\,a_1)(fg'-f'g)-a_2M^2g\\&-(a_5Re\,\lambda)g-(a_1Fr\,Re)g^2\end{aligned} \tag{7.32}$$

$$\begin{aligned}\mathcal{N}_{.} =&\left(\frac{1}{Re\,Pr}\right)\left[a_4\{(1+\epsilon\,\theta)\theta''+\epsilon\,\theta'^2\}+Rd\,\theta''\right]+2(a_3)f\theta'\\&+\left(\frac{Ec}{Re\,A^2}\right)(a_5+We)\left[12f'^2+A^2(f''^2+g'^2)\right]\end{aligned} \tag{7.33}$$

When q equals 0 and q equals 1,

$$f(\eta,0)=f_0(\eta), f(\eta,1)=f(\eta)$$

$$g(\eta,0)=g_0(\eta), g(\eta,1)=g(\eta)$$

$$\theta(\eta,0)=\theta_0(\eta), \theta(\eta,1)=\theta(\eta) \tag{7.34}$$

The formulation of $f(\eta;q), g(\eta;q)$, and $\theta(\eta;q)$ through a Taylor series expand yields the following outcomes:

$$f(\eta;q)=f_0(\eta)+\sum_{n=1}^{\infty}f_n(\eta)q^n$$

$$g(\eta;q) = g_0(\eta) + \sum_{n=1}^{\infty} g_n(\eta) q^n$$

$$\theta(\eta;q) = \theta_0(\eta) + \sum_{n=1}^{\infty} \theta_n(\eta) q^n \tag{7.35}$$

For the series (7.35) to achieve convergence at $q=1$, it is essential to assign suitable values to the parameters $\hbar_f$, $\hbar_g$, and $\hbar_\theta$. The equations governing the deformation of mth order can be articulated as follows:

$$\begin{aligned}
&\mathcal{L}_f\left[f - \chi_m f_{m-1}\right] = \hbar_f \mathcal{R}_m^f \\
&\mathcal{L}_g\left[g_m - \chi_m g_{m-1}\right] = \hbar_g \mathcal{R}_m^g \\
&\mathcal{L}_\theta\left[\theta_m - \chi_m \theta_{m-1}\right] = \hbar_\theta \mathcal{R}_m^\theta
\end{aligned} \tag{7.36}$$

where χ_m is defined as

$$\chi_m = 0, m \le 1 = 1, m > 1$$

and

$$\begin{aligned}
\mathcal{R}_m^f(\eta) = {} & (a_5 + We) f_{m-1}^{iv} + 2(Rea_1)\left[\sum_{j=0}^{m-1}\left(f_{m-1-j} f_j''' + g_{m-1-j} g_j'\right)\right] \\
& - a_2 M^2 f_{m-1}'' - (a_5 Re\lambda) f_{m-1}' - 2(a_1 FrRe) \sum_{j=0}^{m-1} f_{m-1-j}' f_j''
\end{aligned} \tag{7.37}$$

$$\begin{aligned}
\mathcal{R}_m^g(\eta) = {} & (a_5 + We) g_{m-1}'' + 2(Rea_1)\left[\sum_{j=0}^{m-1}\left(f_{m-1-j} g_j' - g_{m-1-j} f_j'\right)\right] \\
& - a_2 M^2 g_{m-1} - (a_5 Re\lambda) g_{m-1} - (a_1 FrRe) \sum_{j=0}^{m-1} g_{m-1-j} g_j
\end{aligned} \tag{7.38}$$

$$\begin{aligned}
\mathcal{R}_m^\theta = {} & \frac{1}{Re\,Pr}\left[a_4\left\{\theta_{m-1}'' + \epsilon \sum_{j=0}^{m-1} \theta_{m-1-j} \theta_j'' + \epsilon \sum_{j=0}^{m-1} \theta_{m-1-j}' \theta_j'\right\} + Rd\ \theta_{m-1}''\right] \\
& + 2a_3 \sum_{j=0}^{m-1} f_{m-1-j} \theta' + \frac{Ec}{Re\,A^2}(a_5 + We) \\
& \left[12 \sum_{j=0}^{m-1} f_{m-1-j}' f_j' + A^2\left(\sum_{j=0}^{m-1} f_{m-1-j}'' f_j'' + \sum_{j=0}^{m-1} g_{m-1-j}' g_j'\right)\right]
\end{aligned} \tag{7.39}$$

7.7 THE CONVERGENCE OF HAM

Liao (2010) presented the notion of reducing errors in convergent series solutions, which we can calculate using the average squared residuals representation:

$$E_{f_m} = \frac{1}{L}\sum_{i=0}^{L}\left[\mathcal{N}_f\left(\sum_{k=0}^{m} f_k\left(i\Delta\bar{t}\right)\right)\right]^2$$

$$E_{g_m} = \frac{1}{L}\sum_{i=0}^{L}\left[\mathcal{N}_g\left(\sum_{k=0}^{m} g_k\left(i\Delta\bar{t}\right)\right)\right]^2$$

$$E_{\theta_m} = \frac{1}{L}\sum_{i=0}^{L}\left[\mathcal{N}_\theta\left(\sum_{k=0}^{m} \theta_k\left(i\Delta\bar{t}\right)\right)\right]^2 \tag{7.40}$$

With the present setup, where $\Delta\bar{t} = \frac{1}{L}$ and $L = 20$, we may choose suitable values for the auxiliary parameters $\left(\hbar_f, \hbar_g, \hbar_\theta\right)$ using this approach. The parameters are essential in establishing the convergence zone that is being evaluated while creating $\hbar$ -curves. Figure 7.2(a–c) shows the $\hbar$-curves for the mth-order approximation. The graphs depict the range of convergence that is considered acceptable: $-2.25 \le \hbar_f \le 0.25$, $-2 \le \hbar_g \le 1$, $-3.5 \le \hbar_\theta \le 0.5$. The curves representing $f(0.1)$, $g(0.1)$, and $\theta(0.1)$ are shown. The graph of $f(0.1), g(0.1)$ and $\theta(0.1)$ has a parallel segment that extends into a certain area. This phenomenon arises due to the uniqueness of the solution, where all convergent series of $f(0.1), g(0.1)$, and $\theta(0.1)$ derived from different values of ħ converge to a unique value. The category to which these graphs pertain is the permissible domain of ħ, commonly known as ħ-curves. The optimal values for $\hbar_f$, $\hbar_g$, and $\hbar_\theta$ in the present research are –0.94, –0.94, and –1.91, respectively.

The HAM is employed in the present study owing to its strong capability to tackle highly nonlinear problems without relying on small parameter assumptions. Moreover, HAM offers a built-in mechanism to control convergence and provides highly accurate analytic approximations, making it particularly suited for the complex flow dynamics and variable thermal conditions considered in this investigation.

7.8 VALIDATION

A comparison between the results of this study and those of Al-Mdallal et al.'s (2021) work reveals a strong agreement in the outcomes, as Table 7.3 illustrates. The strong alignment between the two kinds of

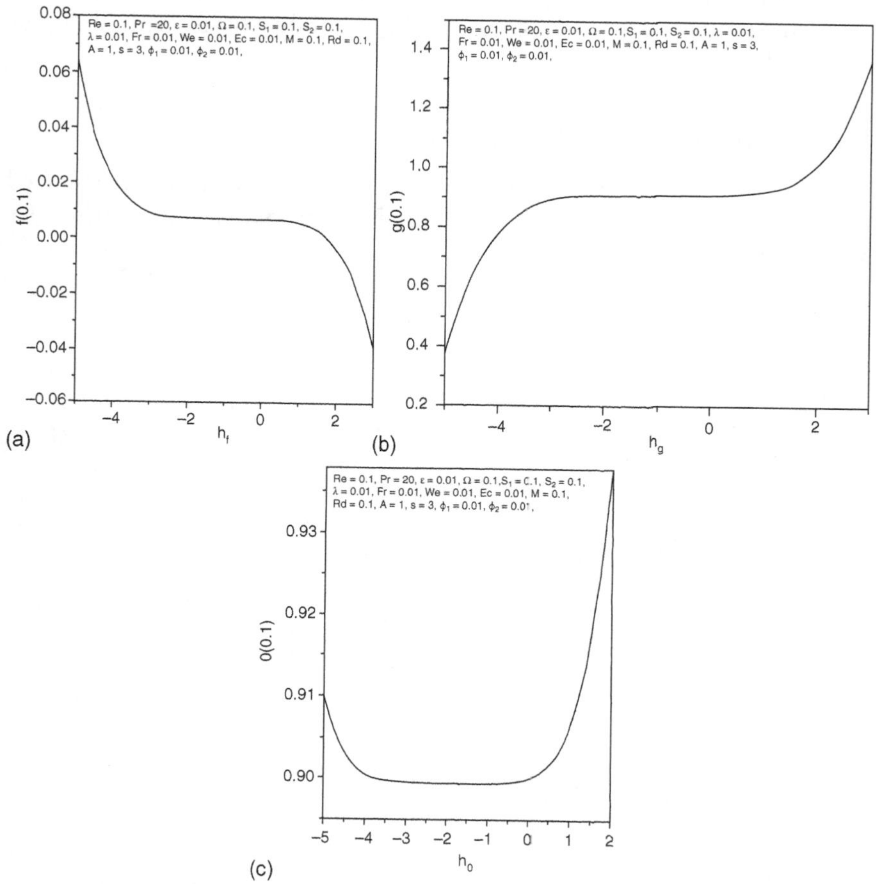

Figure 7.2 (a) ħ-curves for *f*; (b) ħ-curves for *g*; (c) ħ-curves for *θ*.

Table 7.3 Comparison of $-g'(0)$ when $Re = 1, Pr = 1, \epsilon = 0, a_1 = a_2 = \lambda = Fr = We = Ec = M = Rd = \phi_1 = \phi_2 = 0$

Ω	$-h'(0)$ *(Al-Mdallal et al., 2021)*	$-h'(0)$ *(Present work)*
−1.0	2.00095	2:00095
−1.8	1.80259	1:80259
−1.3	1.30442	1.30442
0.0	1.00428	1.00428
0.5	0.50261	0.50261

research results underscores the robustness and reinforces the reliability of the present investigation, affirming the credibility of its methodologies and results.

7.9 RESULTS AND DISCUSSION

The significance of several parameters in the N-N scenario may be recognized in Figures 7.3–7.8. The curves are plotted to approximate the 14th-order approximation.

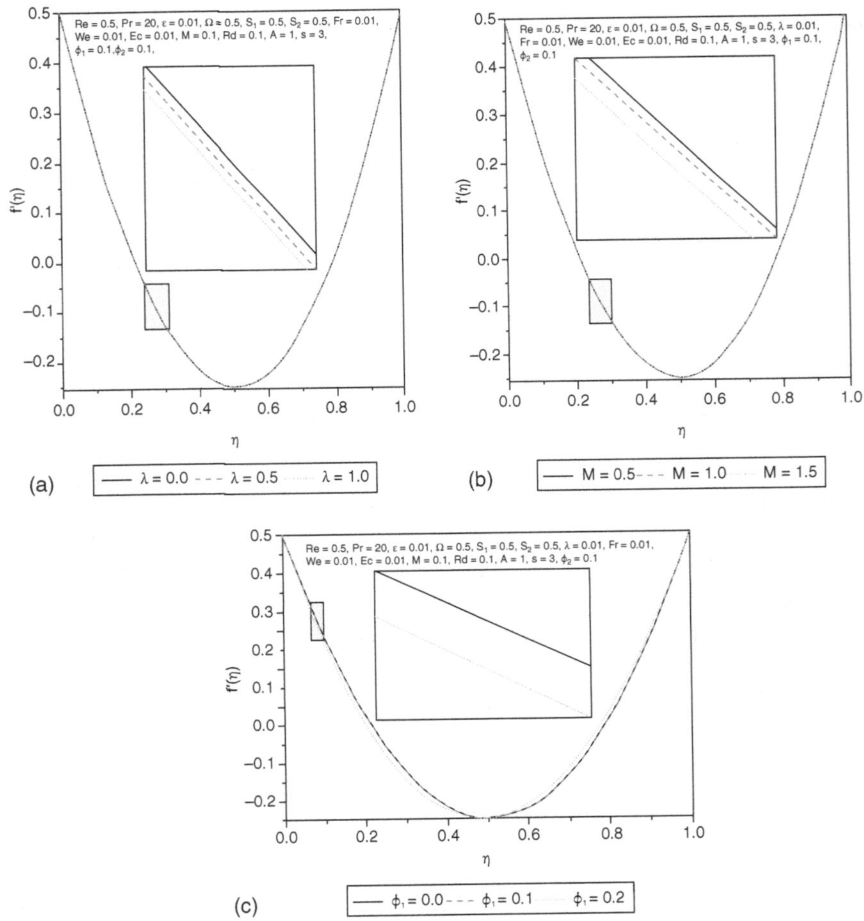

Figure 7.3 (a) Impact of λ on $f'(\eta)$; (b) Impact of M on $f'(\eta)$; (c) Impact of ϕ_1 on $f'(\eta)$.

Radial velocity: Figure 7.3(a–c) illustrates how different critical factors affect the radial velocity $f'(\eta)$.

Evidently, the radial velocity reduces as the porosity parameter λ hikes, as seen in Figure 7.3(a). Porous materials provide a greater degree of retardation. Consistently, the force exerts its influence to slow down the $f'(\eta)$ of the fluid.

An illustration of the connection between $f'(\eta)$ and M is shown in Figure 7.3(b). An elevation of M leads to a reduction in $f'(\eta)$. The engagement of an MF with an electrically conductive fluid results in the generation of the Lorentz force (LF). This, consequently, leads to an increase in flow resistance, which subsequently results in a decrease in $f'(\eta)$. Figure 7.3(c) demonstrates a scaling down in the $f'(\eta)$ as the titanium dioxide concentration increases. Through the presence of nanoparticles, the viscosity is enhanced, which ultimately results in a decline in velocity. These findings have practical significance in magnetohydrodynamic flow control, thermal management systems, and material processing industries, where precise regulation of fluid motion under magnetic influence is essential.

Tangential velocity: Figure 7.4(a) reveals a decrease in tangential velocity corresponding to an increase in the porosity parameter λ. Flow experiences more resistance in porous materials, and bigger pore sizes often correlate with increased porosity, resulting in heightened flow resistance. A representation of the thorough relationship between M and $g(\eta)$ is shown in Figure 7.4(b). When *M* increases, the tangential velocity decreases, which is the outcome of the increase in *M*. The LF is produced when a MF is present in a fluid that is electrically conducting. This causes the flow resistance to rise, which in turn causes the velocity to decrease.

As the titanium dioxide's concentration, ϕ_1 rises in the presence of cobalt ferrite $CoFe_2O_4$, $g(\eta)$ decreases, as seen in Figure 7.4(c) to illustrate this phenomenon. The viscosity rises through the presence of nanoparticles, which ultimately results in a decrease in velocity. These findings are highly relevant for designing porous media-based heat exchangers, magnetic flow control systems, and nanoengineered cooling technologies where controlling fluid resistance and flow behavior is critical.

Axial velocity: This section investigates the effect of various dimensionless variables on the axial velocity, as seen in Figure 7.5(a–c). The link between the axial velocity and λ is clearly seen in Figure 7.5(a), which exhibits an inverse relationship, a reduction in the velocity relative to the tangential direction. The flow encounters increased resistance when interacting with a porous material, and a rise in the porosity parameter typically signifies the existence of larger pore sizes. Figure 7.5(b) illustrates the connection regarding the relationship between *M* and the axial velocity $f(\eta)$. The axial velocity drops with a larger value of *M*. The Lorentz force is produced when

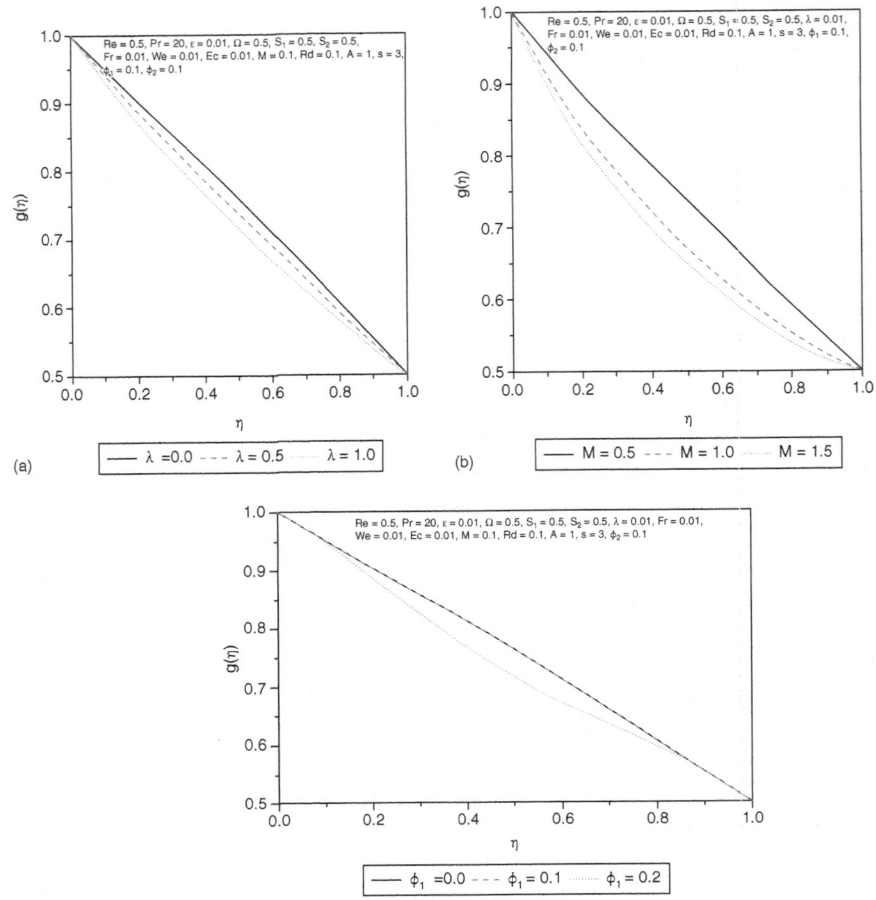

Figure 7.4 (a) Impact of λ on $g(\eta)$; Fi(b) Impact of M on $g(\eta)$; (c) Impact of ϕ_1 on $g(\eta)$.

an MF is present in a fluid that conducts electricity. This, therefore, leads to an elevation in the resistance of fluid dynamics and a subsequent decline in axial velocity. These observations are pivotal for optimizing designs in porous media cooling systems, magnetohydrodynamic flow control devices, and advanced fluidic applications where managing axial momentum under external fields is essential.

Temperature profile: Figure 7.6(a–d) shows how several fluid parameters affect the temperature profile, $\theta(\eta)$. As the Eckert number (Ec) increases, there is a corresponding enhancement in the temperature of the fluid, as seen in Figure 7.6(a). Ec precisely calculates heat dissipation in a system by

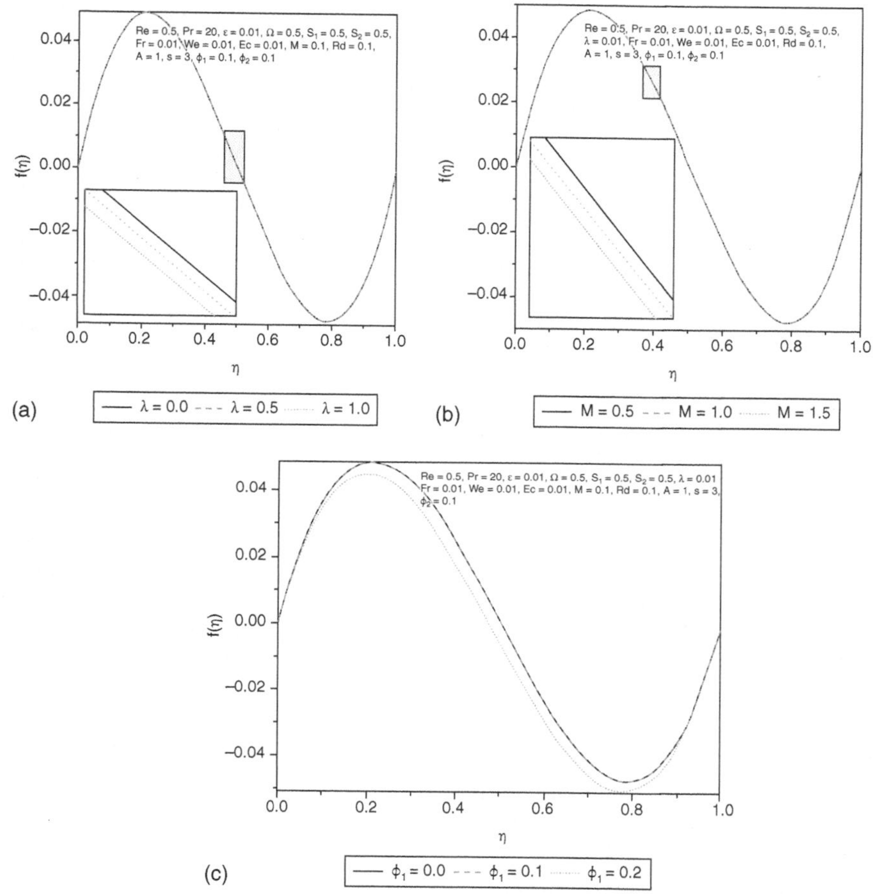

Figure 7.5 (a) Impact of λ on $f(\eta)$; (b) Impact of M on $f(\eta)$; (c) Impact of ϕ_1 on $f(\eta)$.

correlating kinetic energy with enthalpy changes. Consequently, the kinetic energy increases as the Ec value rises, resulting in an escalation of friction within the fluid.

Figure 7.6(b) shows that for larger values of the shape factor *s*, the temperature profile decreases. Greater shape factor *s* often correlates to more elongated or intricate geometries, resulting in an increased surface area to volume ratio. Elaborate geometries may facilitate the even distribution of heat over the material. Increased shape factors of nanoparticles enhance heat dispersion, hence reducing the formation of hot spots and promoting a more homogeneous distribution of temperature. This efficient dispersion mechanism may reduce the highest temperatures inside the system, resulting in a reduction of the overall temperature profile.

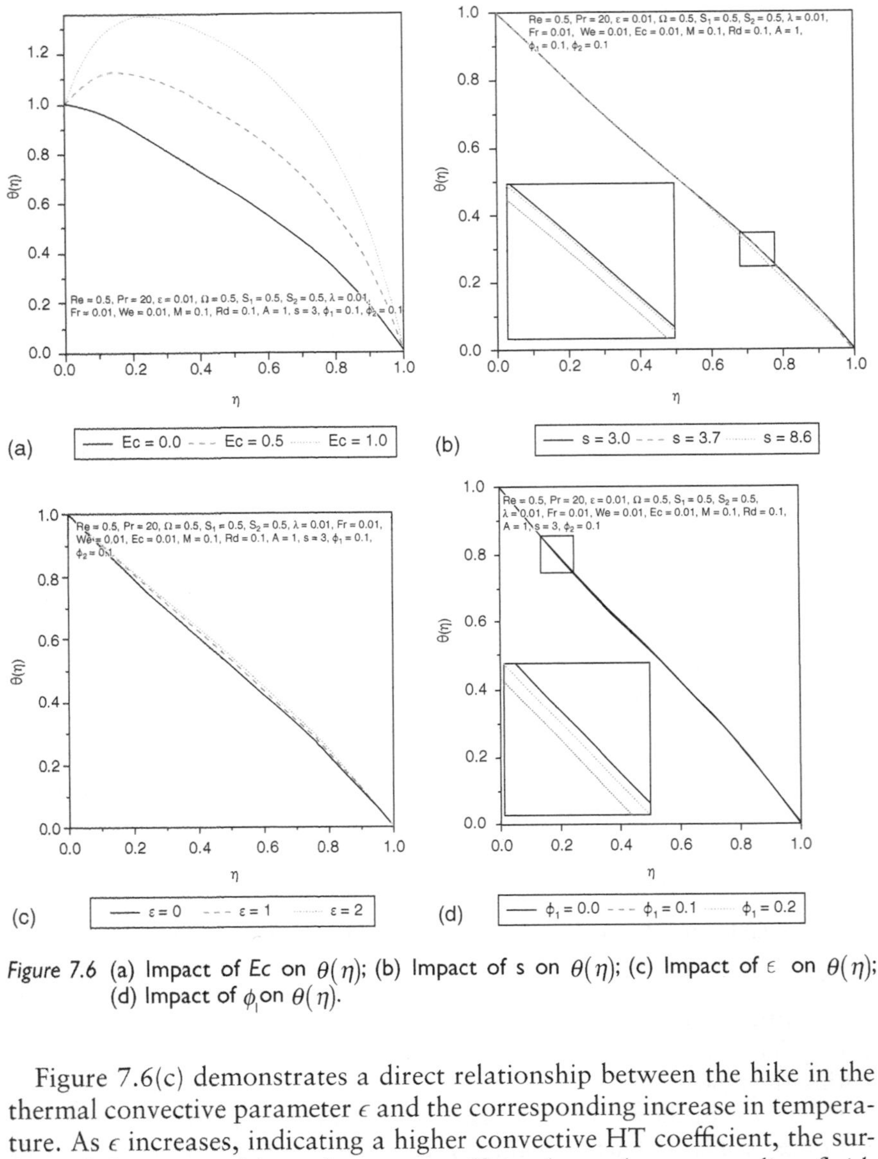

Figure 7.6 (a) Impact of Ec on $\theta(\eta)$; (b) Impact of s on $\theta(\eta)$; (c) Impact of ε on $\theta(\eta)$; (d) Impact of ϕ_1 on $\theta(\eta)$.

Figure 7.6(c) demonstrates a direct relationship between the hike in the thermal convective parameter ϵ and the corresponding increase in temperature. As ϵ increases, indicating a higher convective HT coefficient, the surface of the material loses heat more efficiently to the surrounding fluid. This efficient heat loss requires the surface to be at a higher temperature to maintain the same heat flux from the interior of the material. Consequently, $\theta(\eta)$ near the surface is elevated, and the temperature gradient becomes steeper, resulting in an elevated temperature profile in proximity to the surface. This phenomenon mirrors practical applications such as cooling of electronic components, thermal management in aerospace structures, and advanced heat exchanger designs, where enhanced convective heat transfer demands higher surface temperatures to maintain system stability and performance.

Figure 7.6(d) illustrates that the $\theta(\eta)$ rises with an increase in ϕ_1. The enhancement of TC and HT characteristics in NF is the underlying reason for this phenomenon. The inclusion of cobalt ferrite ($CoFe_2O_4$) in the base fluid, together with a rising concentration of titanium dioxide (TiO_2) nanoparticles, improves the thermal conductivity of the HNF. Titanium dioxide is recognized for its exceptional thermal conductivity, and when it is disseminated in a fluid, it greatly enhances the overall ability to transmit heat. Increasing the volume percentage of TiO_2 enhances the nanofluid's capacity to transmit heat from the heated surface to the bulk fluid. The enhanced thermal conductivity facilitates a more effective heat transfer mechanism, leading to an elevated temperature profile. The presence of both $CoFe_2O_4$ and higher concentrations of TiO_2 enhances the convective heat transfer, resulting in an elevated temperature distribution throughout the fluid. The synergistic impact of $CoFe_2O_4$ and increased levels of TiO_2 enhances convective heat transfer while simultaneously achieving a consistent temperature distribution. This uniformity is essential for various applications, including high-performance cooling systems, thermal energy storage solutions, and sophisticated industrial heat exchangers.

Skin Friction: Figure 7.7(a–d) demonstrates the impact of critical fluid factors on skin friction. Figure 7.7(a) and (b) demonstrate that the skin friction for the bottom disk increases in both the radial and tangential directions as the values of λ and M rise. The coefficient of skin friction quantifies the level of flow resistance caused by the surface friction between the fluid and the disk. In the event that the porosity parameter λ is increased, it signifies a higher level of porosity in the medium, enabling enhanced fluid infiltration and interaction within the porous structure. This contact enhances the drag force exerted on the fluid, thereby increasing the coefficient of skin friction. Similarly, a shoot-up in the MF parameter M signifies a more intense MF exerted on the fluid, leading to the occurrence of magnetohydrodynamic (MHD) processes. The LF, which emerges from the interplay between the MF and electric currents within the fluid, introduces supplementary resistance to the flow dynamics. The increased resistance results in a higher skin friction coefficient in both the tangential and radial directions on the lower disk. Therefore, both the higher porosity and magnetic field characteristics result in increased frictional resistance, which subsequently elevates the C_f.

Figure 7.7(c) and (d) demonstrates that the skin friction coefficient for the bottom disk rises in both the radial direction and the tangential direction when the values of the Weissenberg number We and M increase. The fluid flow may be elucidated by considering the influences of viscoelasticity and

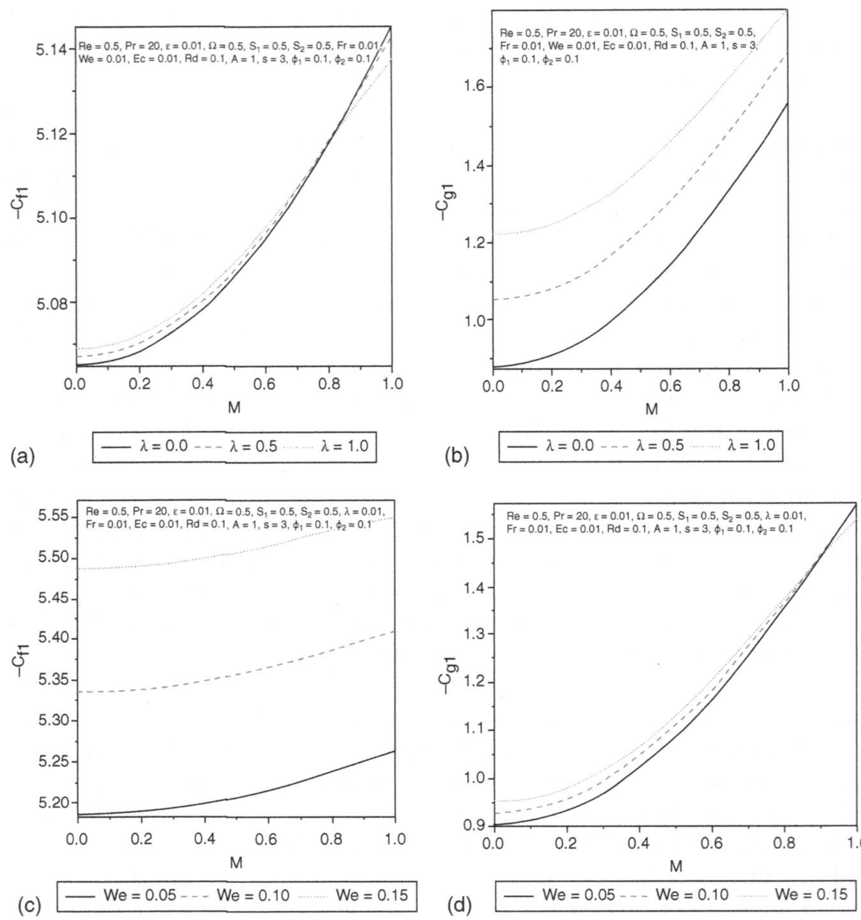

Figure 7.7 (a) Impact of λ and M on -C_{f1}; (b) Impact of λ and M on -C_{g1}; (c) Impact of We and M on -C_{f1}; (d) Impact of We and M on -C_{g1}.

MHD. The Weissenberg number (We) is a dimensionless number that quantifies the ratio of elastic effects to viscous effects in a viscoelastic fluid. A rise in the value of We implies a stronger presence of elastic effects, leading to a higher level of resistance to deformation. The heightened resistance amplifies the frictional forces at the boundary among the fluid and solid states, leading to an elevation of the C_f in both the radial direction and tangential direction. M quantifies the influence of an externally applied MF in an electrically conductive fluid. As M grows, the influence of the LF (the force exerted by the MF on the charges in the fluid) becomes more pronounced. This force exerts its influence in the inverse direction to the fluid velocity, resulting in more resistance and hence causing a rise in C_f. The synergistic impact of the increased viscoelastic characteristics (attributable to higher We) and the amplified magnetic resistance (due to higher M) results in an elevation of C_f in both the tangential and radial orientations on the bottom disk.

Nusselt number: The characteristics of the fluid and the conditions of its flow have a significant impact on the Nusselt number. This method is frequently employed to examine the transfer of thermal energy across a range of technological applications, such as cooling systems, pipelines, and heat exchangers.

Figure 7.8(a) and (b) demonstrates that the Nusselt number experiences a rise in proximity to both disks when larger values of the radiation parameter (Rd) and the MF parameter M. The Rd measures the influence of thermal radiation on the process of HT. A higher Rd value indicates that radiative HT is becoming more prominent. This improvement in radiative HT results in an increased temperature differential between the surfaces of the disk, thereby raising the Nusselt number. The MF parameter (M) quantifies the effect of an external MF on the fluid. An elevation in the value of M leads to a more significant MHD effect. These effects provide a Lorentz force that helps stabilize the boundary layer and improve convective HT by lowering the thermal boundary layer's thickness. Consequently, the larger values of the radiation parameter and magnetic field parameter increase the efficiency of heat transmission, leading to an enhanced Nusselt number in the vicinity of the disks. Such improvements are critical in high-temperature industrial processes, magnetic control systems, and thermal management technologies where maximizing heat dissipation is essential for operational reliability.

Moreover, Figure 7.8(c) and (d) illustrates that the Nusselt number exhibits an increase when the concentration of TiO_2 volume percentage and M are bigger, particularly in the vicinity of both disks. An elevation in the volume percentage of TiO_2 enhances the TC of the HNF, resulting in improved heat transfer efficiency. As the thermal conductivity increases, the fluid becomes more effective at transferring heat. This allows the fluid to remove heat from the disks more efficiently. As a result, the Nusselt number increases. Moreover, an increased MF parameter (M) signifies a more pronounced MHD phenomena, whereby the LF produced by the MF interacts with the fluid flow. This interaction enhances the stability of the boundary layer and promotes the passage of heat by convection by decreasing the boundary layer's thickness. The combination of increased thermal conductivity resulting from greater TiO_2 concentration and better convective heat transfer owing to stronger MHD effects lead to a substantial rise in the Nusselt number in the vicinity of the disks. The synergistic effect of enhanced thermal conductivity due to higher TiO_2 concentrations and improved convective transport driven by stronger MHD phenomena leads to a significant rise in the Nu around the disks. Such mechanisms are highly relevant to real-world applications,

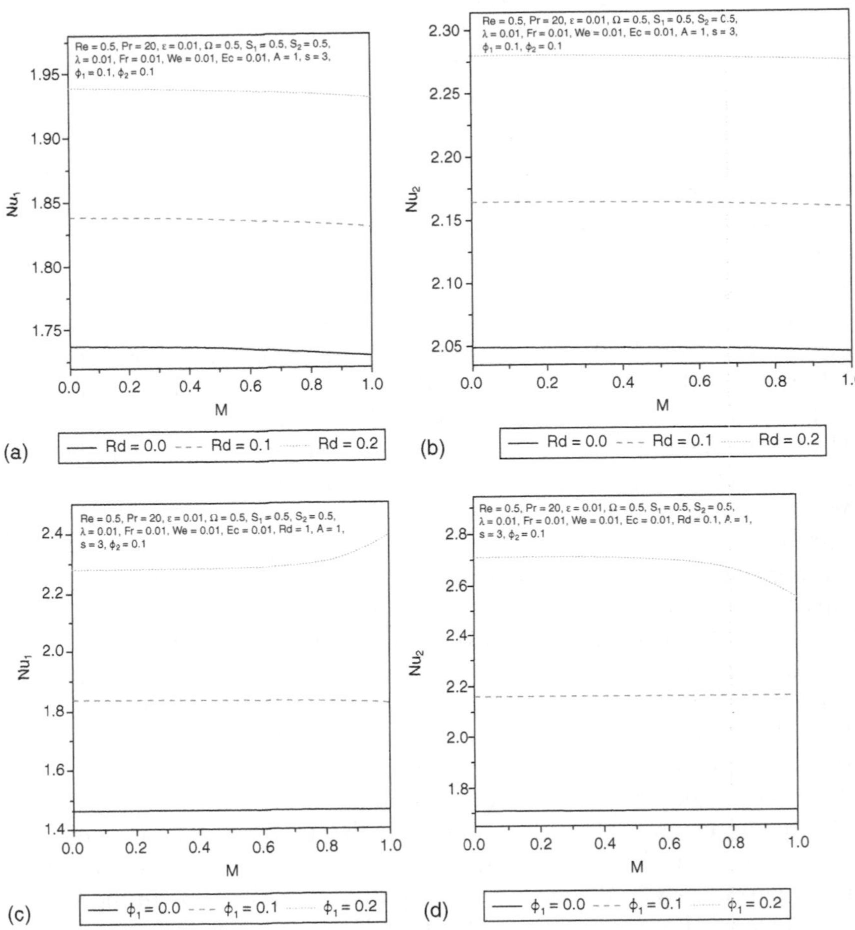

Figure 7.8 (a) Impact of Rd and M on Nu_1; (b) Impact of Rd and M on Nu_2; (c) Impact of ϕ_1 and M on Nu_1; (d) Impact of ϕ_1 and M on Nu_2.

including cooling of rotating machinery, magnetic control of industrial heat exchangers, and thermal management in energy systems where maximizing heat removal is critical for operational efficiency.

7.10 CONCLUSION

This chapter presents a detailed examination of the transmission of heat properties of a Ree–Eyring hybrid nanofluid in a complex system circulating between two rotating disks. The flow takes place inside a medium that follows the Darcy–Forchheimer law and has a thermal conductivity that varies. The results illustrate the potential of hybrid nanofluids to enhance heat transfer across various engineering applications. The current mathematical model of the issue is resolved using a methodology called the HAM. The findings obtained provide good comparisons with the previous study conducted by Al-Mdallal et al. (2021). The results of this research have practical consequences for several engineering applications, including the cooling of electronic devices, industrial heat exchangers, and various systems where enhanced thermal performance is essential. To summarize, the simulations have resulted in the following conclusions:

- Radial velocity decreases for higher porosity and magnetic field parameter values. Additionally, it shows an inverse relation with the titanium dioxide concentration.
- As the porosity and magnetic field parameters rise, the tangential velocity decreases.
  Furthermore, it demonstrates a negative correlation with the volume fraction concentration of titanium dioxide nanoparticles.
- As the porosity and magnetic field parameters rise, the axial velocity decreases. Moreover, it has an inverse relationship with the volume fraction concentration of titanium dioxide nanoparticles.
- The skin friction coefficient for the bottom disk escalates in the radial and tangential directions for larger values of the porosity parameter, Weissenberg number, and magnetic field parameters.
- The Nusselt number for both disks increases with the elevation of the magnetic field parameter, the radiation parameters, and the titanium dioxide volume fraction concentration.

Future studies may build upon this foundation by integrating the influences of slip boundary conditions, temperature-dependent viscosity, or the phenomenon of nanoparticle aggregation. Additionally, experimental validation of the model could further solidify the practical relevance of the findings for real-world thermal management systems.

REFERENCES

Ali, F., Zaib, A., Faizan, M., Zafar, S. S., Alkarni, S., Shah, N. A., & Chung, J. D. (2024). Heat and mass exchanger analysis for Ree–Eyring hybrid nanofluid through a stretching sheet utilizing the homotopy perturbation method. *Case Studies in Thermal Engineering*, *54*, 104014.

Al-Mdallal, Q. M., Renuka, A., Muthtamilselvan, M., & Abdalla, B. (2021). Ree–Eyring fluid flow of Cu–water nanofluid between infinite spinning disks with an effect of thermal radiation. *Ain Shams Engineering Journal*, *12*(3), 2947–2956.

Almusawi, B. A., &Abdulhadi, A. M. (2021). Heat transfer analysis and magnetohydrodynamics effect on peristaltic transport of Ree–Eyring fluid in rotating frame. *Iraqi Journal of Science*, 2714–2725.

Alzahrani, F., & Khan, M. I. (2022). Entropy generation and Joule heating applications for Darcy Forchheimer flow of Ree–Eyring nanofluid due to double rotating disks with artificial neural network. *Alexandria Engineering Journal*, *61*(5), 3679–3689.

Baithalu, R., & Mishra, S. R. (2023a). On the free convection of magneto-micropolar fluid in association with thermal radiation and chemical reaction and optimized heat transfer rate using response surface methodology. *Modern Physics Letters B*, *37*(33), 2350171.

Baithalu, R., & Mishra, S. R. (2023b). Response surface methodology in optimizing heat transfer rate for homogeneous–heterogeneous reactions with thermal radiation effect on the micropolar nanofluid. *Numerical Heat Transfer, Part A: Applications*, 1–20.

Choi, S. U.-S. (1998). *Nanofluid Technology: Current Status and Future Research*. www.osti.gov/servlets/purl/11048

Devi, S. P. A., & Devi, S. S. U. (2016). Numerical investigation of hydromagnetic hybrid Cu--Al_2O_3/water nanofluid flow over a permeable stretching sheet with suction. *International Journal of Nonlinear Sciences and Numerical Simulation*, *17*(5), 249–257.

Enamul, S., & Ontela, S. (2024). Magnetohydrodynamic Darcy–Forchheimer flow of non-Newtonian second-grade hybrid nanofluid bounded by double-revolving disks with variable thermal conductivity: Entropy generation analysis. *Hybrid Advances*, 6(May), 100226. https://doi.org/10.1016/j.hybadv.2024.100226

Eyring, H. (1936). Viscosity, plasticity, and diffusion as examples of absolute reaction rates. *The Journal of Chemical Physics*, *4*(4), 283–291.

Gowda, R. J. P., Kumar, R. N., Aldalbahi, A., Issakhov, A., Prasannakumara, B. C., Rahimi-Gorji, M., &Rahaman, M. (2021). Thermophoretic particle deposition in time-dependent flow of hybrid nanofluid over rotating and vertically upward/downward moving disk. *Surfaces and Interfaces*, *22*, 100864.

Gul, T., Kashifullah, Bilal, M., Alghamdi, W., Asjad, M. I., &Abdeljawad, T. (2021). Hybrid nanofluid flow within the conical gap between the cone and the surface of a rotating disk. *Scientific Reports*, *11*(1), 1180.

Hamid, A., Khan, M., & others. (2018). Unsteady mixed convective flow of Williamson nanofluid with heat transfer in the presence of variable thermal conductivity and magnetic field. *Journal of Molecular Liquids*, *260*, 436–446.

Hamilton, R. L., & Crosser, O. K. (1962). Thermal conductivity of heterogeneous two-component systems. *Industrial & Engineering Chemistry Fundamentals*, *1*(3), 187–191.

Hosseinzadeh, K., Roghani, S., Asadi, A., Mogharrebi, A., & Ganji, D. D. (2021). Investigation of micropolar hybrid ferrofluid flow over a vertical plate by considering various base fluid and nanoparticle shape factor. *International Journal of Numerical Methods for Heat & Fluid Flow*, *31*(1), 402–417.

Ijaz, N., Zeeshan, A., & Bhatti, M. M. (2018). Peristaltic propulsion of particulate non-Newtonian Ree–Eyring fluid in a duct through constant magnetic field. *Alexandria Engineering Journal*, *57*(2), 1055–1060.

Jagadesh, V., Sreenadh, S., Ajithkumar, M., Sucharitha, G., & Lakshminarayana, P. (2024). Convective peristaltic pumping of MHD Ree–Eyring nanofluid in a chemically reacting flexible divergent channel with activation energy and radiation. *Radiation Effects and Defects in Solids*, 1–16.

Jamshed, W., Baleanu, D., Nasir, N. A. A. M., Shahzad, F., Nisar, K. S., Shoaib, M., Ahmad, S., & Ismail, K. A. (2021). The improved thermal efficiency of Prandtl–Eyring hybrid nanofluid via classical Keller box technique. *Scientific Reports*, *11*(1), 23535.

Khan, M. I., Ahmad Khan, S., Hayat, T., Faisal Javed, M., & Waqas, M. (2020). Entropy generation in radiative flow of Ree–Eyring fluid due to due rotating disks. *International Journal of Numerical Methods for Heat & Fluid Flow*, *30*(4), 1839–1865.

Khashi'ie, N. S., Arifin, N. M., Sheremet, M., & Pop, I. (2021). Shape factor effect of radiative Cu-Al_2O_3/H_2O hybrid nanofluid flow towards an EMHD plate. *Case Studies in Thermal Engineering*, *26*, 101199.

Li, S., Shaheen, N., Ramzan, M., Kadry, S., & Saleel, C. A. (2024). Thermophoretic particle deposition on double-diffusive Ree–Eyring fluid flow across two deformable rotating disks with Hall current and ion slip. *Journal of Magnetism and Magnetic Materials*, *589*, 171547.

Liao, S. (2010). An optimal homotopy-analysis approach for strongly nonlinear differential equations. *Communications in Nonlinear Science and Numerical Simulation*, *15*(8), 2003–2016.

Loganathan, K., Alessa, N., Jain, R., Ali, F., &Zaib, A. (2022). Dynamics of heat and mass transfer: Ree–Eyring nanofluid flow over a Riga plate with bioconvention and thermal radiation. *Frontiers in Physics*, *10*, 974562.

Mishra, S. R., Sharma, R. P., & Swain, L. (2024). Illustration of Joule dissipation on the time-dependent stagnation point flow of nanofluid through a porous surface. *International Journal of Modern Physics B*, *38*(5), 2450077.

Musawi, B. A., &Abdulhadi, A. M. (2023). Influence of heat transfer analysis and (MHD) on peristaltic transport of Ree–Eyring fluid in rotating frame. *AIP Conference Proceedings*, *2414*(1), 040076.

Ogunseye, H. A., Tijani, Y. O., & Sibanda, P. (2020). Entropy generation in an unsteady Eyring–Powell hybrid nanofluid flow over a permeable surface: A Lie group analysis. *Heat Transfer*, *49*(6), 3374–3390.

Ontela, S., Musala, V., Ahmed, S. E., & Thumma, T. (2023). Optimization and sensitivity analysis on axisymmetric motile microorganism flow of visco-elastic nanofluid over a spinning circular disk with a central composite model. *Numerical Heat Transfer, Part A: Applications*, 1–26.

PurnaChandar Rao, D., Thiagarajan, S., & Srinivasa Kumar, V. (2021). Darcy--Forchheimer flow of Ree--Eyring fluid over an inclined plate with chemical reaction: a statistical approach. *Heat Transfer*, *50*(7), 7120–7138.

PurnaChandar Rao, D., Thiagarajan, S., & Srinivasa Kumar, V. (2022). Significance of quadratic thermal radiation on the bioconvective flow of Ree–Eyring fluid through an inclined plate with viscous dissipation and chemical reaction: Non-Fourier heat flux model. *International Journal of Ambient Energy*, *43*(1), 6436–6448.

Rajashekhar, C., Vaidya, H., Manjunatha, G., Dodo, Y. A., Prasad, K. V, Gnaba, H., Amat, R. H. J. C., & Khan, M. I. (2023). Mathematical modeling and simulation of peristaltic activity in Ree–Eyring fluid flow through non-uniform complaint channel: Different varying conditions. *Journal of the Indian Chemical Society*, *100*(3), 100954.

Raju, S. S. K., Babu, M. J., & Raju, C. S. K. (2021). Irreversibility analysis in hybrid nanofluid flow between two rotating disks with activation energy and cross-diffusion effects. *Chinese Journal of Physics*, *72*, 499–529.

Ramzan, M., Riasat, S., Aljurbua, S. F., Ghazwani, H. A. S., & Mahmoud, O. (2022). Hybrid nanofluid flow induced by an oscillating disk considering surface catalyzed reaction and nanoparticles shape factor. *Nanomaterials*, *12*(11), 1794.

Rashad, A. M., Nafe, M. A., &Eisa, D. A. (2023). Heat generation and thermal radiation impacts on flow of magnetic Eyring--Powell hybrid nanofluid in a porous medium. *Arabian Journal for Science and Engineering*, *48*(1), 939–952.

Ree, T., & Eyring, H. (1955). Theory of non-Newtonian flow. i. solid plastic system. *Journal of Applied Physics*, *26*(7), 793–800.

Seth, G. S., Kumar, R., & Bhattacharyya, A. (2018). Entropy generation of dissipative flow of carbon nanotubes in rotating frame with Darcy–Forchheimer porous medium: A numerical study. *Journal of Molecular Liquids*, *268*, 637–646.

Shafiq, A., Çolak, A. B., & Sindhu, T. N. (2023). Optimization of the numerical treatment of the Darcy--Forchheimer flow of Ree--Eyring fluid with chemical reaction by using artificial neural networks. *International Journal for Numerical Methods in Fluids*, *95*(1), 176–192.

Shawky, H. M. (2009). Pulsatile flow with heat transfer of dusty magnetohydrodynamic Ree–Eyring fluid through a channel. *Heat and Mass Transfer*, *45*, 1261–1269.

Shoaib, M., Raja, M. A. Z., Sabir, M. T., Awais, M., Islam, S., Shah, Z., & Kumam, P. (2021). Numerical analysis of 3-D MHD hybrid nanofluid over a rotational disk in presence of thermal radiation with Joule heating and viscous dissipation effects using Lobatto IIIA technique. *Alexandria Engineering Journal*, *60*(4), 3605–3619.

SK, E., & Ontela, S. (2024). Entropy analysis of Darcy–Forchheimer flow of Reiner–Rivlin $A1_2O_3$–Cu/engine oil based hybrid nanofluid between two rotating disks. *Numerical Heat Transfer; Part A: Applications*, 1–30. https://doi.org/10.1080/10407782.2024.2382926

Vergne, F., Chaomleffel, J. P., &Dalmaz, G. (1990). Elastohydrodynamic lubrication of point contacts with a Ree–Eyring fluid: Film thickness and traction. *Journal of Rheology*, *34*(8), 1357–1371.

Wang, S., Cusano, C., & Conry, T. F. (1991). *Thermal Analysis of Elastohydrodynamic Lubrication of Line Contacts Using the Ree–Eyring Fluid Model. Journal of Tribology* Apr 1991, *113*(2), 232–242. https://doi.org/10.1115/1.2920611

Zafar, S. S., Zaib, A., Ali, F., Alduais, F. S., Bossly, A. Al, & Saeed, A. (2024). Second law analysis on Ree–Eyring nanoliquid and Darcy Forchheimer flow through a significant stratification in the gyrotactic microorganism. *International Journal of Numerical Methods for Heat & Fluid Flow*, *34*(2), 494–519.

Zahir, H., Gul, J., Inc, M., Alqahtani, R. T., & others. (2024). Impact of fractional magnetohydrodynamic and hall current on Ree–Eyring fluid flow by using radial basis function method. *Alexandria Engineering Journal*, *88*, 210–215.

Chapter 8

Illustration of *n*th order chemical reaction on second-grade nanofluid flow past a Riga sensor plate with the influence of viscous dissipation

Abhishek Sharma and Ram Prakash Sharma

8.1 INTRODUCTION

Numerous technical applications gain advantages from nanofluid utilization. Researchers and engineers find nanofluid characteristics highly relevant. Various sectors, including solar, thermal fluid systems, power, and nuclear reactors, may benefit from advancements in nanofluid technologies. In recent years, substantial studies have been undertaken by many scientists on the diverse uses of nanofluids. Mohan Krishna et al. (2017) explored the conductive nanofluid flow through a slender needle. They utilized Sakiadis and Blasius boundary layer flow in a slender needle configuration. Felicita et al. (2023) applied Runge–Kutta (R-K) method to explore the impact of Brownian and thermophores effect in nanofluid flow across a microchannel. Deepthi et al. (2024) inspected the nanofluid flow over an extending sheet utilizing the Buongiorno nanofluid model. They observed that the Brownian parameter is responsible for enhanced thermal distribution across the domain. Ratha et al. (2024) investigated the impact of radiating and dissipating heat on nanofluid flow past a permeable sheet using in-built bvp4c program in MATLAB. Their findings indicated that nanofluids played a significant role in raising the rate of thermal transfer. Sharma et al. (2024) focused on the influence of chemical reactions on the nanofluid flow past a curved geometry. It was shown in their research the chemical reaction strength enhanced the rate of mass transfer.

The examination of non-Newtonian fluid dynamics has attracted minimal focus within the field of fluid mechanics, even though these fluids have vital applications in industry and technology. The study of such fluids presents considerable challenges to engineers and mathematicians. These types of fluids find several applications in different fields, including drilling operations, food production, and bioengineering. Among the viscoelastic liquid models, the second-grade (SG) fluid model is particularly noteworthy. Over the past few years, research on non-Newtonian liquids has intensified due to their superior functional characteristics. Jamshed et al. (2021) reported the impact of dissipative heat on second-grade nanofluid

DOI: 10.1201/9781003642169-8

flow past a permeable surface using R-K method. Sunthrayuth et al. (2022) inspected the characteristics of thermal transfer in a SG fluid with the impact of nanoparticles agglomeration. In their study, they observed that the particle without agglomeration has higher heat transfer. Dhlamini et al. (2023) illustrated the flow characteristics of second-grade nanofluid in the presence of microorganisms past an extending sheet. They observed that the random movement of microbes due to Brownian motion leads to a growth in thermal distribution. Gangadhar et al. (2023) scrutinized the radiative flow of SG fluid past a nonlinear stretching surface. Within the context of a moving cylinder, Cham and Mustafa (2023) investigated the slip flow of a second-grade fluid. In their study, it is observed that rising the SG fluid parameter enhances streamwise momentum.

Gailitis and Lielausis, researchers at the Physics Institute in Riga, introduced an electromagnetic control apparatus characterized by a span-wise configuration of electrodes and magnets. The Riga plate is the name of such an actuator. The Riga plate uses external magnetic and electric field to create a Lorentz force that is parallel to the wall. It is made up of electrodes and permanent magnets. Hakeem et al. (2020) investigated flow and thermal properties of radiative nanofluid through a Riga surface using R-K method. Their findings indicated that the modified Hartmann number positively influenced the thermal transfer rate. Sharma et al. (2023) focused on the radiative flow of Jeffry fluid over a Riga surface, incorporating radiation and melting condition. Madhukesh et al. (2024) examined aggregated nanofluid flow past a Riga surface in the presence of thermal radiation employing the R-K method. The study revealed that an increase in the radiation resulted in better thermal transfer. Upreti et al. (2024) analyzed impact of dissipative heat on the flow of nanofluid fluid over Riga surface. They observed that a higher modified Hartmann number led to an increased momentum distribution. Panda et al. (2024) examined the transient behavior of hybrid nanofluid flow over a Riga channel while considering the influence of dissipative heating.

In fluid flow problems, statistical analysis is emerging as a method for optimizing thermal and mass transport. The response surface methodology (RSM), based on numerical experiments. Mackolil and Mahanthesh (2021) optimized the thermal rate in flow of hybrid nanofluid via expanding surface using RSM. Reddy et al. (2023) conducted a sensitivity analysis in hybrid nanofluid flow within an extending plate using RSM combined with analysis of variance (ANOVA). Reshmi et al. (2024) utilized RSM to optimize the thermal transfer rate in nanofluid flow through a permeable extending sheet. They observed that the correlation obtained from RSM predicts the thermal transfer rate with 99 percent accuracy. Sharma and Sharma (2025) optimized the thermal transport rate in nanofluid flow past a curved surface using statistical method called RSM. Galal et al. (2024) considered nanofluid within a vertical channel with dissipative heat and radiation to explore the sensitivity investigation for the rate of thermal transfer.

The objectives of the present study are as follows:

- To analyze the thermal and mass flux boundary conditions on the flow of a second-grade nanofluid past an extending Riga plate, considering the effects of dissipative heat and chemical reaction of *n*th order.
- To reformulate the controlling equations into non-dimensional form using similarity transformation and solve them numerically with the Runge–Kutta fourth-order technique.
- To develop a statistical model using response surface methodology (RSM) for predicting and optimizing skin friction coefficients.

The novelty of the current study is as follows:

- The proposed study adopts a novel approach by combining analysis of variance with RSM to optimize skin friction coefficients, allowing for an effective examination of multi-dimensional parameters.
- The study is extended by considering the effectiveness of dissipative heat, which plays a crucial role in applications involving energy loss due to friction.
- The introduction of *n*th-order chemical reaction, and thermal and mass flux conditions introduce a new perspective on nanoparticle–fluid interactions.

8.2 MATHEMATICAL FORMULATION

The investigation focuses on the steady-state, incompressible, two-dimensional second-grade nanofluid flow past a stretching Riga plate within a coordinate system (x, y), as presented in Figure 8.1. The Riga plate moves along the x-axis with an extending velocity described by $\left[u_w = ax\right]$ under the condition that a is positive.

At the surface, heat and mass flux boundary conditions are considered and the ambient values of concentration $\left[C_\infty\right]$ and temperature $\left[T_\infty\right]$. Based on boundary layer approximations, the governing expressions are formulated as follows (Jamshed et al., 2021; (Hakeem et al., 2020).

$$\frac{\partial u}{\partial x} = -\frac{\partial v}{\partial y} \tag{8.1}$$

$$\rho_f\left(v\frac{\partial u}{\partial y} + u\frac{\partial u}{\partial x}\right) - \alpha_1\left(v\frac{\partial^3 u}{\partial y^3} + u\frac{\partial^3 u}{\partial x \partial y^2} + \frac{\partial u}{\partial x}\frac{\partial^2 u}{\partial y^2} - \frac{\partial u}{\partial y}\frac{\partial^2 u}{\partial x \partial y}\right) = \mu_f\frac{\partial^2 u}{\partial y^2} + \frac{M_0 J_0 \pi}{8}\exp\left\{-\frac{\pi}{c}y\right\}, \tag{8.2}$$

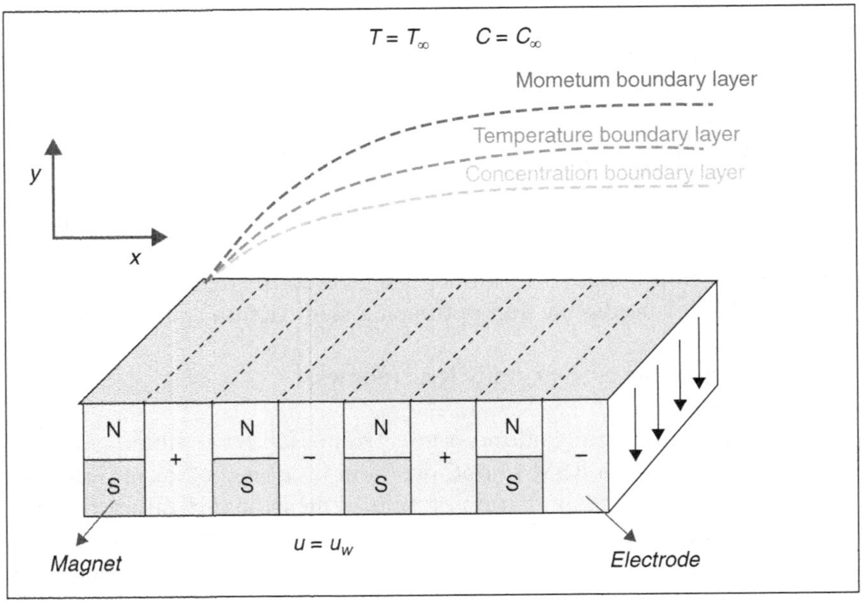

Figure 8.1 Flow geometry of problem.

$$\left(\rho C_p\right)_f\left(v\frac{\partial T}{\partial y}+u\frac{\partial T}{\partial x}\right)=+\mu_{nf}\left(\frac{\partial u}{\partial y}\right)^2+\kappa_f\frac{\partial^2 T}{\partial y^2}+\tau\left[\frac{D_T}{T_\infty}\left(\frac{\partial T}{\partial y}\right)^2+D_B\frac{\partial T}{\partial y}\frac{\partial C}{\partial y}\right] \tag{8.4}$$

$$v\frac{\partial C}{\partial y}+u\frac{\partial C}{\partial x}=D_b\frac{\partial^2 C}{\partial y^2}-K_r\left(C-C_\infty\right)^n+\frac{D_T}{T_\infty}\frac{\partial^2 T}{\partial y^2} \tag{8.5}$$

With suitable boundary conditions (Beg et al., 2009; Ganesan and Loganathan, 2002)

$$\left.\begin{aligned} &\left. u=u_w,\ v=0,\ -\kappa\frac{\partial T}{\partial y}=q_w,\left(q_w=E_1\left(x/l\right)^2\right),\right\} \\ &\left.-D_B\frac{\partial C}{\partial y}=h_w,\left(h_w=E_2\left(x/l\right)^2\right),\right\}_{as\,y=0} \\ &\left. u\to 0,\ \frac{\partial u}{\partial y}\to 0,\ T\to T_\infty, C\to C_\infty,\ \right\}_{as\ \ y\to\infty} \end{aligned}\right. \tag{8.6}$$

The following transformations, based on similarity principles, are introduced in Kumari et al. (2024):

$$u = axf'(\eta),\ v = -\sqrt{a\upsilon}f(\eta),\ \Phi(\eta) = \frac{C - C_\infty}{C_w - C_\infty},$$
$$\Theta(\eta) = \frac{T - T_\infty}{T_w - T_\infty},\ \eta = \sqrt{\frac{a}{\upsilon}}y, \tag{8.7}$$

Using the .above similarity transformations Eq. (8.1) satisfied and rest equations transform as follows:

$$f''' + K2f'f''' - Kff^{iv} + ff'' - Kf''^2 - f'^2 + \mathrm{M}\exp(d\eta) = 0 \tag{8.8}$$

$$\Theta'' + \Pr f\Theta' + \Pr\left(\mathrm{N}_b\Phi'\Theta' + \mathrm{N}_t\Theta'^2 + Ecf''^2\right) = 0 \tag{8.9}$$

$$\Phi'' + Sc\left(f\Phi' - K^*\Phi^n\right) + \frac{\mathrm{N}_t}{\mathrm{N}_b}\Theta'' = 0 \tag{8.10}$$

with

$$\left.\begin{aligned} &f'(\eta)|_{\eta\to\infty} \to 0, f''(\eta)|_{\eta\to\infty} \to 0, \Theta(\eta)|_{\eta\to\infty} \to 0, \Phi(\eta)|_{\eta\to\infty} \to 0 \\ &f(\eta)|_{\eta=0} = 0, f'(\eta)|_{\eta=0} = 1, \Theta'(\eta)|_{\eta=0} = -1,\ \Phi'(\eta)|_{\eta=0} = -1 \end{aligned}\right\} \tag{8.11}$$

Dimensionless parameters are as follows:

$$\left.\begin{aligned} &d = \frac{-\pi}{c}\sqrt{\frac{\upsilon_f}{a}},\ \Pr = \frac{\upsilon_f}{\alpha_f},\ \mathrm{N}_b = \frac{\tau D_B\left(C_w - C_\infty\right)}{\upsilon_f},\ \mathrm{N}_t = \frac{\tau D_T\left(T_w - T_\infty\right)}{\upsilon_f T_\infty} \\ &Sc = \frac{\upsilon_f}{D_B}, K^* = \frac{Kr\left(C_w - C_\infty\right)^{n-1}}{a}, Ec = \frac{u_w^2}{C_p\left(T_w - T_\infty\right)},\ \mathrm{M} = \frac{M_0 J_0 \pi}{\rho_f 8a^2 x},\ K = \frac{\alpha_1 a}{\mu_f} \end{aligned}\right\rangle \tag{8.12}$$

The skin friction in the x direction is stated as follows for numerical results:

$$\mathbb{C}f_x = \frac{\tau_w}{\rho_f u_w^2} = \frac{2}{\sqrt{\mathrm{Re}_x}}\left[f'' + 3Kf'f''\right]_{\eta=0} \tag{8.13}$$

Table 8.1 comparison of $\left[\mathbb{C}f_x\right]$ with specific value of [*K*] in the absence of $\left[M_r, Ec, \mathbb{N}_b \,\&\, \mathbb{N}_t\right]$

[*K*]	*Massoudi and Maneschy (2004)*	*Present*
0.01	−1.02489	−1.0248812
0.1	−1.23453	−1.2394987
1.0	−2.83024	−2.8284248

where

$$\tau_w = \left[\mu_f \frac{\partial u}{\partial y} + \rho_f \alpha_1 \left(2\frac{\partial u}{\partial x}\frac{\partial u}{\partial y} + u\frac{\partial^2 u}{\partial x \partial y}\right)\right]_{y=0} \tag{8.14}$$

8.3 METHODOLOGY AND VALIDATION OF RESULTS

The solution is obtained by applying a hybrid numerical approach that combines the shooting technique with the fourth-order approximation of R-K method to analyze the nonlinear equations (Eqs. 8.8–8.10) and the boundary condition (Eq. 8.11). To implement this method, the nonlinear coupled system of equations (8.8)–(8.10) is transmuted into a first-order system by defining a new variable:

$$\begin{bmatrix} f \\ f' \\ f'' \\ f''' \end{bmatrix} = \begin{bmatrix} \aleph_1 \\ \aleph_2 \\ \aleph_3 \\ \aleph_4 \end{bmatrix}, \begin{bmatrix} \Theta \\ \Theta' \end{bmatrix} = \begin{bmatrix} \aleph_5 \\ \aleph_6 \end{bmatrix}, \begin{bmatrix} \Phi \\ \Phi' \end{bmatrix} = \begin{bmatrix} \aleph_7 \\ \aleph_8 \end{bmatrix}, \tag{8.15}$$

Applying the above-mentioned variables, Eqs. (8.8)–(8.11) are converted into

$$f^{iv} = \frac{\aleph_4 + \aleph_1\aleph_3 - \aleph_2^{\,2} + K2\aleph_2\aleph_4 - K\aleph_3^{\,2} - \mathrm{H}\aleph_2 + \mathrm{M}\exp\{-\vartheta\eta\}}{K\aleph_1} \tag{8.16}$$

$$\Theta'' = -\Pr\aleph_1\aleph_6 - \Pr\left(\mathbb{N}_b\aleph_8\aleph_6 + \mathbb{N}_t\aleph_6^{\,2} + Ec\aleph_2^{\,2}\right) \tag{8.17}$$

$$\Phi'' = -Sc\left(\aleph_1\aleph_8 - K^*\aleph_7^{\,n}\right) - \frac{\mathbb{N}_t}{\mathbb{N}_b}\Theta'' \tag{8.18}$$

Boundary conditions are

$$\left.\begin{array}{l} \aleph_1(0) = 0, \aleph_2(0) = 1, \aleph_6(0) = -1,\ \aleph_8(0) = -1 \\ \aleph_2(\infty) \to 0, \aleph_3(\infty) \to 0, \aleph_5(\infty) \to 0, \aleph_7(\infty) \to 0 \end{array}\right\} \tag{8.19}$$

The convergence criterion for the numerical method is set to achieve accuracy up to the fifth decimal place with a step size of $[\eta] = 0.01$. Further, Table 8.1 presents the validation of the numerical results against previous studies (Massoudi and Maneschy, 2024) showing that the current findings exhibit excellent agreement with prior work in the limiting case.

8.4 DISCUSSION OF GRAPHICAL RESULTS

This research examines the radiative second-grade nanofluid flow over a stretching Riga surface while considering various factors influencing heat and solutal transfer, including viscous dissipation, convective conditions, and an *n*th-order chemical reaction. The uniqueness of this research lies in employing a statistical method, that is, ANOVA test for optimizing the skin friction coefficient. Further, the physical impact of the defining terms in the flow phenomena is represented through graphs. The impact of modified Hartman number $[M_r]$ and viscoelastic parameter $[K]$ on nanofluid flow velocity across the surface is depicted in Figure 8.2. The flow in the positive x -direction is reduced due to the action of the surface-parallel Lorentz force. A stronger magnetic field leads to lower wall velocity gradient. However, larger value of $[K]$ parameter rises the velocity profile $[f'(\eta)]$. In second-grade nanofluids, an increase in the fluid parameter counteracts the fluid's viscosity, leading to an enhanced flow rate. In general, a rise in the fluid parameter results in lower viscosity, thereby diminishing flow resistance.

Figure 8.3 presents the outcome of Eckert number $[Ec]$, modified Hartman number $[M_r]$, Brownian parameter $[\mathbb{N}_b]$, and thermophoretic parameters $[\mathbb{N}_t]$ on temperature distribution $[\Theta(\chi)]$. The $[Ec]$ parameter is elevated, and the $[\Theta(\chi)]$ field experiences an increase. From a physical perspective, more heat is generated within the flow domain due to stronger frictional heating. Consequently, a growth in $[Ec]$ leads to an augmentation in the $[\Theta(\chi)]$ profile. A stronger magnetic field leads to increased temperature distribution because the resistive characteristics of the force accumulate energy in the fluid's layer. The accumulated energy progressively intensifies and exceeds the $[\Theta(\chi)]$ profile. The analysis also shows that the thermal fields are directly proportional to $[\mathbb{N}_b]$. This occurs because Brownian motion represents nanoparticles in the fluid, which leads to a rise in heat transfer. Further, the thermal field increases with greater values of $[\mathbb{N}_t]$. In thermophoresis, particles of fluid move from the heated zone to the cooler zone. The migration of heated liquid particles from low-temperature areas to hot surfaces during thermophoresis results in temperature elevation.

Figure 8.4 depicts the impact of power index factor $[n]$, chemical reaction $[K^*]$ and Schmidt number $[Sc]$ on solutal profile $[\Phi(\eta)]$. The exponent variation of the chemical reaction is characterized by the power index $[n]$. As $[n]$ increases, a considerable decrease in nanofluid concentration is observed while the positive values of the $[K^*]$ indicate its destructive nature,

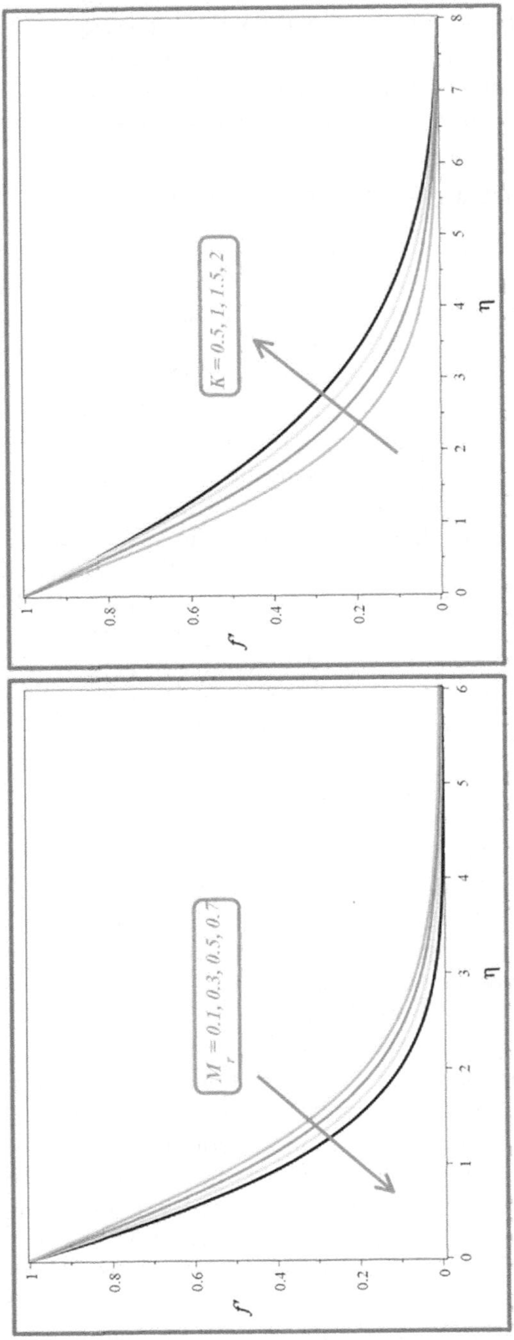

Figure 8.2 Impact of $[M_r]$ and $[K]$ on $[f'(\eta)]$.

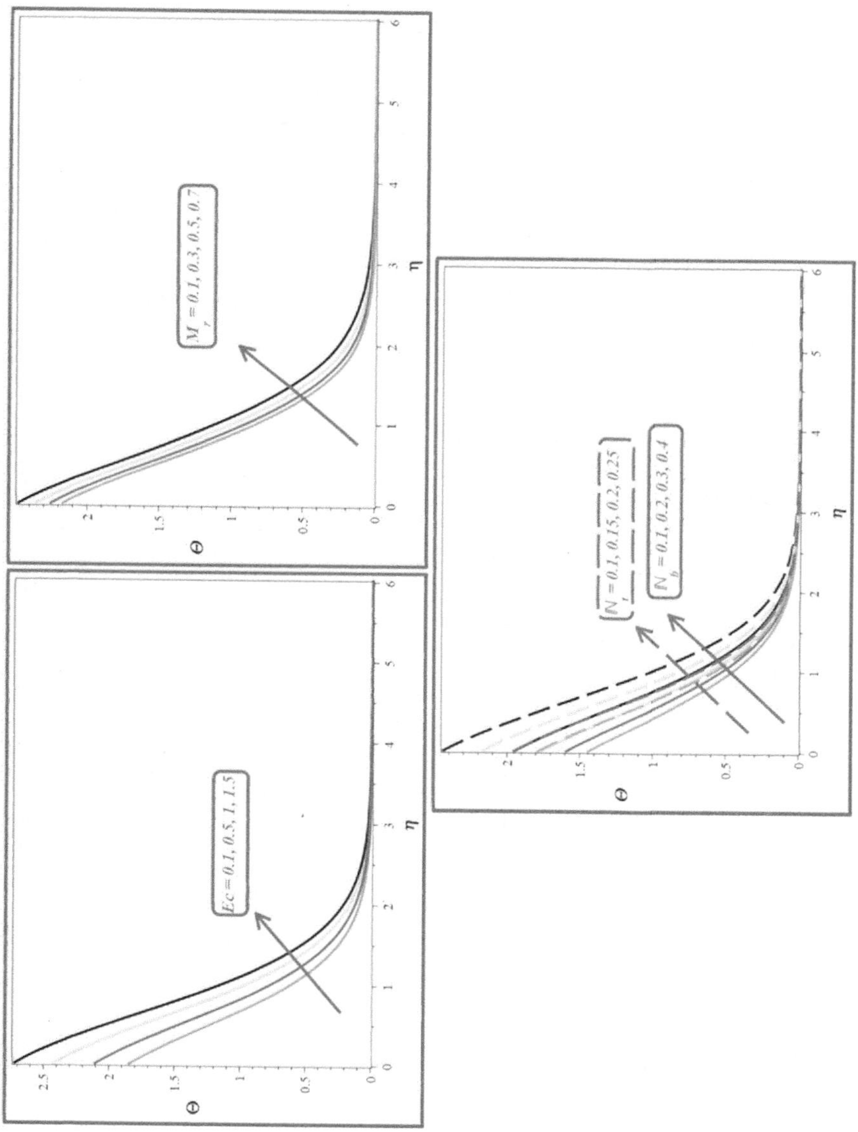

Figure 8.3 Impact of $\left[Ec, M_r, N_t\right]$ and $\left[N_b\right]$ on $\left[\theta(\eta)\right]$.

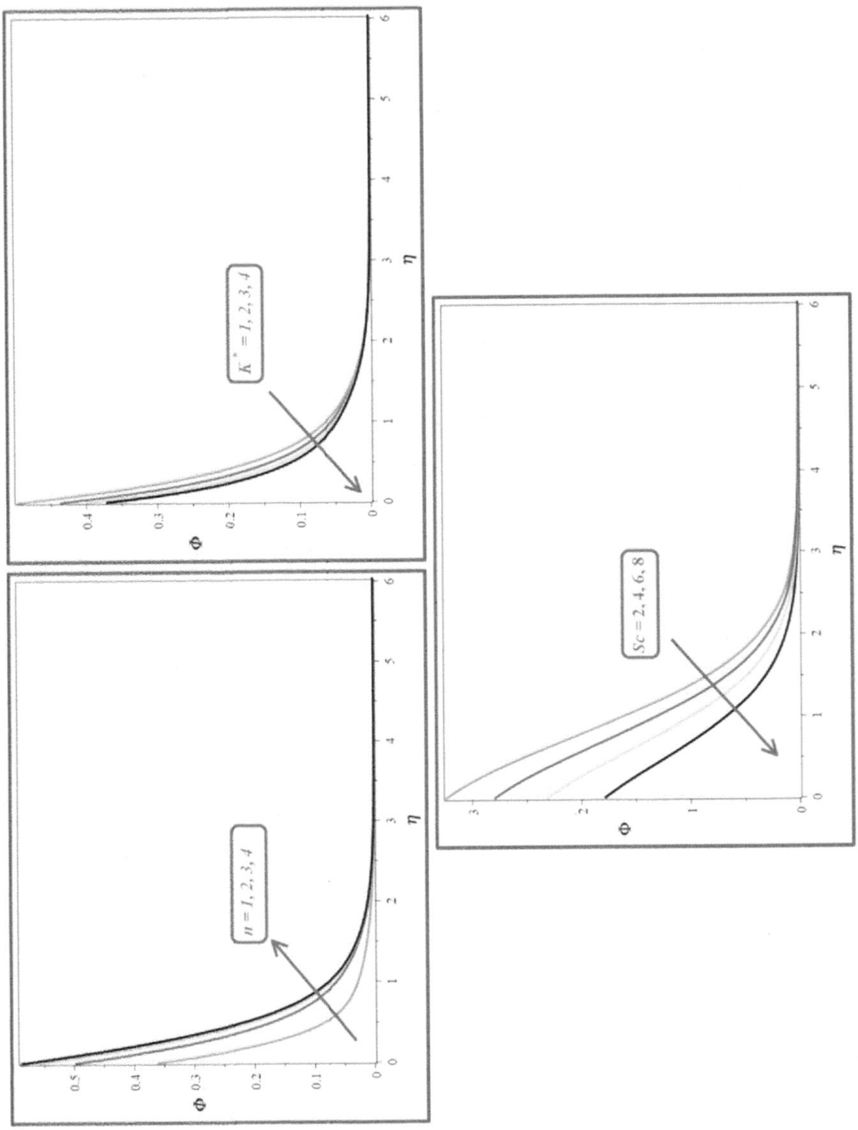

Figure 8.4 Impact of $[n, K^*]$ and $[Sc]$ on $[\theta(\eta)]$.

Table 8.2 Experimental runs

$[M_r]$	$[K]$	$[d]$	$[Cf_x]$
0.6	1	2	-0.41202
0.3	1	2	-0.37530
0.3	1	2	-0.37530
0.1	0.5	3	-0.40767
0.1	0.5	1	-0.42634
0.3	1	1	-0.46156
0.3	1	2	-0.37530
0.5	0.5	3	-0.45443
0.1	1	2	-0.34545
0.3	1.5	2	-0.31050
0.3	1	2	-0.37530
0.3	1	2	-0.37530
0.3	1	2	-0.53338
0.5	1.5	3	-0.33522
0.3	1	2	-0.37530
0.5	0.5	1	-0.55061
0.3	1	3	-0.51841
0.5	1.5	1	-0.38637
0.1	1.5	1	-0.32265
0.1	1.5	3	-0.31266

further contributing to the decline in concentration. However, as the $[Sc]$ increases, molecular diffusion decreases, leading to lower species concentration. Consequently, higher values of Sc result in reduced concentration. This reduction in concentration weakens the effects of concentration buoyancy.

8.5 STATISTICAL APPROACH (RSM AND ANOVA)

This chapter presents an optimization of the skin friction $\left[\mathbb{C}f_x\right]$ coefficient using a statistical approach called RSM, while analyzing the impact of key factors such as modified Hartman number $\left[M_r\right]$, viscoelastic parameter $[K]$, and width of electrode $[d]$. During computation, these constraints fall within the designated range, that is, $0.1 \le M_r \le 0.6$, $0.5 \le K \le 1.5$ and $1 \le d \le 3$. Table 8.2 provides the range of influential factors for simulating the $\left[\mathbb{C}f_x\right]$, along with the distribution of 20 unique runs with different parameter settings. Additionally, the model is validated using analysis of variance (ANOVA), incorporating surface and contour plots. The standard methodology, along with the simulated results for various factors, is briefly explained. However, the response corresponding to the specified components is expressed using the multivariate classic model as

$$\mathbb{C}f_x = -0.5703 - 0.578\, M_r + 0.2003\, K + 0.1332\, d - 0.03403\, d^2$$

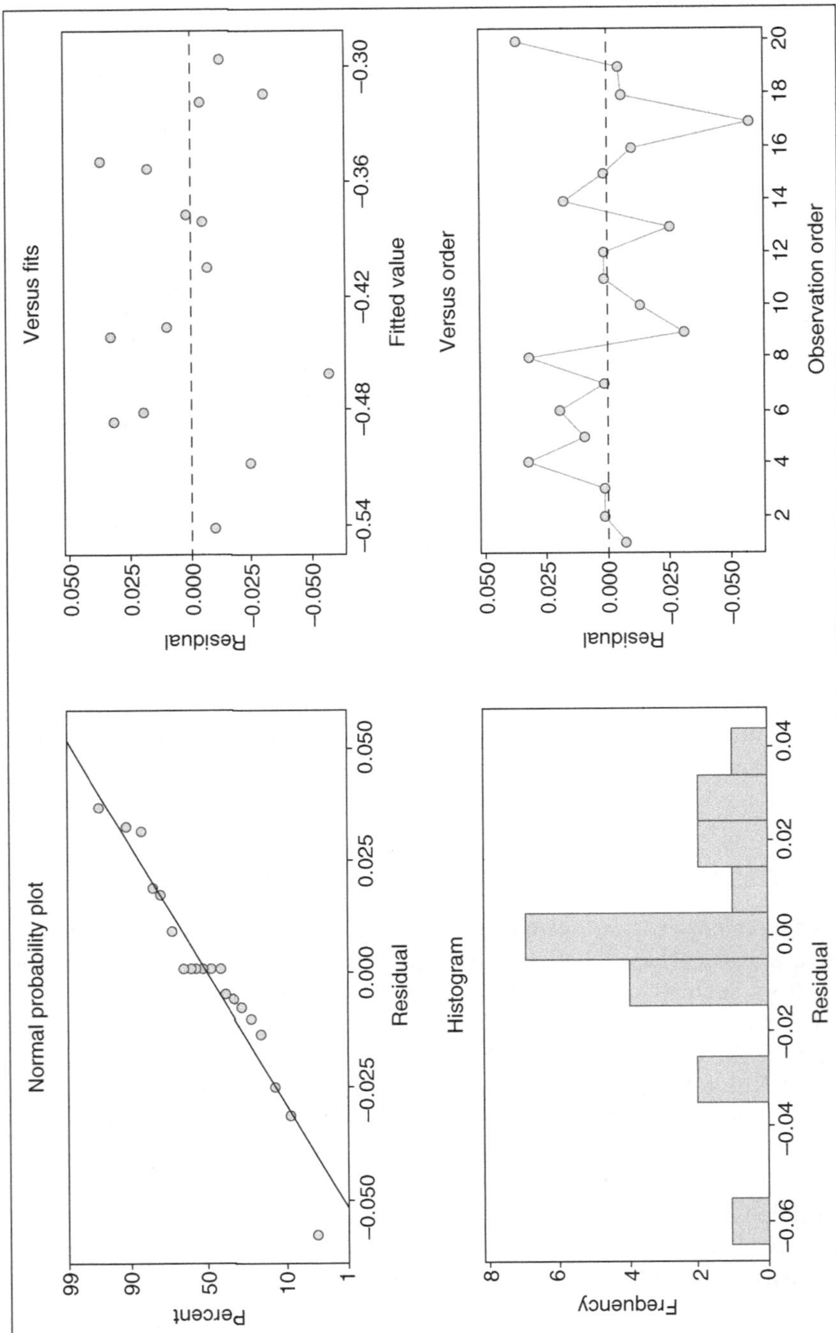

Figure 8.5 Residual plots for shear rate.

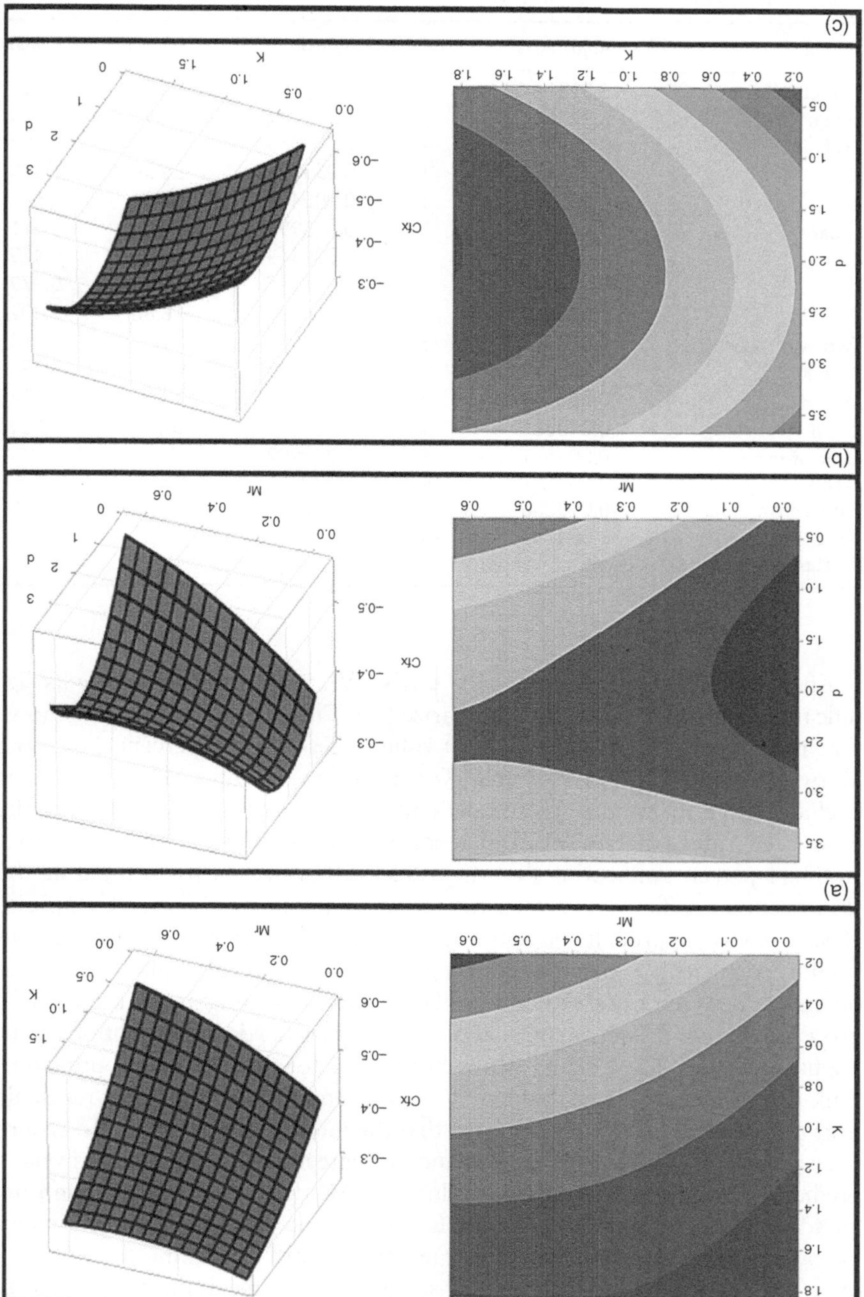

Figure 8.6 (a–c) Contour and surface plots for shear rate.

Table 8.3 ANOVA table for shear rate

Source	*Adjusted SS*	*DF*	*Adjusted MS*	*F-Value*	*P-Value*
Model	8.5231e-02	9	9.4700e-03	9.96	0.001
Linear	6.4236e-02	3	2.1412e-02	22.51	0.000
$[M_r]$	0.009984	1	0.009984	10.50	0.009
K	0.053779	1	0.053779	56.54	0.000
d	0.000473	1	0.000473	0.50	0.497
Square	0.017978	3	0.005993	6.30	0.011
M_r^2	0.000514	1	0.000514	0.54	0.479
K^2	0.001247	1	0.001247	1.31	0.279
d^2	0.016038	1	0.016038	16.86	0.002
Two-way interaction	0.003018	3	0.001006	1.06	0.410
$M_r \times K$	0.000898	1	0.000898	0.94	0.354
$M_r \times d$	0.001760	1	0.001760	1.85	0.204
$K \times d$	0.000361	1	0.000361	0.38	0.552
Lack-of-fit	0.009511	5	0.001902		
Error	0.009511	10	0.000951		
Pure error	0.000000	5	0.000000		
Total	0.094742	19			

SS: sum of square; DF: degree of freedom; MS: mean of square.

After designing the model for $\left[\mathbb{C}f_x\right]$, ANOVA is applied to evaluate its significance, and the results are summarized in Table 8.3. The analysis shows that the model is significant if the p-value is less than 5 percent. Since the all-two-way interaction $\left[M_r^2\right]$ and $\left[K^2\right]$ terms have p-value above 0.05, it is excluded from the model. The model's adequacy is assessed by evaluating the R-squared value, that is, 91.22 percent and adjusted R-squared value, that is, 83.31 percent, indicating that the model is a perfect fit for computing the skin friction coefficient. Further, model adequacy in the Central Composite Design (CCD) approach for optimizing shear rate, characterized by the skin friction coefficient, is determined using residual plots. These plots depict the difference between observed and predicted values, helping assess the model's precision. Figure 8.5 shows normal plots that illustrate data distributions from the fitting values. These figures show histograms for the same distribution and rank data distributions related to $\left[\mathbb{C}f_x\right]$. In normal probability plot, residuals near the straight line strongly suggest that the model is appropriate. To ensure the assumption of constant variance and evaluate model fit, the residual versus predicted plot is analyzed. If the residuals are randomly scattered, it confirms an adequate correlation between the factors and the Nusselt number. Finally, histogram of residuals visually illustrates their distribution, and the presence of a bell-shaped curve signifies the significance of the influencing factors. Meanwhile, the residuals versus factor level plot evaluates the influence of each factor. The random scattering of residuals confirms the model's appropriate fit.

The contour and surface plots in Figure 8.6(a–c) visually depict the pattern of $[\mathbb{C}f_x]$, showcasing the combined influence of the interaction terms. The influence of $[K]$ and $[M_r]$ is visualized through contour and surface plots in Figure 8.6(a), assuming $[d]$ is zero. It is observed that the $[K]$ factor contributes effectively to enhancing the shear rate. The plots in Figure 8.6(b) represents the effects of $[M_r]$ and magnetism $[d]$ considering a zero value for $[K]$ factor. The figure illustrates a significant improvement due to $[d]$, whereas $[M_r]$ factor reduces the $[\mathbb{C}f_x]$ response. Figure 8.6(c) displays the contour and surface plots showing the variation in $[K]$ and $[d]$ when the $[M_r]$ factor is set to zero. The findings suggest that these factors are beneficial in improving the shear rate.

8.6 CONCLUSION

This research examines the radiative second-grade nanofluid flow over an extending Riga surface while considering various factors influencing heat and solutal transfer, including viscous dissipation, convective conditions, and an *n*th-order chemical reaction. Further, a statistical model is formulated by integrating RSM to predict and optimize the performance of coefficient of the skin friction. Finally, the major conclusions drawn from the study are as follows:

- The study's numerical comparison with earlier research confirms a good correlation, reinforcing the accuracy and convergence of the applied technique.
- The solutal profile of the second-grade fluid decreases due to the Schmidt number and chemical reaction parameter.
- The velocity profile is enhanced by the viscoelastic parameter, while a significant reduction is observed due to the modified Hartmann number.
- The analysis conducted via RSM demonstrates considerable advancements in shear rate performance and flow dynamics under varying thermal and hydrodynamic conditions

REFERENCES

Bég, O.A., Bakier, A.Y., Prasad, V.R., 2009. Numerical study of free convection magnetohydrodynamic heat and mass transfer from a stretching surface to a saturated porous medium with Soret and Dufour effects. *Computational Materials Science* 46, 57–65.

Cham, A., & Mustafa, M., 2023. Exploring the dynamics of second-grade fluid motion and heat over a deforming cylinder or plate affected by partial slip conditions. *Arabian Journal for Science and Engineering* 49, 1505–1514.

Deepthi, V.V.L., Narla, V.K., Raju, R.S., 2024. Double diffusive effects on nanofluid flow toward a permeable stretching surface in presence of thermophoresis and Brownian motion effects: A numerical study. *Numerical Methods for Partial Differential Equations* 40, e23086.

Dhlamini, M., Zondo, K., Mondal, H., Sibanda, P., Shaw, S., Motsa, S.S., Mkhatshwa, M.P., Magodora, M., 2023. Second-grade bioconvection flow of a nanofluid with slip convective boundary conditions. *Pramana* 97, 197. https://doi.org/10.1007/s12043-023-02676-0

Felicita, A., Venkatesh, P., Gireesha, B., Krishnamurthy, M., 2023. Slip and convective flow of Williamson nanofluid influenced by Brownian motion and thermophoresis mechanism in a horizontal microchannel. *Proceedings of the Institution of Mechanical Engineers, Part N: Journal of Nanomaterials, Nanoengineering and Nanosystems* 239, 59–69. https://doi.org/10.1177/23977914231177340

Galal, A.M., Mishra, S.R., Baithalu, R., Kallel, M., Panda, S., Chamkha, A.J., Ali, R., Afshar, S.R., 2024. Three-dimensional flow of nanofluid through a vertical channel considering KKL model: Hydrothermal analysis via response surface methodology. *Journal of the Taiwan Institute of Chemical Engineers* 168, 105931.

Ganesan, P., &Loganathan, P., 2002. Heat and mass flux effects on a moving vertical cylinder with chemically reactive species diffusion. *Journal of Engineering Physics and Thermophysics* 75, 899–909.

Gangadhar, K., Sree, T.S., Thumma, T., 2023. Impact of Arrhenius energy and irregular heat absorption on generalized second grade fluid MHD flow over nonlinear elongating surface with thermal radiation and Cattaneo–Christov heat flux theory. *Modern Physics Letters B* 38, 2450077. https://doi.org/10.1142/s0217984924500775

Hakeem, A.K.A., Ragupathi, P., Saranya, S., Ganga, B., 2020. Three Dimensional Non-linear Radiative Nanofluid Flow over a Riga Plate. *DOAJ (Directory of Open Access Journals)* 6, 1012–1029.

Jamshed, W., Mishra, S.R., Pattnaik, P.K., Nisar, K.S., Devi, U., Prakash, M., Shahzad, F., Hussain, M., Vijayakumar, V., 2021. Features of entropy optimization on viscous second grade nanofluid streamed with thermal radiation: A Tiwari and Das model. *Case Studies in Thermal Engineering* 27, 101291.

Kumari, P.S., Ibrahim, S.M., Lorenzini, G., 2024. Chemical reaction on Williamson nanofluid's radiative MHD dissipative stagnation point flow over an exponentially inclined stretching surface with multi-slip effects. *Frontiers in Heat and Mass Transfer* 22, 1839–1863.

Mackolil, J., & Mahanthesh, B., 2021. Optimization of heat transfer in the thermal Marangoni convective flow of a hybrid nanomaterial with sensitivity analysis. *Applied Mathematics and Mechanics* 42, 1663–1674.

Madhukesh, J.K., Paramesh, S.O., Prasanna, G.D., Prasannakumara, B.C., Khan, M.I., Abdullaev, S., Rasool, G., 2024. Impact of magnetized nanoparticle aggregation over a Riga plate with thermal radiation in water-Al_2O_3 based nanofluid flow. *ZAMM—Journal of Applied Mathematics and Mechanics/Zeitschrift für Angewandte Mathematik und Mechanik*, e202300270. https://doi.org/10.1002/zamm.202300270

Massoudi, M., & Maneschy, C.E., 2004. Numerical solution to the flow of a second grade fluid over a stretching sheet using the method of quasi-linearization. *Applied Mathematics and Computation* 149, 165–173. https://doi.org/10.1016/s0096-3003(02)00963-3

Mohan Krishna, P., Sharma, R.P., Sandeep, N., 2017. Boundary layer analysis of persistent moving horizontal needle in Blasius and Sakiadis magnetohydrodynamic radiative nanofluid flows. *Nuclear Engineering and Technology* 49, 1654–1659.

Panda, S., Baithalu, R., Mishra, S.R., 2024. Melting heat impact on the time-dependent squeezing hybrid nanofluid flow over a Riga surface with non-uniform heat source/sink. *Mechanics of Time-Dependent Materials* 29, 4.

Ratha, P.K., Tripathy, R.S., Mishra, S.R., 2024. On the impact of temperature and space dependent heat source/sink with Richardson number for the MHD convection flow of nanofluid. *International Journal of Applied and Computational Mathematics* 10, 166.

Reddy, C.A., Thumma, T., Goud, J.S., Panda, S., 2023. Thermal and sensitivity analysis on hydromagnetic CuO-Ag-H_2O nanofluid radiative flow over an elongating convective thermal surface: RSM-CCD model. *Journal of Thermal Analysis and Calorimetry* 148, 12195–12210.

Reshmi, V.P., Nagaraja, K.V., Shankara, 2024. Heat transfer analysis of TiO_2-H_2O based nanofluid flow across a porous stretching sheet with heat source/sink using response surface methodology. *Journal of Physics Conference Series* 2818, 012010.

Sharma, A., &Sharma, R.P., 2025. Statistical and numerical analysis of MHD nanofluid (TiO_2/EG) flow on a nonlinear curved stretching surface with heat source: A RSM approach. *Journal of Applied Mathematics and Mechanics/Zeitschrift für Angewandte Mathematik und Mechanik* 105, e202400513.

Sharma, B.K., Kumar, A., Mishra, N.K., Albaijan, I., Fernandez-Gamiz, U., 2023. Computational analysis of melting radiative heat transfer for solar Riga trough collectors of Jeffrey hybrid-nanofluid flow: A new stochastic approach. *Case Studies in Thermal Engineering* 52, 103658.

Sharma, R.P., Sharma, A., Madhukesh, J.K., Prasannakumara, B.C., Nagaraja, K.V., Kumar, R., 2024. Significance of homogeneous–heterogeneous reaction on MHD nanofluid flow over a curvilinear stretching surface. *Pramana* 98, 164.

Sunthrayuth, P., Abdelmohsen, S.A.M., Rekha, M.B., Raghunatha, K.R., Abdelbacki, A.M.M., Gorji, M.R., Prasannakumara, B.C., 2022. Impact of nanoparticle aggregation on heat transfer phenomena of second grade nanofluid flow over melting surface subject to homogeneous–heterogeneous reactions. *Case Studies in Thermal Engineering* 32, 101897.

Upreti, H., Prakash, J., Usman, N., Pandey, A.K., Tripathi, D., 2024. Thermal analysis of 3D Darcy–Forchheimer flow of SWCNT–MWCNT/sodium alginate on Riga plate. *Journal of Thermal Analysis and Calorimetry* 149, 3891–3911.

Chapter 9

Thermo-bioconvective transport in a lid-driven porous cavity with oxytactic microorganisms under magnetic field and Joule heating effects

Sahin Ahmed, Anil Nangkar, and Bikash Das

9.1 INTRODUCTION

Convective transport in enclosures has long been a subject of fundamental and applied interest in thermal sciences, particularly due to its relevance in areas such as electronic cooling, nuclear waste disposal, geophysics, and biomedical engineering. Among the various enclosure types, the lid-driven square cavity remains an established problem that has been extensively studied for both natural and mixed convection flows (De Vahl Davis, 1983; Cheng, 2011). The moving lid conveys momentum to the fluid, leading to complex recirculating patterns that are sensitive to thermal and solutal boundary conditions, as well as to external influences like magnetic fields or porous structures.

Porous media are universal in nature and engineering systems, appearing in insulation materials, biological tissues, and subsurface reservoirs. The modelling of fluid flow through porous media often begins with Darcy's law, which relates the velocity field to the pressure gradient and permeability. However, Darcy's model is valid primarily for slow, viscous-dominated flows. At moderate to high velocities, inertial effects become significant, and the Darcy–Forchheimer model is employed to capture the nonlinear resistance to flow (Vafai & Tien, 1981; Nield & Bejan, 2006). Early investigations by Walker and Homsy (1978) laid the groundwork for understanding convection in porous cavities, while Ingham and Pop (2005) compiled extensive theoretical and numerical treatments of transport in porous systems. Subsequent studies further explored how variations in Darcy and Forchheimer numbers influence convective structures and heat transfer characteristics (Saeid & Pop, 2005; Basak, Roy, Paul, & Pop, 2006).

Incorporating magnetic fields into cavity flows introduces magnetohydrodynamic (MHD) effects, where the interaction of the magnetic field with the electrically conducting fluid gives rise to Lorentz forces. These forces can dampen convective currents and recreate flow patterns depending on

 DOI: 10.1201/9781003642169-9

the field's intensity and orientation. Moreover, Joule heating, resulting from resistive dissipation of electric currents, can substantially affect temperature distributions and flow dynamics (Zhang, Li, & Chen, 2015). Several studies have examined MHD flows in porous media, demonstrating how magnetic field strength and orientation influence thermal behaviour (Bakar, Karimipour, & Roslan, 2016; Swain, Parida, Kar, & Senapati, 2020; Abbas, Jeelani, & Alharthi, 2022). Sugilal, Wattal, and Iyer (2005) analysed convective patterns in a uniformly Joule-heated pool, highlighting how internal heat generation modifies buoyancy-driven flows. Yashkun et al. (2021) considered Joule heating in hybrid nanofluids, indicating its critical role in mixed convection regimes.

A further level of complexity arises when the fluid medium contains motile microorganisms, particularly in biological or environmental contexts. Bioconvection refers to the convective motion induced by the collective behaviour of microorganisms, typically driven by mechanisms like gyrotaxis (orientation due to gravity and viscous torques) or oxytaxis (movement along oxygen gradients). In the case of oxytactic microorganisms, their upward movement toward higher oxygen concentrations creates density gradients that can initiate bioconvective instabilities. Sheremet and Pop (2014) were among the first to study thermo-bioconvection in porous square cavities with oxytactic microorganisms, showing how these organisms affect flow and transport characteristics. Ahmed, Oztop, Mansour, and Abu-Hamdeh (2018) extended this work to include MHD effects, revealing how magnetic fields suppress microorganism-induced convection. Hussain, Aly, and Öztop (2022) investigated magneto-bioconvection in a lid-driven cavity with a streamlined obstacle, emphasizing the effects of oxytactic bacteria in altering flow symmetry. More recently, Chowdhury and Rao (2023) analysed thermo-bioconvection in a tilted square porous cavity heated sinusoidally, uncovering how cavity orientation and heating patterns influence microorganism distribution.

This study introduces a novel configuration of oxygen-driven bioconvection in a lid-driven porous cavity filled with oxytactic microorganisms. In contrast to traditional models with uniform wall conditions, the present setup applies oxygen only at the top wall and introduces microorganisms exclusively in the central third of the bottom wall, capturing more realistic conditions seen in biological and microfluidic systems. The effects of an inclined magnetic field and Joule heating are considered simultaneously, a combination not extensively explored in previous bioconvective studies. Furthermore, the porous medium is modelled using the Darcy–Forchheimer framework to account for both viscous and inertial resistance. The system is numerically solved using the finite element method (FEM) via COMSOL Multiphysics 6.2, enabling a detailed analysis of coupled heat, mass, and microorganism transport phenomena.

This framework raises several key research questions:

- How does a localized source of microorganisms alter bioconvective circulation in comparison to uniformly distributed sources?
- What are the individual and combined effects of magnetic field strength, Joule heating, and porous resistance on thermal and oxygen transport?
- How do variations in Reynolds number and bioconvection Lewis number influence the efficiency of microbial migration and oxygen uptake in the lower region of the cavity?
- How does the combination of Joule heating and microorganism diffusivity influence heat and oxygen transport near the low-oxygen region?

9.2 MATHEMATICAL FORMULATION

The present study investigates a two-dimensional, steady, incompressible flow of a Newtonian fluid within a square cavity of side length L, saturated with a homogeneous and isotropic Darcy–Forchheimer porous medium. The cavity is subjected to a uniform magnetic field B_0 applied at an angle of $\lambda = \pi/4$, and the effects of Joule heating are included. The top wall of the cavity moves horizontally at a constant velocity U_0 and is maintained at a low temperature T_c and high oxygen concentration (c_0), while the bottom wall is heated T_h and oxygen-deficient (c_{min}). The vertical walls are adiabatic and impermeable to both oxygen and microorganisms. Microorganisms are introduced only along the central one-third $(L/3)$ segment of the bottom wall, representing a localized biological source (n_0). These microorganisms exhibit oxytactic behaviour, migrating upward toward regions of higher oxygen concentration. The flow is assumed to be governed by the Boussinesq approximation, and all thermophysical properties are taken as constant throughout the domain. The flow configuration of the proposed model is sketched in Figure 9.1.

The governing equations, incorporating fluid flow, heat transfer, oxygen transport, and microorganism concentration, are given by:

$$\frac{\partial \bar{u}}{\partial \bar{x}} + \frac{\partial \bar{v}}{\partial \bar{y}} = 0, \tag{9.1}$$

$$\rho\left(\bar{u}\frac{\partial \bar{u}}{\partial \bar{x}} + \bar{v}\frac{\partial \bar{u}}{\partial \bar{y}}\right) = \left\{\begin{array}{c} -\dfrac{\partial p}{\partial \bar{x}} + \mu\left(\dfrac{\partial^2 \bar{u}}{\partial \bar{x}^2} + \dfrac{\partial^2 \bar{u}}{\partial \bar{y}^2}\right) - \dfrac{\mu}{k}\bar{u} \\ -\dfrac{\rho F}{\sqrt{K}} u\sqrt{u^2 + v^2} + \sigma B_0^2\left(\bar{v} sin\lambda cos\lambda - \bar{u} sin^2\lambda\right) \end{array}\right\}, \tag{9.2}$$

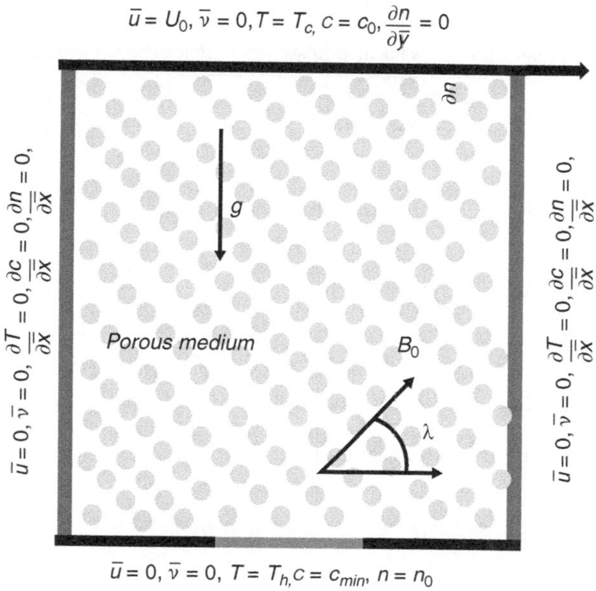

Figure 9.1 Model geometry.

$$\rho\left(\bar{u}\frac{\partial \bar{v}}{\partial \bar{x}} + \bar{v}\frac{\partial \bar{v}}{\partial \bar{y}}\right) = \left\{ \begin{array}{c} -\frac{\partial p}{\partial \bar{y}} + \mu\left(\frac{\partial^2 \bar{v}}{\partial \bar{x}^2} + \frac{\partial^2 \bar{v}}{\partial \bar{y}^2}\right) - \frac{\mu}{k}\bar{v} - \frac{\rho F}{\sqrt{K}} v\sqrt{u^2 + v^2} \\ +\sigma B_0^2\left(\bar{u} sin\lambda cos\lambda - \bar{v} cos^2\lambda\right) + \rho g\beta\left(T - T_c\right) - \gamma\Delta\rho g n \end{array} \right\}, \tag{9.3}$$

$$\rho c_p\left(\bar{u}\frac{\partial T}{\partial \bar{x}} + \bar{v}\frac{\partial T}{\partial \bar{y}}\right) = \kappa\left(\frac{\partial^2 T}{\partial \bar{x}^2} + \frac{\partial^2 T}{\partial \bar{y}^2}\right) + \sigma B_0^2 \bar{v}^2, \tag{9.4}$$

$$\bar{u}\frac{\partial c}{\partial \bar{x}} + \bar{v}\frac{\partial c}{\partial \bar{y}} = D_c\left[\frac{\partial^2 c}{\partial \bar{x}^2} + \frac{\partial^2 c}{\partial \bar{y}^2}\right] - \delta n, \tag{9.5}$$

$$\frac{\partial}{\partial \bar{x}}\left[\bar{u}n + \left(\frac{bW_c}{\Delta c}\right)\frac{\partial c}{\partial \bar{x}} n - D_n\frac{\partial n}{\partial \bar{x}}\right] + \frac{\partial}{\partial \bar{y}}\left[\bar{v}n + \left(\frac{bW_c}{\Delta c}\right)\frac{\partial c}{\partial \bar{x}} n - D_n\frac{\partial n}{\partial \bar{y}}\right] = 0, \tag{9.6}$$

subject to the following constraints:

Cavity walls	*Temperature*	*Velocity profiles*	*Concentration profiles*
Bottom wall	$T = T_h$	$\bar{u} = \bar{v} = 0$	$c = c_{min},\ n = n_0;\ \frac{L}{3} \le x \le \frac{2L}{3}$
Top wall	$T = T_c$	$\bar{u} = U_0, \bar{v} = 0$	$c = c_0, \frac{\partial n}{\partial \bar{y}} = 0$
Right wall	$\frac{\partial T}{\partial \bar{x}} = 0$	$\bar{u} = \bar{v} = 0$	$\frac{\partial c}{\partial \bar{x}} = 0, \frac{\partial n}{\partial \bar{x}} = 0$
Left wall	$\frac{\partial T}{\partial \bar{x}} = 0$	$\bar{u} = \bar{v} = 0$	$\frac{\partial c}{\partial \bar{x}} = 0, \frac{\partial n}{\partial \bar{x}} = 0$

Using these dimensionless parameters:

$$X = \frac{\bar{x}}{L},\ Y = \frac{\bar{y}}{L},\ U = \frac{\bar{u}}{U_0},\ V = \frac{\bar{v}}{U_0},\ P = \frac{pL^2}{\rho},$$
$$\theta = \frac{T - T_c}{T_h - T_c},\ C = \frac{c - c_{min}}{c_0 - c_{min}},\ N = \frac{n}{n_0}.$$

the above equations reduce to their non-dimensional forms:

$$\frac{\partial U}{\partial X} + \frac{\partial V}{\partial Y} = 0, \tag{9.7}$$

$$U\frac{\partial U}{\partial X} + V\frac{\partial U}{\partial Y} = \left\{ \begin{array}{c} -\frac{\partial P}{\partial X} + \frac{1}{Re}\left[\frac{\partial^2 U}{\partial X^2} + \frac{\partial^2 U}{\partial Y^2}\right] - \frac{U}{Re.Da} \\ -Fr.U.\sqrt{U^2 + V^2} + \frac{Ha^2}{Re}\left[V\sin\lambda\cos\lambda - U\sin^2\lambda\right] \end{array} \right\}, \tag{9.8}$$

$$U\frac{\partial U}{\partial X} + V\frac{\partial U}{\partial Y} = \left\{ \begin{array}{c} -\frac{\partial P}{\partial Y} + \frac{1}{\mathrm{Re}}\left[\frac{\partial^2 V}{\partial X^2} + \frac{\partial^2 V}{\partial Y^2}\right] - \frac{V}{\mathrm{Re.Da}} - Fr.V.\sqrt{U^2 + V^2} \\ +\frac{Ha^2}{Re}\left[U\sin\lambda\cos\lambda - V\cos^2\lambda\right] + Ri.\theta - Ri.Rb.N \end{array} \right\}, \tag{9.9}$$

$$U\frac{\partial \theta}{\partial X} + V\frac{\partial \theta}{\partial Y} = \frac{1}{\mathrm{Re.Pr}}\left[\frac{\partial^2 \theta}{\partial X^2} + \frac{\partial^2 \theta}{\partial Y^2}\right] + J.V^2, \tag{9.10}$$

$$U\frac{\partial C}{\partial X}+V\frac{\partial C}{\partial Y}=\frac{1}{\text{Le.Pr.Re}}\left[\frac{\partial^2 C}{\partial X^2}+\frac{\partial^2 C}{\partial Y^2}\right]-\frac{\sigma_1}{\text{Le.Pr.Re}}N, \tag{9.11}$$

$$\begin{aligned}U\frac{\partial N}{\partial X}+V\frac{\partial N}{\partial Y}+\frac{Pe}{\text{Lb.Pr.Re}}\left[\frac{\partial}{\partial X}\left(N\frac{\partial C}{\partial X}\right)+\frac{\partial}{\partial Y}\left(N\frac{\partial C}{\partial Y}\right)\right]\\=\frac{1}{\text{Lb.Pr.Re}}\left[\frac{\partial^2 N}{\partial X^2}+\frac{\partial^2 N}{\partial Y^2}\right],\end{aligned} \tag{9.12}$$

where,

$$\text{Pr}=\frac{\nu}{\alpha},\ \text{Re}=\frac{U_0L}{\nu},\ \text{Ha}=B_0L\sqrt{\frac{\sigma}{\mu}},\ \text{Da}=\frac{K}{L^2},\ J=\frac{\sigma L U_0 B_0^2}{\rho C_p \Delta T},$$
$$\text{Gr}=\frac{g\beta L^3\left(T_h-T_c\right)}{\nu^2},\ \text{Ri}=\frac{Gr}{Re^2},\ \text{Fr}=\frac{FU_0}{\sqrt{K}},\ \text{Le}=\frac{\alpha}{D_c}\text{Rb}=\frac{\gamma\Delta\rho n_0}{\rho\beta\Delta T},$$
$$\text{Lb}=\frac{\alpha}{D_n},\ \text{Pe}=\frac{bW_c}{D_n},\ \sigma_1=\frac{\delta n_0 L^2}{D_c\Delta c}.$$

Subject to the constraints

Cavity walls	*Temperature*	*Velocity profiles*	*Concentration profiles*
Bottom wall	$\theta=1$	$U=V=0$	$C=1\,,N=1 for\frac{L}{3}\le X\le\frac{2L}{3}$
Top wall			$C=0,\frac{\partial N}{\partial Y}=0$
Right wall	$\frac{\partial\theta}{\partial X}=0$	$U=V=0$	$\frac{\partial C}{\partial X}=0,\frac{\partial N}{\partial X}=0$
Left wall	$\frac{\partial\theta}{\partial X}=0$	$U=V=0$	$\frac{\partial C}{\partial X}=0,\frac{\partial N}{\partial X}=0$

9.3 PARAMETERS OF INTEREST

(a) Nusselt Number

The local Nusselt number along the isothermal walls:

$$\text{Nu}(X)=-\left.\frac{\partial\theta}{\partial Y}\right|_{Y=0,1}.$$

Then, the average Nusselt number along the isothermal walls is

$$\mathrm{Nu}_{\mathrm{avg}} = \int_0^1 \mathrm{Nu}(X)\,dX.$$

(b) Sherwood Number

The local Sherwood number along the middle $L/3$ wall:

$$\mathrm{Sh}(X) = -\frac{\partial C}{\partial Y}.$$

Then, the average Sherwood number along the middle $L/3$ wall is

$$\mathrm{Sh}_{\mathrm{avg}} = \int_{\frac{L}{3}}^{\frac{2L}{3}} \mathrm{Sh}(X)\,dX.$$

9.4 NUMERICAL METHODOLOGY

To analyse the coupled behaviour of fluid flow, heat transfer, oxygen distribution, and microorganism transport, the governing equations (9.1 – 9.6) were first non-dimensionalized using suitable scaling parameters based on characteristic length, velocity, and temperature. This ensures consistency with the incompressibility condition and facilitates parametric analysis across a range of physical scenarios. The resulting system of non-dimensional partial differential equations (9.7 – 9.12) was solved using the FEM in COMSOL Multiphysics 6.2, which offers healthy multiphysics coupling capabilities. The model incorporates the laminar flow interface coupled with the heat transfer in fluids and transport of diluted species modules to account for momentum, thermal, oxygen, and microorganism fields, respectively. The Darcy–Forchheimer drag was included to account for inertial resistance due to the porous medium.

A structured mesh was generated, with localized refinement near the walls and boundaries where sharp gradients in velocity, temperature, and concentration were anticipated. The computational model employed a steady-state solver, and convergence was ensured by monitoring residual norms with relative tolerance set below 10^{-6}. Boundary and initial conditions were defined according to the physical configuration of the problem, ensuring accurate representation of flow, thermal, and bioconvective behaviour.

9.4.1 Mesh grid test

A mesh independence study was carried out to ensure that the numerical results are not influenced by the discretization size. The analysis was

Table 9.1 Grid independence analysis

Mesh refinement level	*No. of elements*	*Nusselt Number at the bottom wall*	*Sherwood Number at the bottom wall*
Extremely Coarse	209	0.44579	−0.33069
Extra Coarse	360	0.47722	−0.33251
Coarser	544	0.49022	−0.33339
Coarse	1005	0.50728	−0.33447
Normal	1505	0.51396	−0.33486
Fine	2510	0.52677	−0.33562
Finer	6617	0.52677	−0.33562

performed for a representative set of governing parameters, which were kept constant during this test. Various levels of mesh refinement were used, ranging from extremely coarse to finer, and the number of elements varied from 209 to 6,617. The corresponding average Nusselt and Sherwood numbers at the bottom wall were used as benchmark outputs to evaluate mesh sensitivity. As shown in Table 9.1, the results begin to stabilize at the fine mesh level with 2,510 elements, where both the Nusselt and Sherwood numbers converge to consistent values. Further refinement (finer mesh with 6,617 elements) yielded identical results, confirming mesh independence. Based on this observation, the fine mesh was selected for all subsequent simulations to balance accuracy and computational efficiency.

A mesh independence study was carried out to ensure that the numerical results are not influenced by the discretization size. The analysis was performed for a representative set of governing parameters, which were kept constant during this test. Various levels of mesh refinement were used, ranging from extremely coarse to finer, and the number of elements varied from 209 to 6,617. The corresponding average Nusselt and Sherwood numbers at the bottom wall were used as benchmark outputs to evaluate mesh sensitivity. As shown in Table 9.1, the results begin to stabilize at the fine mesh level with 2,510 elements, where both the Nusselt and Sherwood numbers converge to consistent values. Further refinement (finer mesh with 6,617 elements) yielded identical results, confirming mesh independence. Based on this observation, the fine mesh was selected for all subsequent simulations to balance accuracy and computational efficiency. Mesh distribution is sketched in Figure 9.2.

9.5 VALIDATION OF THE STUDY

To establish confidence in the numerical framework employed, a validation study was performed by comparing the current simulation results with established benchmark solutions available in the literature. For this comparison, the effects of oxygen transport, microorganism motion, Joule

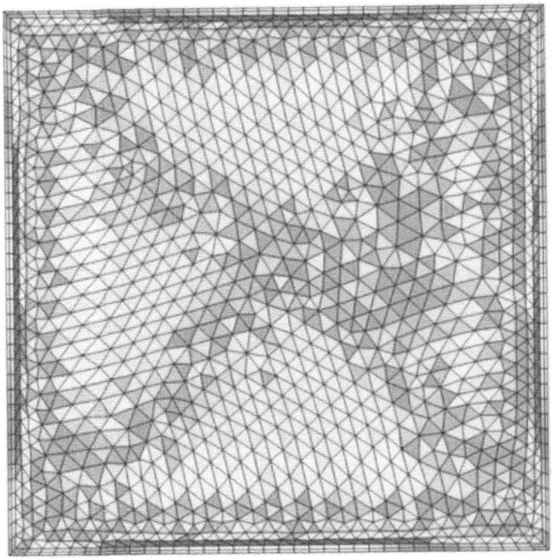

Figure 9.2 Mesh distribution.

heating, and magnetic field were deactivated, effectively reducing the model to its fundamental flow and thermal components. The isothermal contours and streamline patterns obtained under these simplified conditions were compared with the results of Cheng (2011) and Moallemi et al. (1992), as shown in Figures 9.3 and 9.4. The present results demonstrate excellent agreement with these references, thereby validating the correctness of the flow model, numerical formulation, and solution methodology. This consistency confirms that the finite element-based approach used in this study is healthy and reliable for simulating the extended model with additional physical effects.

9.6 RESULTS AND DISCUSSION

The present boundary value problem governed by the non-dimensional equations is solved numerically using FEM to investigate the influence of key non-dimensional parameters: Reynolds number (Re), Hartmann number (Ha), bioconvection Rayleigh number (Rb), bioconvection Lewis number (Lb), Joule heating parameter (J), and Forchheimer number (Fr). The parameter ranges considered in this study are: $\mathrm{Re} = (10-1000)$, $\mathrm{Pr} = (0.7, 7, 14)$, $\mathrm{Ha} = (0-100)$, $\mathrm{Rb} = (0.1-10)$, $\mathrm{Lb} = (0.01-1)$, $J = (0.01-10)$, and $\mathrm{Fr} = (0.001-1)$. The results are presented in terms of streamline patterns, isotherms, oxygen concentration contours, and

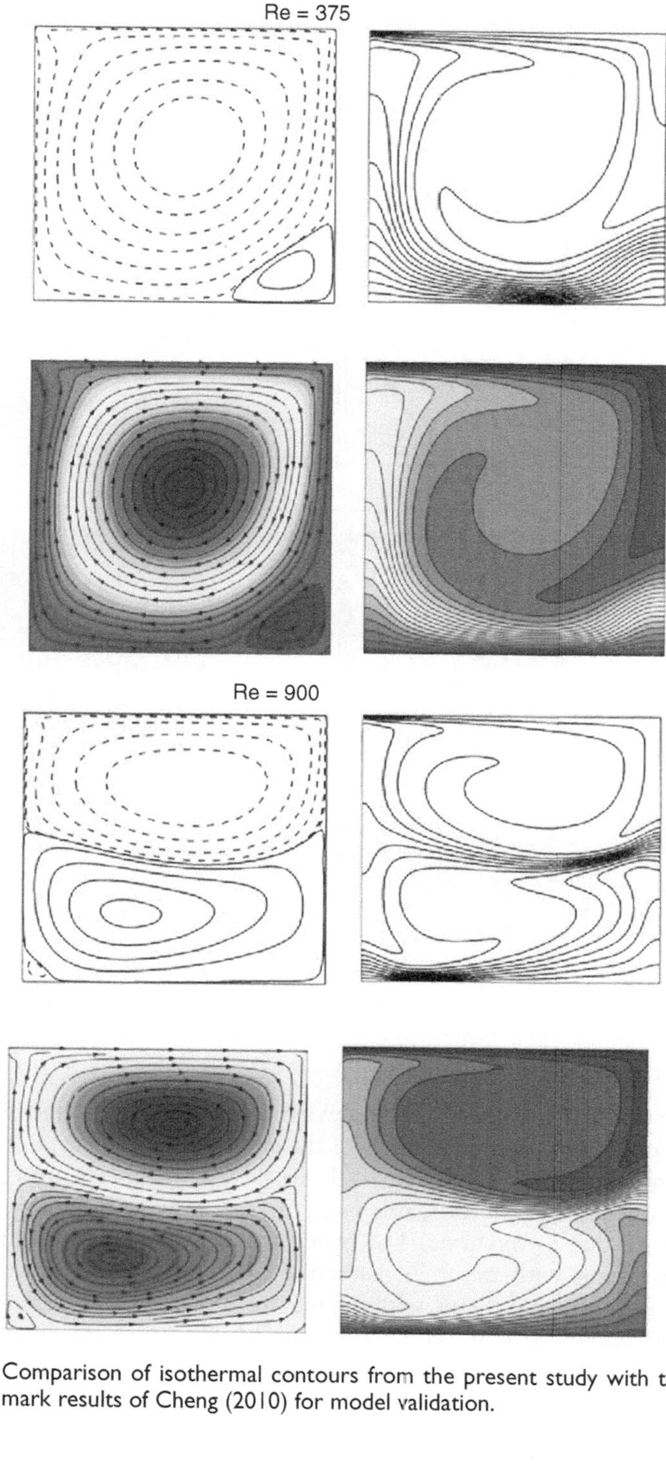

Figure 9.3 Comparison of isothermal contours from the present study with the benchmark results of Cheng (2010) for model validation.

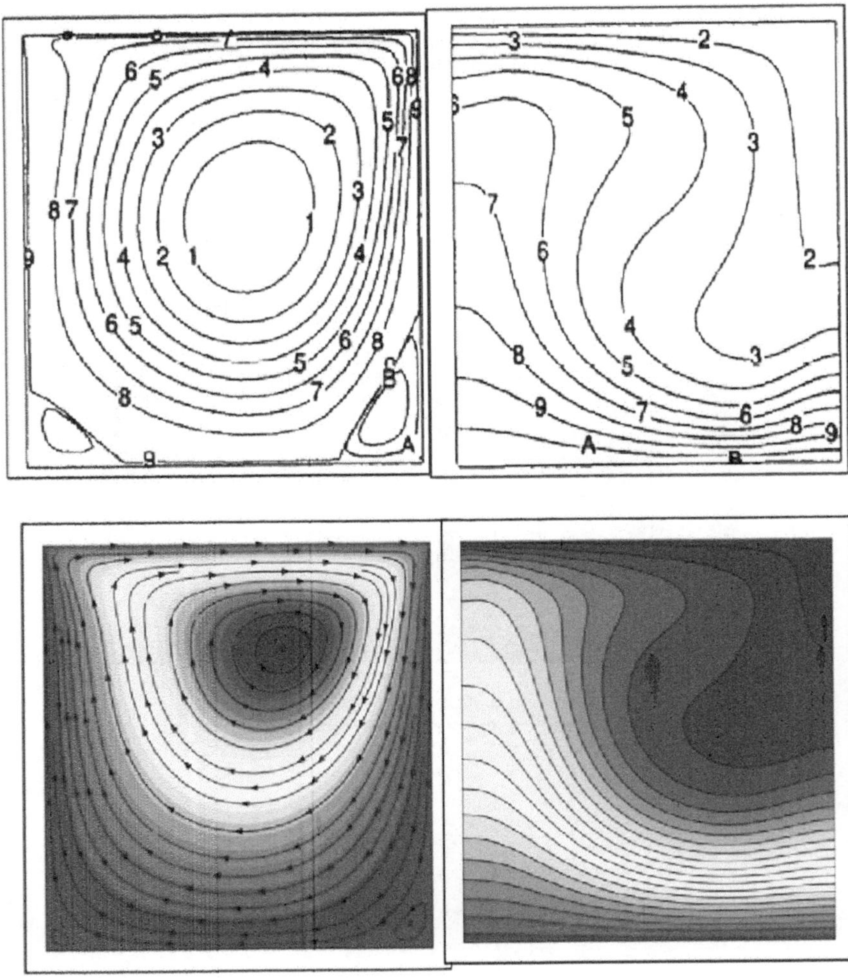

Figure 9.4 Comparison of streamline contours from the present study with the benchmark results of Moallemi et al. (1992) for model validation.

microorganism distributions, along with average Nusselt and Sherwood numbers computed along the walls of the cavity.

As shown in Figure 9.5, at Re = 10, the flow is symmetric and diffusion-dominated, enabling microorganisms to migrate upward smoothly from the bottom wall due to oxytaxis. The concentration contours exhibit a stratified distribution with minimal distortion. In contrast, at Re = 1,000, the flow becomes more difficult with strong recirculation and secondary vortices. This enhanced convective transport leads to a more heterogeneous

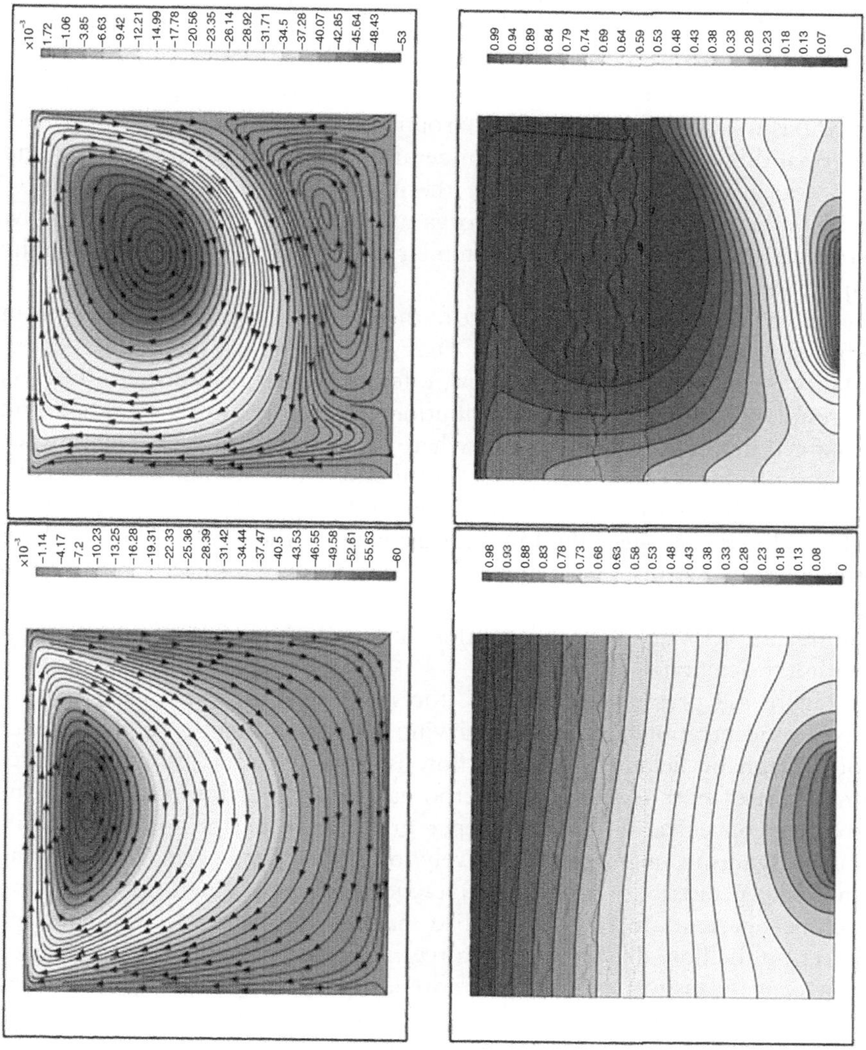

Figure 9.5 Streamlines (Top) and Microorganism concentration contours (Bottom) for Re = 10 and Re = 1000.

distribution of microorganisms. Specifically, the concentration contours indicate areas of high microorganism accumulation near regions of flow separation and reattachment, as well as within recirculation zones. These observations align with findings from studies on lid-driven cavity flows, where higher Reynolds numbers introduce complex flow structures that significantly influence scalar transport phenomena.

As depicted in Figure 9.6 at a lower Prandtl number (Pr = 0.7), the thermal diffusivity is relatively high, resulting in a more stratified and uniform microorganism concentration profile. The contours appear smooth and horizontally layered, indicating steady upward migration due to oxytaxis in a thermally relaxed environment. In contrast, for Pr = 7, where thermal diffusivity is lower, the temperature gradients become steeper and more localized. This sharpens the thermal boundary layer and enhances its coupling with flow and microorganism transport. The contours show stronger spatial variations and asymmetry, with noticeable distortion in the upper region.

Figure 9.7 illustrates the influence of the Darcy number on microorganism concentration profiles. At a lower Darcy number (Da = 0.01), the permeability of the porous medium is low, offering greater resistance to flow. As a result, the microorganism distribution remains relatively stratified and confined near the bottom region, where they are introduced. The upward migration due to oxytaxis is less pronounced due to weaker convective transport. In contrast, at Da = 1, the medium is significantly more permeable, allowing stronger fluid motion throughout the cavity. This enhances the convective transport of microorganisms, leading to more irregular and asymmetric distribution contours. The flow-driven mixing becomes more dominant, helping in better migration and wider spread of microorganisms within the cavity.

Figure 9.8 presents the oxygen (top row) and microorganism (bottom row) concentration contours for varying Hartmann number. In the absence of a magnetic field (Ha = 0), the flow is more dynamic, allowing greater transport of both oxygen and microorganisms throughout the cavity. The oxygen concentration shows steeper gradients near the top wall, while microorganisms migrate more actively toward the upper regions, following the oxygen distribution due to oxytactic behaviour. When the Hartmann number increases to 100, the applied magnetic field introduces a damping effect on the flow due to Lorentz forces. This suppression leads to reduced convective transport, resulting in more laminated and horizontally layered concentration fields for both oxygen and microorganisms. The oxygen remains concentrated near the top wall, and microorganisms exhibit limited vertical migration, remaining mostly near the source region at the bottom wall. These trends demonstrate the inhibiting influence of magnetic fields on bioconvective transport by reducing flow strength and mixing efficiency.

Figure 9.9 further illustrates the effect of the bioconvection Lewis number on microorganism concentration contours for $Lb = 0.01, 0.1,$ and 1. As Lb increases, a clear suppression in the upward migration of microorganisms is observed. At a low Lewis number (Lb = 0.01), the microorganism motility diffusion dominates, resulting in a more uniform and extended distribution across the cavity. The concentration contours are widely spread, indicating

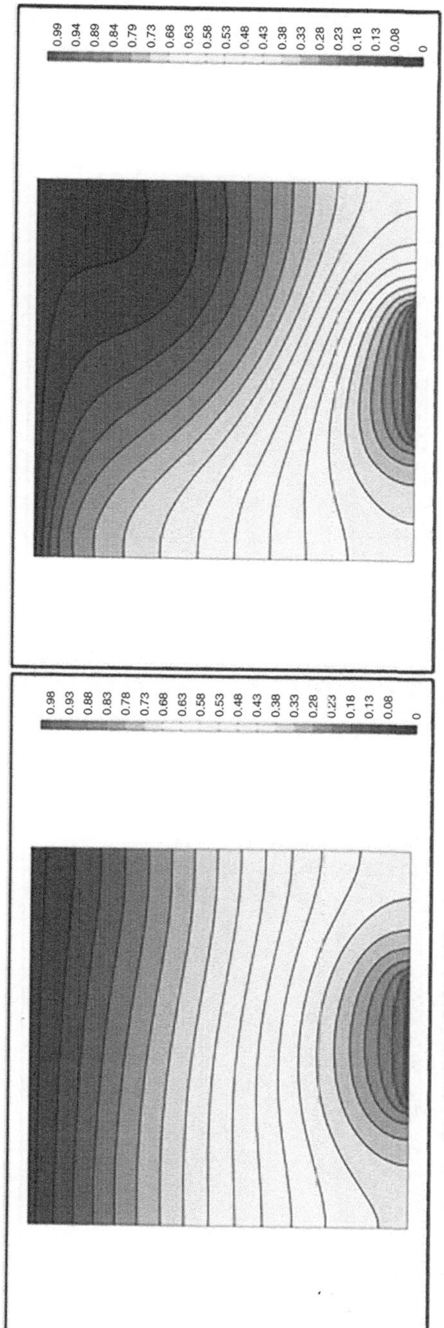

Figure 9.6 Microorganism concentration contours for Pr = 0.7 (left) and Pr = 7 (right).

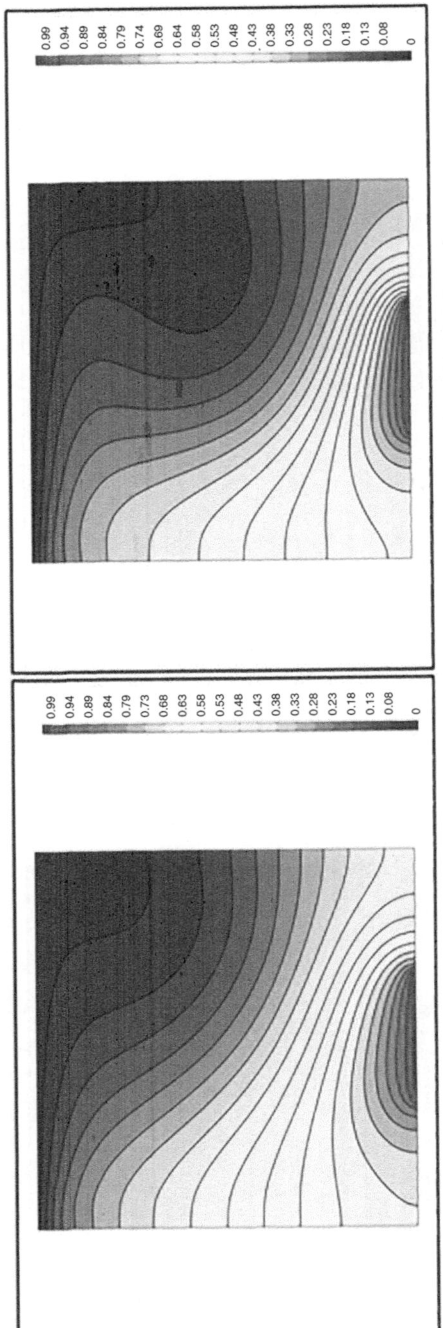

Figure 9.7 Microorganism concentration contours for Da = 0.01 and Da = 1.

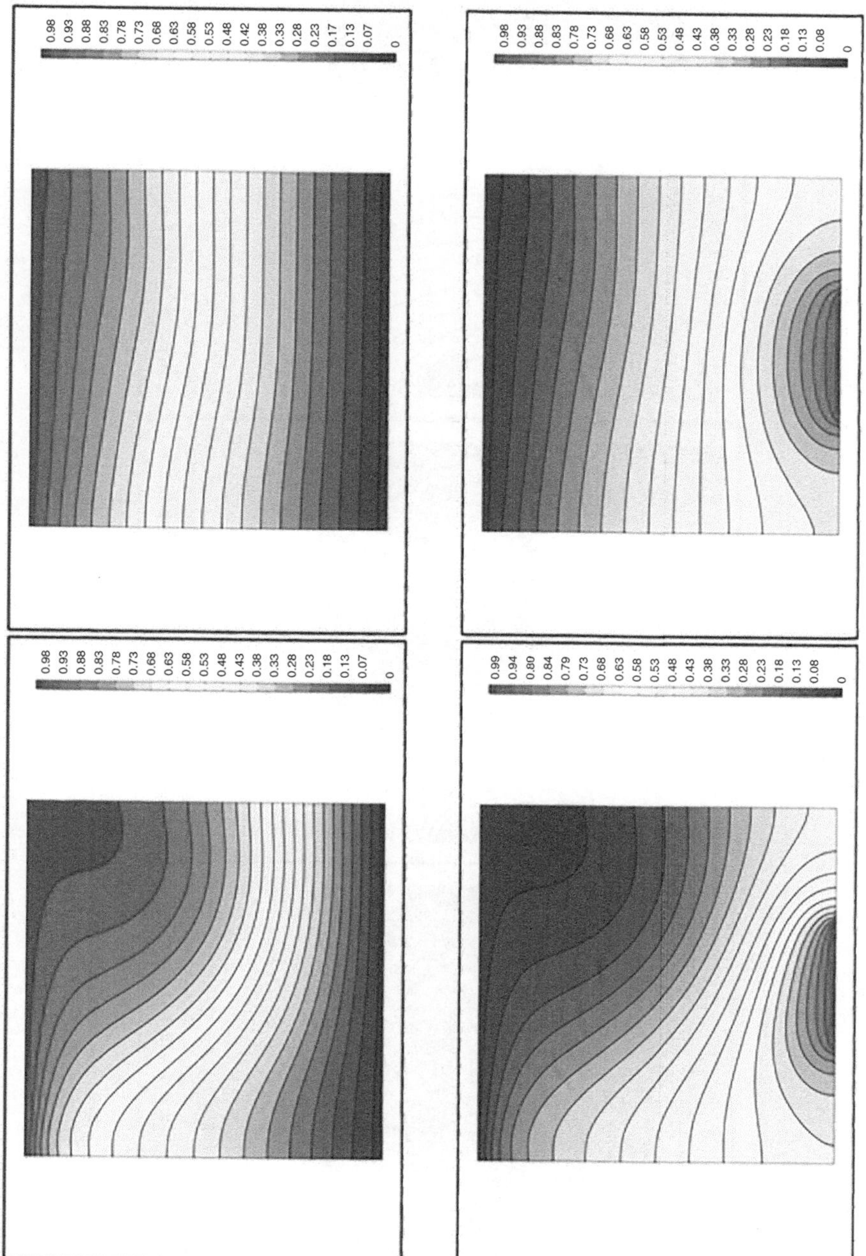

Figure 9.8 Oxygen and Microorganism concentration contours for Ha = 0and Ha = 100.

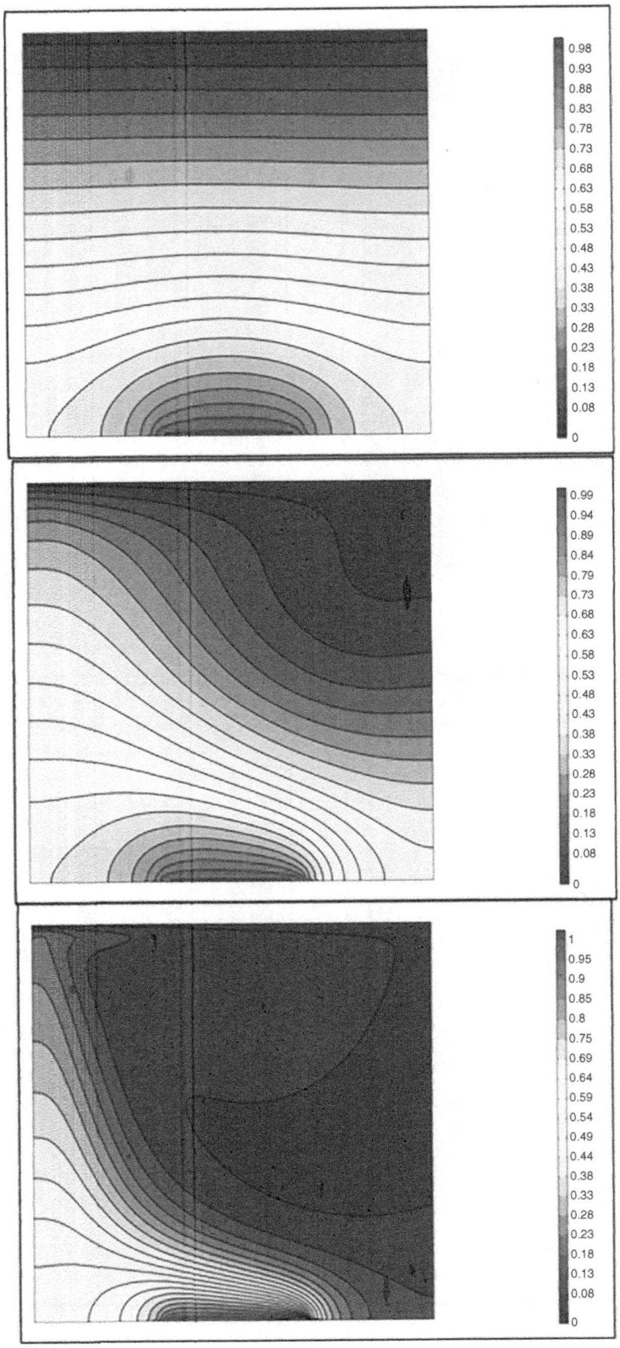

Figure 9.9 Oxygen and Microorganism concentration contours for Lb = 0.01,0.1 and 1.

effective upward transport under the influence of oxytaxis. As Lb increases to 0.1, the distribution becomes moderately confined, with a noticeable reduction in vertical spread. This trend continues for Lb = 1, where the contours are highly concentrated near the lower part of the cavity. The reduced microorganism dispersion at higher Lb reflects the decreasing influence of motile diffusivity, leading to stronger localization near the source region. These results confirm that higher bioconvection Lewis numbers limit the diffusive transport of microorganisms, confining their distribution to the lower zone of the cavity despite the presence of oxygen gradients.

Figure 9.10 illustrates the variation of the average Nusselt number (evaluated at the bottom $L/3$ region) with respect to the Hartmann number (Ha) for different values of the bioconvection Lewis number (Lb). A consistent decline in the Nusselt number is observed with increasing Ha, indicating a suppression of convective heat transfer due to the influence of the magnetic field. As the Lorentz force becomes stronger at higher Ha, the fluid motion is damped, resulting in weaker convection and consequently lower thermal transport. Among the different Lb values, the differences in Nusselt number are marginal but noticeable at lower Ha, as highlighted in the zoomed inset. Lower Lb corresponds to slightly higher heat transfer, suggesting that weaker diffusion of microorganisms (lower Lb) leads to more vigorous bioconvective activity, thereby enhancing fluid mixing and

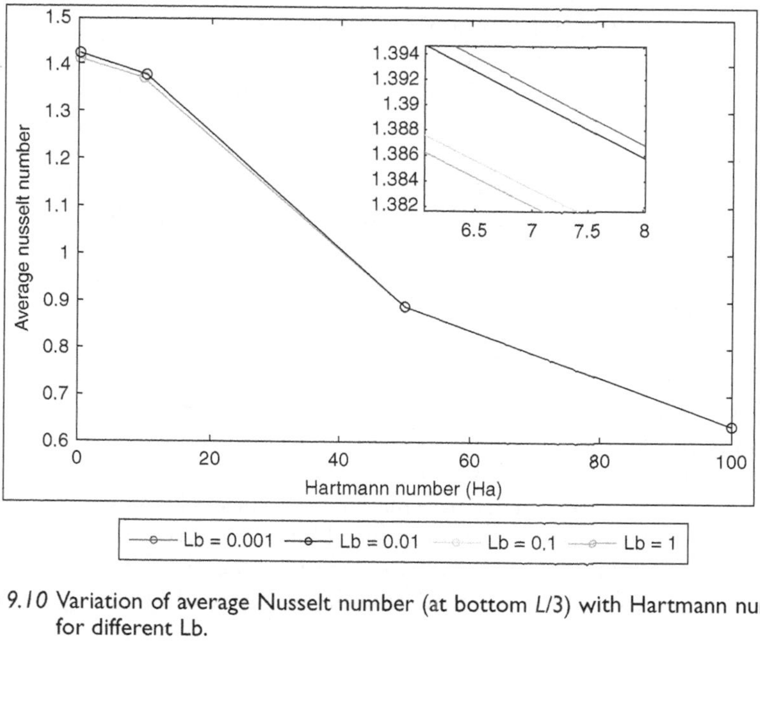

Figure 9.10 Variation of average Nusselt number (at bottom *L*/3) with Hartmann number for different Lb.

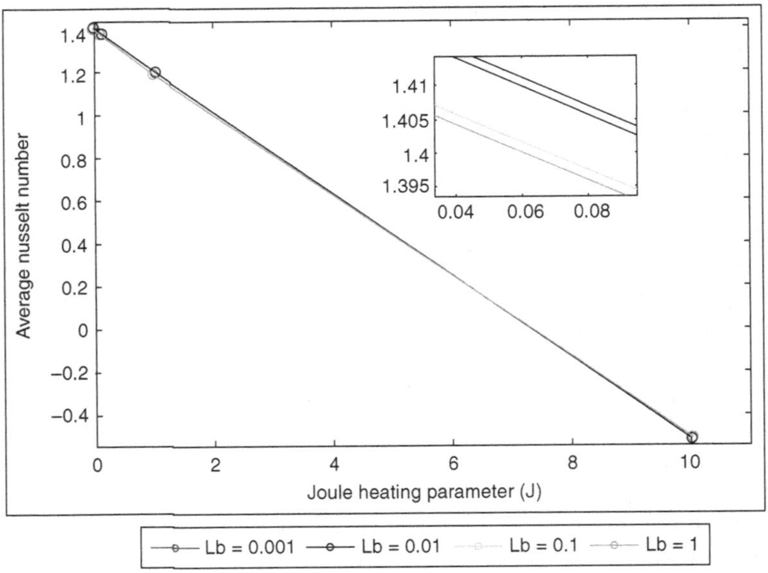

Figure 9.11 Variation of average Nusselt number (at bottom L/3) with Joule heating parameter for different Lb.

thermal transport. However, as Ha increases, magnetic damping dominates and diminishes these differences across Lb values.

Figure 9.11 presents the variation of average Nusselt number (measured at the bottom $L/3$ section) with respect to the Joule heating parameter (J) for different bioconvection Lewis numbers (Lb). A sharp decline in the Nusselt number is evident with increasing J, indicating that internal heat generation due to Joule heating significantly weakens thermal convection in the system. As J increases, the accumulation of resistive heat within the fluid elevates the overall thermal energy, thereby reducing the temperature gradient across the cavity. This leads to a suppression of buoyancy-driven flow, which is responsible for convective heat transport, resulting in a continuous reduction of the Nusselt number. In fact, for higher J values, the Nusselt number becomes negative, suggesting the reversal of heat transport direction at the lower region, which may be attributed to dominant conduction opposing the weakened convection. The effect of Lb remains relatively subtle throughout, but at lower J values, the zoomed inset reveals a mild enhancement in heat transfer for smaller Lb values. This behaviour can be linked to stronger microorganism-induced bioconvective mixing when their diffusion is limited. However, as J rises, the Joule heating effect overwhelms bioconvective enhancements, rendering the influence of Lb negligible.

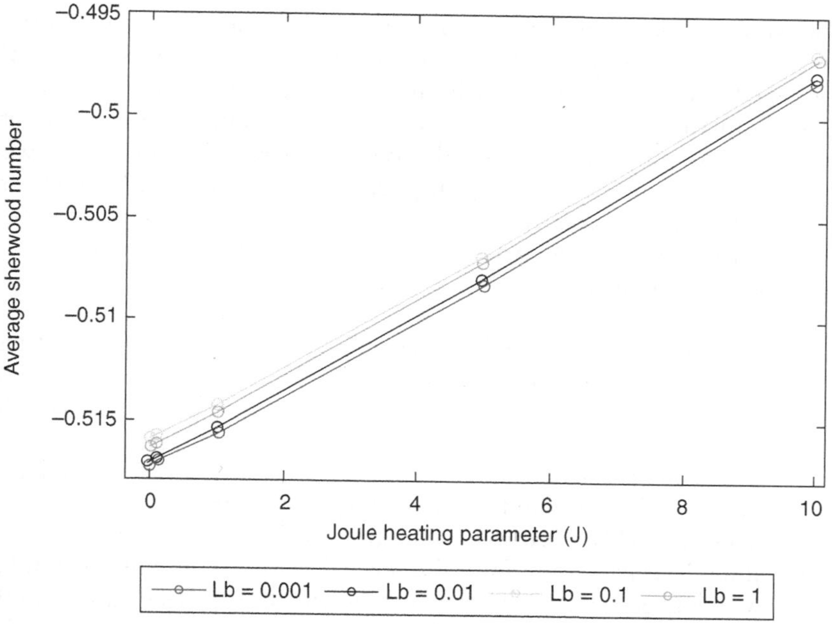

Figure 9.12 Variation of average Sherwood number (at bottom L/3) with Joule heating parameter for different Lb.

Figure 9.12 shows the variation of average Sherwood number at the bottom $L/3$ segment of the cavity with respect to the Joule heating parameter (J) for different bioconvection Lewis numbers (Lb). Contrary to the Nusselt number trend, the Sherwood number demonstrates a gradual increase with increasing J, indicating enhanced mass transfer (oxygen transport) near the bottom wall under the influence of Joule heating. As J increases, resistive heating elevates the local temperature, which can thin the boundary layer and intensify the oxygen gradient across the wall, thereby enhancing diffusive flux. This effect is reflected in the rising Sherwood number. However, the values remain negative throughout, which is characteristic of oxygen transfer occurring against the concentration gradient at the bottom region, where oxygen concentration is minimal. Higher values of Lb (indicating stronger diffusivity of microorganisms) correspond to slightly higher Sherwood numbers, particularly at larger J values. This suggests that enhanced motility of microorganisms (due to higher Lb) possibly improves mixing and creates favorable conditions for more efficient oxygen uptake in the vicinity of the wall, especially under thermal agitation induced by Joule heating.

9.7 CONCLUSIONS AND FUTURE SCOPE

This study explored the thermo-bioconvective behaviour of a lid-driven square porous cavity filled with oxytactic microorganisms under the combined effects of magnetic field and Joule heating. The finite element-based simulations were carried out in COMSOL Multiphysics 6.2, focusing on the bottom $L/3$ segment of the cavity where microorganisms were introduced. Parametric variations in Hartmann number (Ha), Joule heating parameter (J), and bioconvection Lewis number (Lb) were examined to assess their influence on heat and mass transfer characteristics. The following key outcomes were derived:

- Increasing the Reynolds number (Re) strengthens the lid-driven vortex structure, which enhances convective circulation within the cavity and leads to noticeable increases in both microorganism dispersion and oxygen transport near the bottom $L/3$ region.
- Increasing the Hartmann number led to a monotonic reduction in average Nusselt number, confirming the suppressive influence of magnetic fields on convective motion and thermal transport.
- Joule heating significantly diminished the average Nusselt number, with higher values even reversing the direction of heat transfer in the lower cavity region, indicative of dominant internal heat generation and conduction.
- At lower J and Ha, bioconvection Lewis number marginally enhanced heat transfer, particularly when microorganism diffusivity was lower, due to stronger convective mixing.
- Increasing the Darcy number enhances microorganism migration by promoting stronger convective mixing through the porous matrix.
- The average Sherwood number increased with Joule heating, implying stronger oxygen diffusion near the oxygen-depleted zone at the bottom wall.
- Higher values of Lb consistently yielded slightly elevated Sherwood numbers, especially at larger J, highlighting the role of microorganism motility in facilitating oxygen uptake.
- Future investigations may extend the present model by incorporating unsteady effects, non-Newtonian fluid behaviour, and complex geometries to better simulate real-world biomedical and bioengineering systems involving bioconvection and internal heat generation.

REFERENCES

Abbas, A., Jeelani, M. B., & Alharthi, N. H. (2022). Darcy–Forchheimer relation influence on MHD dissipative third-grade fluid flow and heat transfer in porous medium with Joule heating effects: A numerical approach. *Processes, 10*(5), 906.

Ahmed, S. E., Oztop, H. F., Mansour, M. A., & Abu-Hamdeh, N. (2018). Magnetohydrodynamic mixed thermo-bioconvection in porous cavity filled by oxytactic microorganisms. *Thermal Science*, *22*(6 Part B), 2711–2721.

Bakar, N. A., Karimipour, A., & Roslan, R. (2016). Effect of magnetic field on mixed convection heat transfer in a lid-driven square cavity. *Journal of Thermodynamics, 2016*(1), 3487182.

Basak, T., Roy, S., Paul, T., & Pop, I. (2006). Natural convection in a square cavity filled with a porous medium: Effects of various thermal boundary conditions. *International Journal of Heat and Mass Transfer, 49*(7–8), 1430–1441.

Cheng, T. S. (2011). Characteristics of mixed convection heat transfer in a lid-driven square cavity with various Richardson and Prandtl numbers. *International Journal of Thermal Sciences, 50*(2), 197–205.

Chowdhury, S., & Rao, P. S. (2023). Numerical investigation of thermo-bioconvection of oxytactic micro-organism in a tilted square porous cavity heated sinusoidally. *Journal of Nanofluids*, *12*(8), 2285–2296.

De Vahl Davis, G. (1983). Natural convection of air in a square cavity: A benchmark numerical solution. *International Journal for Numerical Methods in Fluids, 3*(3), 249–264.

Hussain, S., Aly, A. M., &Öztop, H. F. (2022). Magneto-bioconvection flow of hybrid nanofluid in the presence of oxytactic bacteria in a lid-driven cavity with a streamlined obstacle. *International Communications in Heat and Mass Transfer*, *134*, 106029.

Ingham, D. B., & Pop, I. (Eds.). (2005). *Transport Phenomena in Porous Media III* (Vol. 3). Elsevier.

Moallemi, M. K., & Jang, K. S. (1992). Prandtl number effects on laminar mixed convection heat transfer in a lid-driven cavity. *International Journal of Heat and Mass Transfer*, *35*(8), 1881–1892.

Nield, D. A., & Bejan, A. (2006). *Convection in Porous Media* (Vol. 3, pp. 629–982). Springer.

Saeid, N. H., & Pop, I. (2005). Non-Darcy natural convection in a square cavity filled with a porous medium. *Fluid Dynamics Research, 36*(1), 35.

Sheremet, M. A., & Pop, I. (2014). Thermo-bioconvection in a square porous cavity filled by oxytactic microorganisms. *Transport in Porous Media*, *103*, 191–205.

Sugilal, G., Wattal, P. K., & Iyer, K. (2005). Convective behavior of a uniformly Joule-heated liquid pool in a rectangular cavity. *International Journal of Thermal Sciences, 44*(10), 915–925.

Swain, B. K., Parida, B. C., Kar, S., & Senapati, N. (2020). Viscous dissipation and Joule heating effect on MHD flow and heat transfer past a stretching sheet embedded in a porous medium. *Heliyon, 6*(10), e05338.

Vafai, K., & Tien, C. L. (1981). Boundary and inertia effects on flow and heat transfer in porous media. *International Journal of Heat and Mass Transfer, 24*(2), 195–203.

Walker, K. L., & Homsy, G. M. (1978). Convection in a porous cavity. *Journal of Fluid Mechanics, 87*(3), 449–474.

Yashkun, U., Zaimi, K., Ishak, A., Pop, I., & Sidaoui, R. (2021). Hybrid nanofluid flow through an exponentially stretching/shrinking sheet with mixed convection and Joule heating. *International Journal of Numerical Methods for Heat & Fluid Flow, 31*(6), 1930–1950.

Zhang, J. K., Li, B. W., & Chen, Y. Y. (2015). The Joule heating effects on natural convection of participating magnetohydrodynamics under different levels of thermal radiation in a cavity. *Journal of Heat Transfer, 137*(5), 052502.

Chapter 10

Heat and mass transfer effects on a stagnation flow considering slip and radiation impact

T. Pradhan, S. Jena, and B. K. Paikaray

10.1 INTRODUCTION

The stagnation-point flow on a stretching surface has drawn abundant attention as it provides various applications in both manufacturing and industrial processes. Wire drawing, aerodynamics, polymer extrusion, electronic colling systems, and microfluidics for particle manipulation. The thermal energy flux and new chemical reaction implemented by Merkin et al. [1] and Khan et al. [2] on the hydromagnetics stagnation-point flow through a stretching cylinder and Joule heating, respectively. Waini et al. [3] scrutinized the heat energy transfer rate of hybrid nano liquid flow to a stagnation point over a shrinking/stretching cylinder where copper and alumina mixed with water were considered.

Bhukta et al. [4] in this study analysed Joulian, Darcy dissipation and viscous flow phenomena on a mouldable surface with Kummer's function. Pattnaik et al. [5] investigated the various uses of MHD liquid flow and thermal energy over a vertical plate with Williamson model. They concluded that the correlation between numerical techniques and methodology gives excellent results. Ahmed [6] studied the numerical paralleling of stagnation-point flow involving thermal energy impact over a horizontal cylinder with stretching velocity. Mahabaleshwar et al. [7] examined the impact of radiation on stagnation-point flow of Casson liquid with nanoparticles over a stretching/shrinking circular cylinder or flat plate. Their study is applicable for electric conductivity, industry, aircraft and other areas.

A numerical analysis studied by Pattanaik et al. [8] on 2D stagnation-point nanoliquid flow past a permeable sheet with the help of RK fourth order method. They conclude that thermal buoyancy boosted the magnitude of nanoliquid velocity. The heterogeneous–homogeneous chemical reaction investigated by Jena et al. [9] on 3D MHD rotational nano liquid flow past a stretching surface with water-based fluid. Pradhan et al. [10] examined the effect of chemical reaction, thermal radiation, and heat source on MHD nanofluid flow on a flat plate. Here, they used the Laplace transfer technique to find the solution for the time-dependent problem.

DOI: 10.1201/9781003642169-10

The impact of chemical reaction, thermal energy, and Joule heating has been investigated by Hayat et al. [11] and Ghadikolaei et al. [12] on 3D flow of Eyring–Powell and Casson fluid with nanoparticles past a porous surface, respectively. The effect of radiation, viscosity, and other characteristics of ferro liquid and non-Newtonian liquid on a shrinking/stretching surface was studied in Refs. [13,14]. The analysis of entropy, stability of nanoliquid and mixed nanoliquid, stagnation nanoliquid and Carreau nano liquid flow past the stretching sheet is presented in Refs. [15–18]. Beg et al. [19] examined the energy transfer on liquid with nanoparticles with the help of a numerical technique. Agrawal flows past a stretching disk with a mixed nanoliquid that has been examined by Waini et al. [20].

10.2 MATHEMATICAL ANALYSIS

Let us consider a 2D steady laminar boundary layer stagnation-point flow for an incompressible fluid flow under effects of radiation and slip, which is driven by shrinking sheet. The corresponding governing equations are as follows:

$$\tilde{u}\partial_x\tilde{u} + \tilde{v}\partial_x\tilde{v} = Ud_xU - \upsilon\partial_y^2\tilde{u} \tag{10.1}$$

$$\tilde{u}\partial_x\tilde{T} + \tilde{v}\partial_y\tilde{T} = \alpha\partial_y^2\tilde{T} - \frac{1}{\rho C_p}\partial_y q_r \tag{10.2}$$

$$\tilde{u}\partial_x\tilde{C} + \tilde{v}\partial_y\tilde{C} = D_B\partial_y^2 C \tag{10.3}$$

The appropriate boundary conditions are

$$\left.\begin{array}{ll} \tilde{u} = cx + L\left(\partial_y\tilde{u}\right), \tilde{v} = 0\,, \tilde{T} = \tilde{T}_w\,, \tilde{C} = \tilde{C}_w & at\, y = 0, \\ \tilde{u} \to U(x) = ax,\ \tilde{T} = \tilde{T}_\infty,\ \tilde{C} = \tilde{C}_\infty & as\, y \to \infty \end{array}\right\} \tag{10.4}$$

where L is the slip length. $c < 0$ and $c > 0$, shrinking and stretching rate, respectively, with $a > 0$ (straining rate parameter). Using the Rosseland approximation, $q_r = -(4\sigma^*/3k^*)\partial_y\tilde{T}^4$. Using Taylor's series expansion for $\tilde{T}^4$ about $\tilde{T}_\infty$ we get, $\tilde{T}^4 = 4\tilde{T}_\infty^3\tilde{T} - 3\tilde{T}_\infty^4$ where higher-order terms are neglected.

So now Equation (10.2) becomes,

$$\tilde{u}\partial_x\tilde{T} + \tilde{v}\partial_y\tilde{T} = \alpha + \alpha\frac{16\sigma^*\tilde{T}_\infty^3}{3kk^*}\partial_y^2\tilde{T}$$

Eqns. (10.1–10.4) undergo the following similarity transformations:

$$\eta=\sqrt{\frac{a}{\upsilon}}y, \tilde{u}=axf'(\eta), \tilde{v}=-\sqrt{a\upsilon}f(\eta), \psi(x,y)=\sqrt{a\upsilon}\,xf(\eta),$$
$$\theta(\eta)=\frac{\tilde{T}-\tilde{T}_{\infty}}{\tilde{T}_{w}-\tilde{T}_{\infty}}, \phi(\eta)=\frac{\tilde{C}-\tilde{C}_{\infty}}{\tilde{C}_{w}-\tilde{C}_{\infty}} \tag{10.5}$$

which can be reduced as:

$$\left.\begin{array}{l} f'''+ff''-f'^{2}+1=0 \\ \frac{1}{\Pr}(1+R)\theta''+f\theta'=0 \\ \frac{1}{Sc}\phi''+f\phi'=0 \end{array}\right\} \tag{10.6}$$

and

$$\left.\begin{array}{llll} f(\eta)=0, f'(\eta)=\lambda+\delta f''(0), & \theta(\eta)=1, & \phi(\eta)=1 & at\ \eta=0 \\ f'(\eta)\to 0, & \theta(\eta)\to 0, & \phi(\eta)\to 0 & as\ \eta\to\infty \end{array}\right\} \tag{10.7}$$

where

$$\Pr=\frac{\upsilon}{\alpha}, Sc=\frac{\upsilon}{D_{B}}, R=\frac{16\sigma^{*}T_{\infty}^{3}}{3kk}, \lambda=\frac{c}{a}, \delta=L\left(\frac{a}{\upsilon}\right)^{1/2} \tag{10.8}$$

10.3 THE PHYSICAL QUANTITIES OF INTEREST

Skin friction coefficient, local Nusselt number, local Sherwood number of the problem are respectively defined as:

$$\tilde{C}_{f}=\frac{\tau_{w}}{\rho U_{w}^{2}}, Nu_{x}=\frac{xq_{w}}{k(\tilde{T}_{w}-\tilde{T}_{\infty})}, Sh_{x}=\frac{xq_{m}}{D_{B}(\tilde{C}_{w}-\tilde{C}_{\infty})}. \tag{10.9}$$

where

$$\tau_{w}=\mu\left(\partial_{y}\tilde{u}\right)_{y=0}, q_{w}=\left(-k\partial_{y}\tilde{T}+q_{r}\right)_{y=0},$$
$$q_{m}=-D_{B}\left(\partial_{y}\tilde{C}\right)_{y=0}, \mathrm{Re}_{x}=\frac{U_{w}x}{\upsilon}$$

A simple calculation gives,

$$\sqrt{\mathrm{Re}_x}\tilde{C}_f = f''(0), Nu_x / \sqrt{\mathrm{Re}_x} = -\left(1 + \frac{4}{3}R\right)\theta'(0),$$
$$Sh_x / \sqrt{\mathrm{Re}_x} = -\phi'(0) \tag{10.10}$$

10.4 RESULTS AND DISCUSSION

Figure 10.1 shows the validation check for velocity, temperature, and concentration profiles and from the figure, it is clear that the problem well agrees the boundary condition which gives an approval for our results. Figures 10.2 and 10.3 present the effect of (δ) on velocity and temperature profiles. Due to active slip parameter, velocity gets enhanced in first solution but in the range $0 \leq \eta \leq 1.5$ and $4.2 \leq \eta < \eta_{max}$, the profile is increasing and for $1.5 \leq \eta \leq 4.2$, it decreases. This behaviour of velocity profile is witnessed in Figure 10.2. Temperature is a decreasing function of velocity slip parameter (δ) for both the solutions which can be confirmed from Figure 10.3. The influence of Prandtl number on both velocity and temperature profiles is depicted in Figures 10.4 and 10.5. Both the profiles are showing opposite behaviour, as expected, for increasing values of Pr. The velocity profile gets decelerated

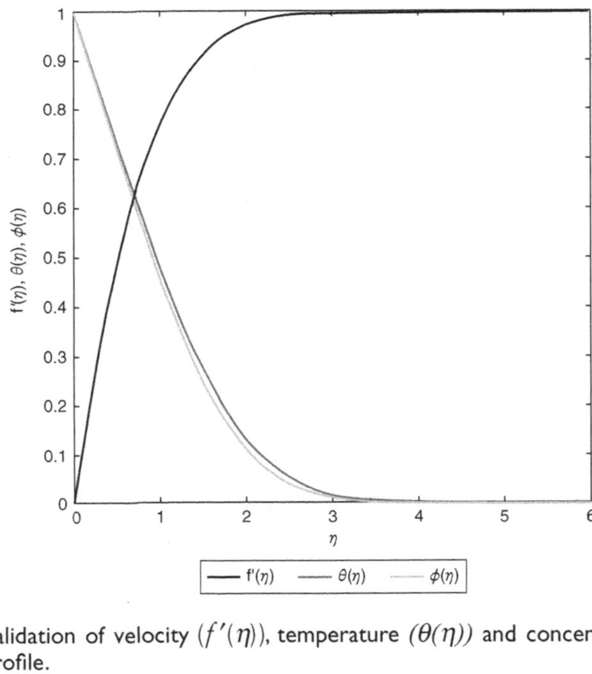

Figure 10.1 Validation of velocity $(f'(\eta))$, temperature $(\theta(\eta))$ and concentration $(\phi(\eta))$ profile.

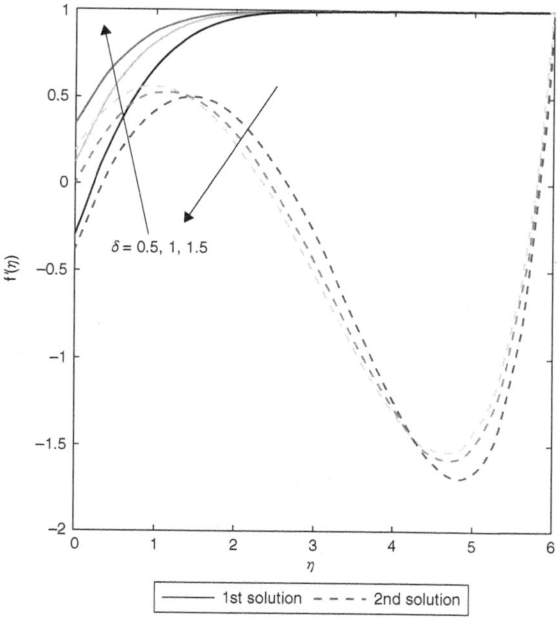

Figure 10.2 Variation of velocity profile $f'(\eta)$ with δ.

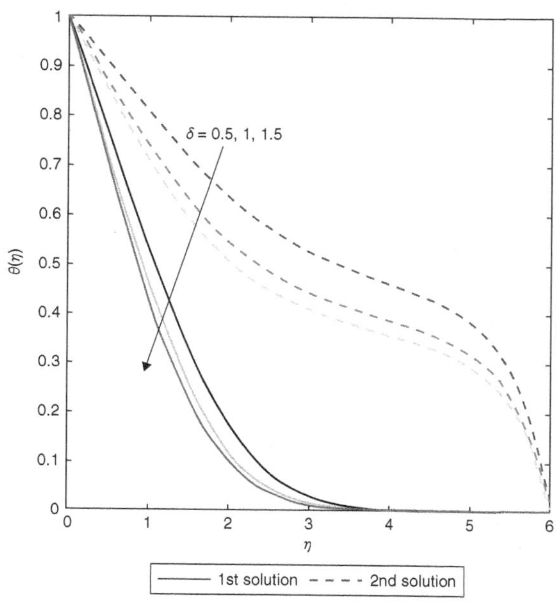

Figure 10.3 Variation of temperature profile $(\theta(\eta))$ with δ.

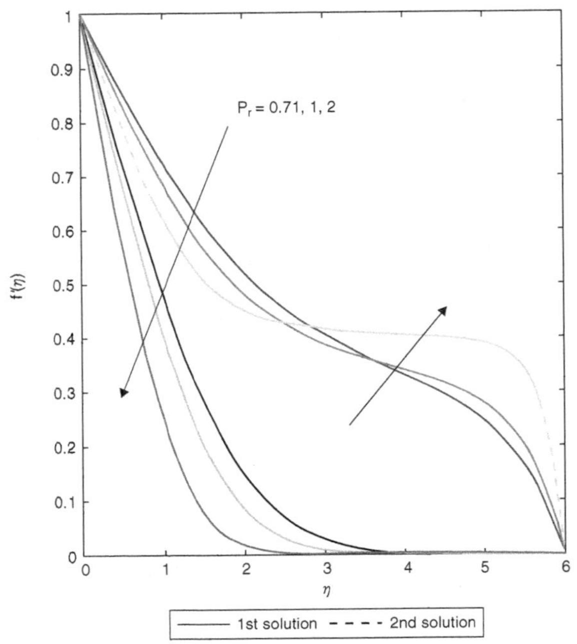

Figure 10.4 Variation of temperature profile $f'(\eta)$ with P_r.

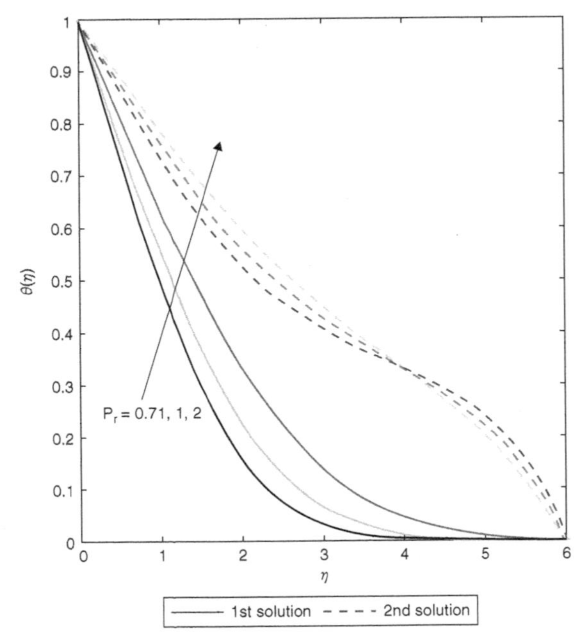

Figure 10.5 Variation of temperature profile $\theta(\eta)$ with *R*.

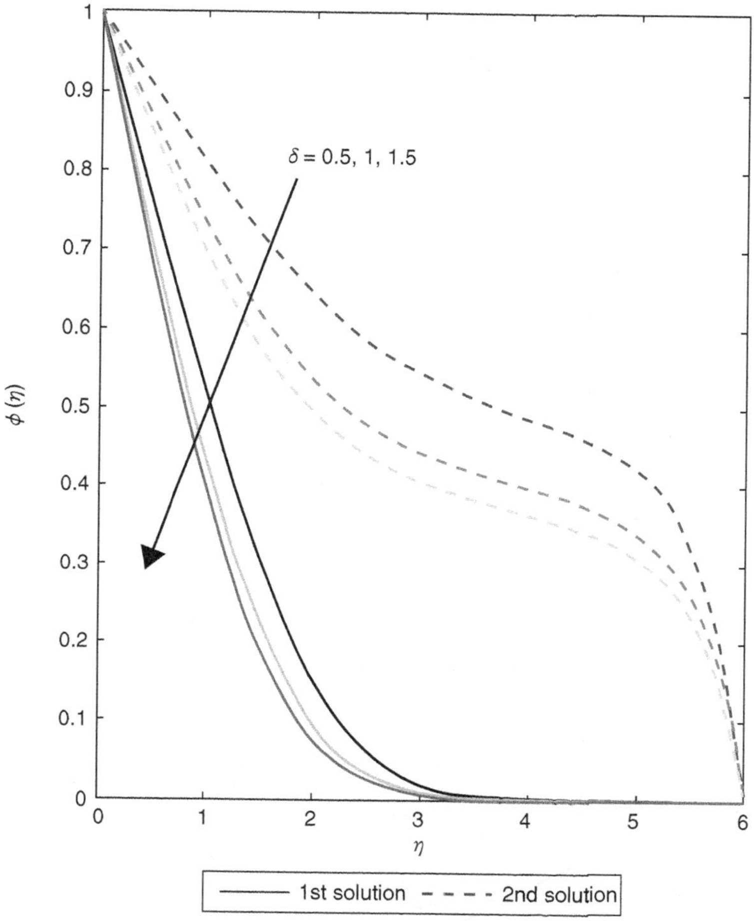

Figure 10.6 Variation of concentration profile $(\phi(\eta))$ with δ.

for increasing values of Prandtl number in both the solutions but second solution shows opposite behaviour away from boundary, that is, for $\eta \geq 3$. Temperature, on the other hand, shows the opposite trend for both the solutions. Figures 10.6 and 10.7 are the evidences of decreasing profile of solutal boundary layer thickness for increasing values of velocity parameter and Schmidt number (S_c). Figure 10.8 presents the graphical behaviour of C_f for both (λ) and (δ). We have taken the '–ve' range for (λ) and '+ve' range for (δ). When δ is increased, C_f increases but λ is to decelerate it. Figures 10.9, 10.10, and 10.11 present the increasing values of both δ and P_r along with '-ve' range of λ increases the Nusselt number coefficient. But it decreases with increasing values of radiation parameter (R). Figures 10.12 and 10.13

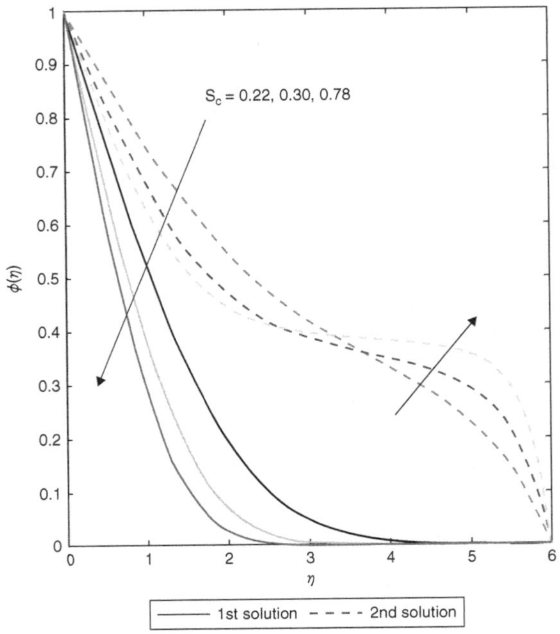

Figure 10.7 Variation of concentration profile $(\phi(\eta))$ with S_c.

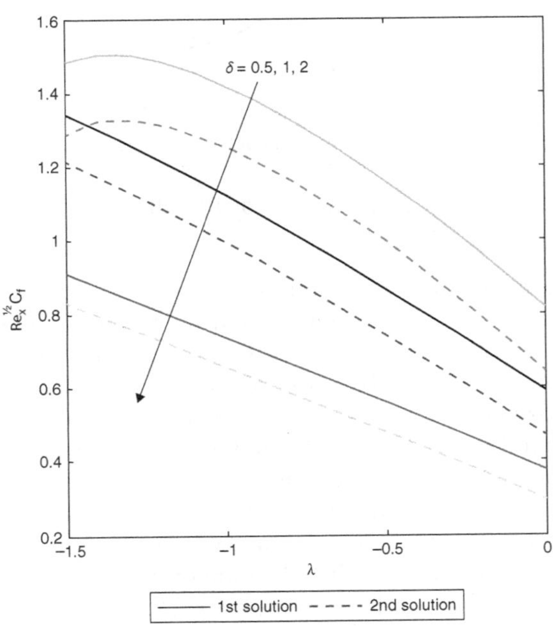

Figure 10.8 Variation of skin friction coefficient with λ for several values of δ.

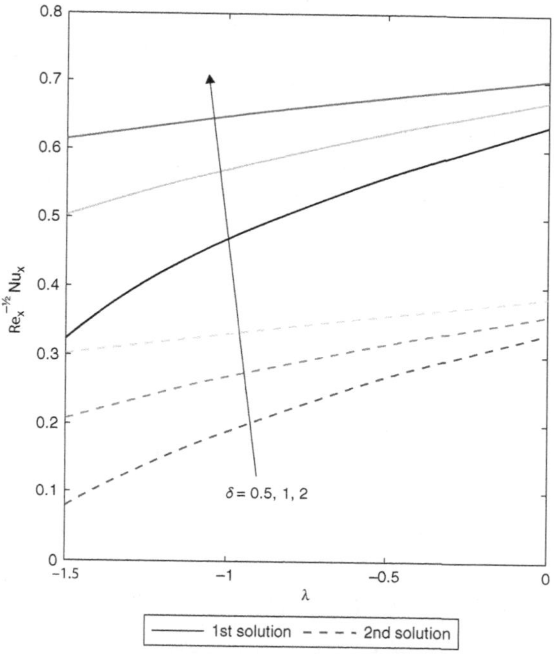

Figure 10.9 Variation of Nusselt Number with λ for different values of δ.

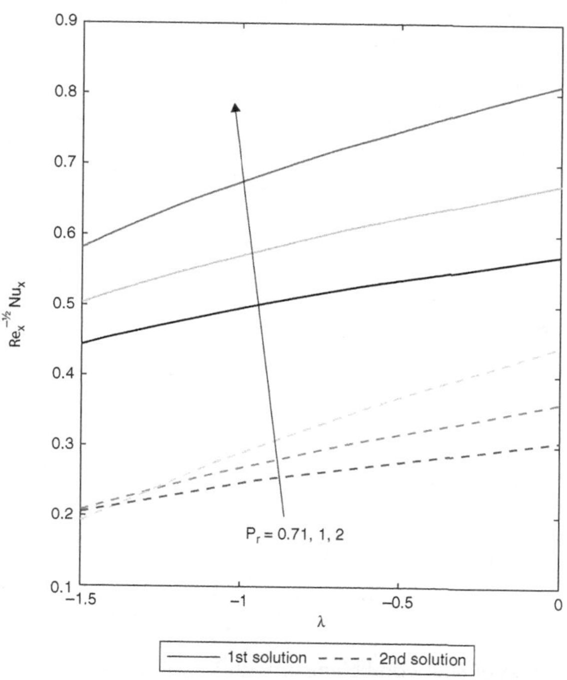

Figure 10.10 Variation of Nusselt Number with λ and P_r.

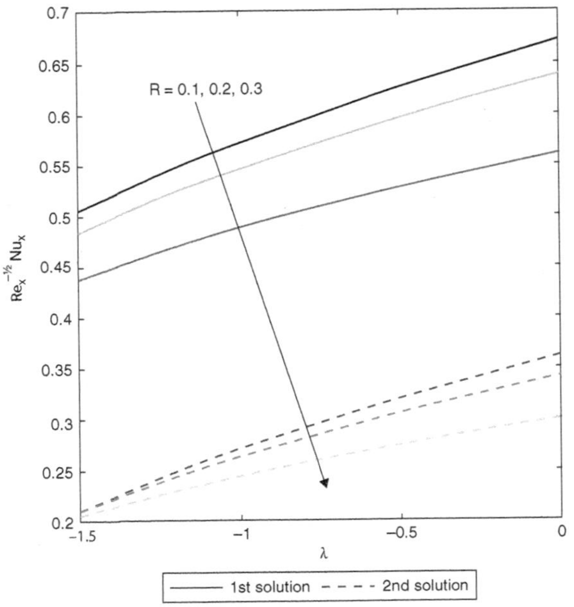

Figure 10.11 Variation of Nusselt Number with λ and *R*.

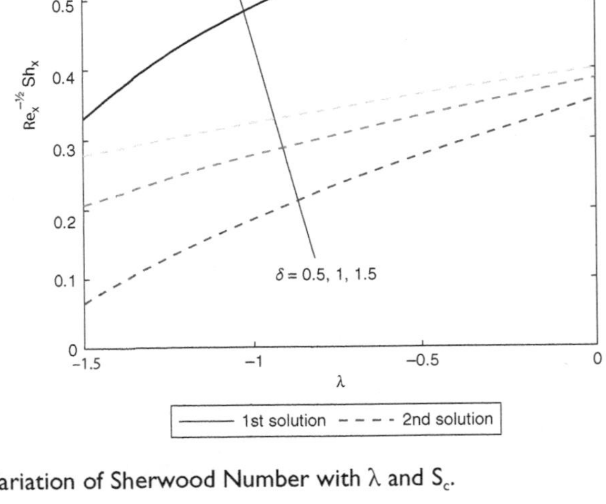

Figure 10.12 Variation of Sherwood Number with λ and S_c.

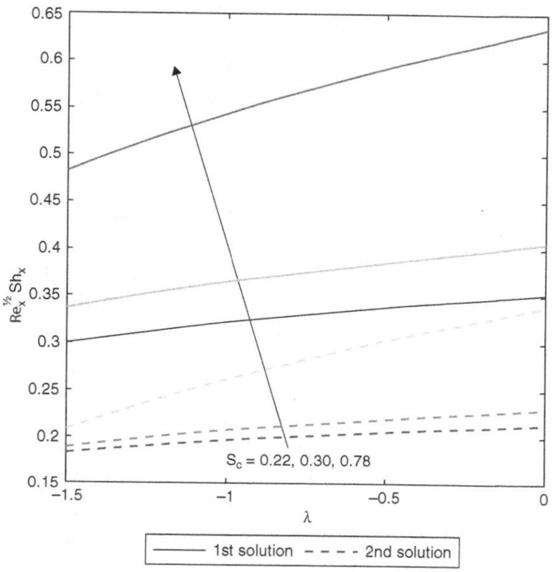

Figure 10.13 Variation of Sherwood Number with λ and S_c.

confirm the accelerated behaviour of Sherwood number with increasing values of both δ and P_r along with '-ve' range of λ.

10.5 CONCLUDING REMARKS

The existence of dual solutions is clearly shown in the figures. The stability analysis is checked through our in-house program using MATLAB software for stability of both the solutions and it is found that the first solution (solid line) is stable and valid physically, while the second solution (dotted line) is unstable.

REFERENCES

[1] Merkin, J. H., Najib, N., Bachok, N., Ishak, A., & Pop, I. (2017). Stagnation-point flow and heat transfer over an exponentially stretching/shrinking cylinder. *Journal of the Taiwan Institute of Chemical Engineers*, 74, 65–72.

[2] Khan, M. I., Hayat, T., Khan, M. I., & Alsaedi, A. (2017). A modified homogeneous-heterogeneous reactions for MHD stagnation flow with viscous dissipation and Joule heating. *International Journal of Heat and Mass Transfer*, 113, 310–317.

[3] Waini, I., Ishak, A., & Pop, I. (2020). Hybrid nanofluid flow towards a stagnation point on a stretching/shrinking cylinder. *Scientific Reports*, 10, 9296.

[4] Bhukta, D., Dash, G. C., Mishra, S. R., & Jena, S. (2021). Analytical estimation of energy dissipations: Viscous, Joulian, and Darcy of viscoelastic fluid flow phenomena over a deformable surface. *Heat Transfer*, 50 (8), 7798–7816.

[5] Pattnaik, P. K., Jena, S., Mishra, S., & Mathur, P. (2023). Application to differential transform method for MHD fluid flow and heat transfer. *Biointerface Research in Applied Chemistry*, 13(2), 110.

[6] Ahmed, B. (2023). Information of stagnation-point flow of Maxwell fluid past symmetrically exponential stretching/shrinking cylinder with prescribed heat flux. *AIP Advances*, 13, 045314.

[7] Mahabaleshwar, U. S., Maranna, T., Mishra, M., Hatami, M., & Sunden, B. (2024). Radiation effect on stagnation point flow of Casson nanofluid past a stretching plate/cylinder. *Scientific Reports*, 14, 1387.

[8] Pattanaik, P C., Jena, S., & Mishra, S. R. (2024). Control of magnetic dissipation and radiation on an unsteady stagnation point nanofluid flow: A numerical approach. *Modern Physics Letters B*, 38(9), 2450039.

[9] Jena, S., Swain, K., Ibrahim, S. M., Sreenivasulu, P., & Lorenzini, G. (2024). Three-dimensional radiative rotating MHD nanofluid flow of over a stretched sheet with homogeneous-heterogeneous chemical reactions. *Journal of Advanced Research in Micro and Nano Engineering*, 21(1), 112–126.

[10] Pradhan, T., Jena, S., & Mishra, S. R. (2025). Laplace transformation technique for free convective time-dependent MHD flow over a vertical porous flat plate with heat sink and chemical reaction: An analytical approach. *Hybrid Advances*, 8, 100363.

[11] Hayat, T., & Nadeem, S. Flow of 3D Eyring-Powell fluid by utilizing Cattaneo-Christov heat flux model and chemical processes over an exponentially stretching surface. *Results in Physics*, 8, 397–403.

[12] Ghadikolaei, S.S., Hosseinzadeh, K., Ganji, D.D., & Jafari, B. (2018). Nonlinear thermal radiation effect on magneto Casson nanofluid flow with Joule heating effect over an inclined porous stretching sheet. *Case Studies in Thermal Engineering*, 12, 176–187.

[13] Jalili, B., Sadighi, S., Jalili, P., & Ganji, D.D. (2019). Characteristics of ferrofluid flow over a stretching sheet with suction and injection. *Case Studies in Thermal Engineering*, 14, 1950121.

[14] Hussain, A., Akbar, S., Sarwar, L., Nadeem, S., & Iqbal, Z. (2019). Effect of time dependent viscosity and radiation efficacy on a non-Newtonian fluid flow. *Heliyon*, 5(2), e01203.

[15] Afridi, M.I., Tlili, I., Goodarzi, M., Osman, M., & Khan, N. A. (2019). Irreversibility analysis of hybrid nanofluid flow over a thin needle with effects of energy dissipation. *Symmetry*, 11(5), 663.

[16] Lund, L.A., Omar, Z., Khan, U., Khan, I., Baleanu, D., Nisar, K.S. (2020). Stability analysis and dual solutions of micropolar nanofluid over the inclined stretching/shrinking surface with convective boundary condition. *Symmetry*, 12(1),74.

[17] Nadeem, S., Amin, A., & Abbas, N. (2020). On the stagnation point flow of nanomaterial with base viscoelastic micropolar fluid over a stretching surface. *Alexandria Engineering Journal*, 59(3), 1751–1760.

[18] Khan, M., Ahmed, J., & Rasheed, Z. (2020). Entropy generation analysis for axisymmetric flow of Carreau nanofluid over a radially stretching disk. *Applied Nanoscience*, 10(12), 5291–5303.

[19] Anwar Beg, O., Zohra, F.T., Uddin, M.J., Ismail, A.I.M., & Sathasivam, S. (2021). Energy conservation of nanofluids from a biomagnetic needle in the presence of Stefan blowing: Lie symmetry and numerical simulation. *Case Studies in Thermal Engineering*, 24, 100861.

[20] Waini, I., Ishak, A., & Pop, I. (2021). Agrawal flow of a hybrid nanofluid over a shrinking disk. *Case Studies in Thermal Engineering*, 25,100950.

Chapter 11

Mathematical model of immiscible Newtonian fluids flow through an anisotropic porous channel

Amit Kumar and Krishna Prasad Madasu

Nomenclature

$\mathbf{V}_i$	Velocity vector $\left(\frac{\mathrm{m}}{\mathrm{s}}\right)$
p_i	Pressure $\left(\frac{\mathrm{N}}{\mathrm{m}^2}\right)$
$\mathbf{K}_j$	Permeability tensor
K_k	Permeability (m^2)
V_0	Characteristics velocity $\left(\frac{\mathrm{m}}{\mathrm{s}}\right)$
u_i	Velocity along x-direction $\left(\frac{\mathrm{m}}{\mathrm{s}}\right)$
Da_i	Darcy number

Greek Symbols

∇	Nabla operator
∇^2	Laplacian operator
μ_i	Dynamic viscosity of fluid
μ_{ei}	Effective viscosity of fluid
σ_1, σ_2	Viscosity ratio $\left(\frac{\mu_1}{\mu_2}\right)$ and $\left(\frac{\mu_3}{\mu_2}\right)$, respectively
λ_i	Ratio of viscosity and effective viscosity $\left(\frac{\mu_i}{\mu_{ei}}\right)$
ψ_i	Anisotropic angle

Suffix

$i=1,2,3$; $j = I, II, III$; $k = 1,2,3,4,5,6$

DOI: 10.1201/9781003642169-11

11.1 INTRODUCTION

The fluids that flow without mixing, like oil and water, industrial wastes in rivers and oceans, contaminated groundwater, and the simultaneous flow of natural gas, oil, and water in petroleum reservoirs, are considered immiscible fluids. The study of immiscible fluid movement is crucial for extracting crude from petroleum reservoirs, filtering contaminated water and understanding transport phenomena. In the petroleum reservoir, oil, gas, and water are trapped inside the pores of reservoir rocks. Some parts of the porous rocks are filled by water, while either gas, oil, or both can occupy the remainder (Allan and Hamdan, 2002; Kamışlı and Öztop, 2008). Another possible scenario is that some parts of the pores are saturated with oil and others with water and gas. Immiscible fluid or multi-layer models like Newtonian–Newtonian or Newtonian–non-Newtonian fluids are used to study such problems. An interface exists between two immiscible fluids, which is crucial for fluid movement. A term viscosity ratio describes it. Also, the porous reservoir rocks have heterogeneous permeability, and adjacent layers need not have the same pore structure. It may vary with depth. Therefore, to gain more insight into the flow, it is necessary to study the fluid's immiscibility and the heterogeneity of porous media.

A porous medium is isotropic when the permeability in all directions is the same. However, every porous medium does not need to have the same permeability in two or more directions; this type of porous medium is referred to as anisotropic. This could be due to uneven arrangement of pore size of a porous material or variations of shape and size of solid matrix. The permeability tensor of the isotropic porous matrix for two-dimensional flow is a diagonal porous matrix. Introducing an anisotropic angle converts a diagonal porous matrix into a non-diagonal porous matrix. This type of porous matrix was first used by Tyvand and Storesletten (1991) to study the convection processes in an anisotropic porous medium with oblique principal axes that form an angle with the horizontal plane. That angle is referred to as an anisotropic angle in many literature (Tyvand & Storesletten, 1991; Degan et al., 2002; Karmakar & Raja Shekhar, 2016). A comprehensive analysis of the significant impact of anisotropic parameters on fluid flow between two porous channels saturated with a Newtonian fluid was conducted by Karmakar & Raja Shekhar (2016). Karmakar (2021) studied a Newtonian fluid through a porous medium having directional permeability between two parallel plates whose upper plate have oscillatory translation. Recently, Couette–Poiseuille flow of immiscible viscous fluids has been studied through two anisotropic porous layers between parallel plates (Jaiswal & Yadav, 2023) and an annular space between two cylinders (Kumar & Madasu, 2024). The strong influence of anisotropic parameters on fluid velocity was observed in the above studies. In this study, we analyse a model comprising three adjacent anisotropic porous medium, each saturated with different fluids, situated between two parallel plates.

The velocity distribution of two immiscible fluids flow between two thin slits is mentioned in a book by Bird et al. (2002). They assumed that the upper layer fluid is less viscous than the lower layer. In the same geometry, Kamışlı and Öztop (2008) used the same velocity profiles to study the entropy generation. The other works that include two immiscible fluids like numerically study the unsteady flow of Newtonian–micropolar fluids between horizontally oriented plates (Devakar & Raje, 2018), immiscible fluids of different viscosity flow through two adjacent layers of the porous channel (Ansari & Deo, 2017), and magnetohydrodynamic (MHD) effect on unsteady flow Newtonian–micropolar model isotropic porous medium between two plates (Devakar & Raje, 2019). However, multi-layer flow is also possible in the reservoir. The effect of the third-layer fluid on flow is not observed when we consider the two-layer model. Allan et al. (2009) investigated the immiscible fluids flow through a porous medium sandwiched between two infinite-depth porous media. The different combinations of Newtonian and non-Newtonian fluids for three-layer immiscible fluids are also studied such as viscous–micropolar–viscous fluids (Umavathi et al., 2008), and micropolar–viscous–micropolar flow between two parallel plates (Umavathi et al., 2014), and nanofluid sandwiched between two viscous fluids in a vertical channel (Lu et al., 2018). Time dependent oscillatory flow and heat transfer of fluids of different viscosities through a porous layer placed in the middle of a horizontal channel investigated by Umavathi et al. (2010). Umavathi and Hemavathi (2019) examined the mixed convective flow of immiscible (viscous–nanofluid-viscous) fluids in the vertical channel in which the middle layer is porous. Rajeev and Mahanthesh (2021) consider different immiscible nanofluids in a vertical channel to investigate the nonlinear convection with viscous dissipation effect. Umavathi and Sutkar (2023) studied the micropolar nanofluid flow between two porous layers saturated with Newtonian fluid. In three adjacent layers of horizontally oriented isotropic porous channels of the same constant permeability between two plates saturated with Newtonian–micropolar-Newtonian fluids, Yadav et al. (2018) investigated the magnetic field impact on flow behaviour. Jaiswal and Yadav (2020) analyse the immiscible fluids flow through a rectangular pipe filled with three isotropic porous layers of different permeability, and interfaces of Newtonian–micropolar fluids are linearly movable. Zaytoon et al. (2016) consider the flow of immiscible fluids through porous layers with variable permeability of the middle layer between two constant permeability layers. In the same horizontal geometry, Yadav et al. (2023) did a detailed analysis of the impact of the MHD on the immiscible Newtonian fluids flowing through a channel containing three adjacent porous mediums whose permeabilities vary linearly. Deepak and Maurya (2022) investigated the MHD effect on micropolar–Newtonian–micropolar fluids flow through a channel contain three adjacent porous medium. The above-three-layer models have significant applications in the extraction of oil from petroleum reservoirs, purification of contaminated groundwater, and other industrial fields where immiscible fluids frequently exist.

The available literature includes several investigations about three immiscible fluids flowing through a porous medium of either constant permeability or permeability as a function. However, we did not find any literature that considers the effect of anisotropic parameters on the three immiscible fluids flowing through three different anisotropic porous mediums in a channel. The primary goal of this work is to analyse the impact of anisotropic parameters and viscosity ratios on the flow of immiscible fluids. We also mention particular cases of the present problem.

11.2 MATHEMATICAL FORMULATION

Figure 11.1 presents a detailed illustration of the present physical model. It consists of two parallel rigid plates filled with three different adjacent porous mediums. The width of each porous medium is constant and named region-I ($h_1 \le y \le 3h$), region-II ($h_0 \le y \le h_1$), and region-III ($0 \le y \le h_0$). The porous medium of each region having anisotropic permeability (assumed) is described by anisotropic parameters (permeability ratio and anisotropic angle). The direction of permeability in region-I is along the principal axis K_1 and K_2, in region-II along the principal axis $K_3$and K_4, and in region-III along the principal axis K_5 and K_6. The principal axis $K_2, K_4,$and K_6 makes an angle ψ_1, ψ_2, and ψ_3 with the x-axis, respectively called anisotropic angles. All three porous media are saturated with Newtonian fluids of different viscosities. In region-I fluid have viscosity μ_1, in region-II μ_2, and in region-III μ_3, and assumed that $\mu_1 \le \mu_2$ and $\mu_3 \le \mu_2$. For the present problem, the steady and incompressible immiscible fluids flow through the porous channel is due to the effect of axial pressure gradient and governed by Brinkman's equation.

The governing equation of each porous region is (Degan et al., 2002; Karmakar & Raja Shekhar, 2016; Karmakar, 2021; Jaiswal & Yadav, 2023)

$$\nabla \cdot \mathrm{V_i} = 0, \tag{11.1}$$

$$\mu_i \mathrm{V_i} + \mathrm{K_j}\left(\nabla p_i - \mu_{ei}\nabla^2 \mathrm{V_i}\right) = 0, \tag{11.2}$$

where $\mathrm{V_i}$ is the velocity vector, μ_i is the viscosity of fluid, μ_{ei} is the effective viscosity of the fluid, p_i is the pressure, and $\mathrm{K_j}$ denotes the asymmetric permeability tensors are given by

$$\mathrm{K_I} = \begin{pmatrix} \mathrm{K_1}\sin^2\psi_1 + \mathrm{K_2}\cos^2\psi_1 & (\mathrm{K_2} - \mathrm{K_1})\sin\psi_1\cos\psi_1 \\ (\mathrm{K_2} - \mathrm{K_1})\sin\psi_1\cos\psi_1 & \mathrm{K_2}\sin^2\psi_1 + \mathrm{K_1}\cos^2\psi_1 \end{pmatrix}, \tag{11.3}$$

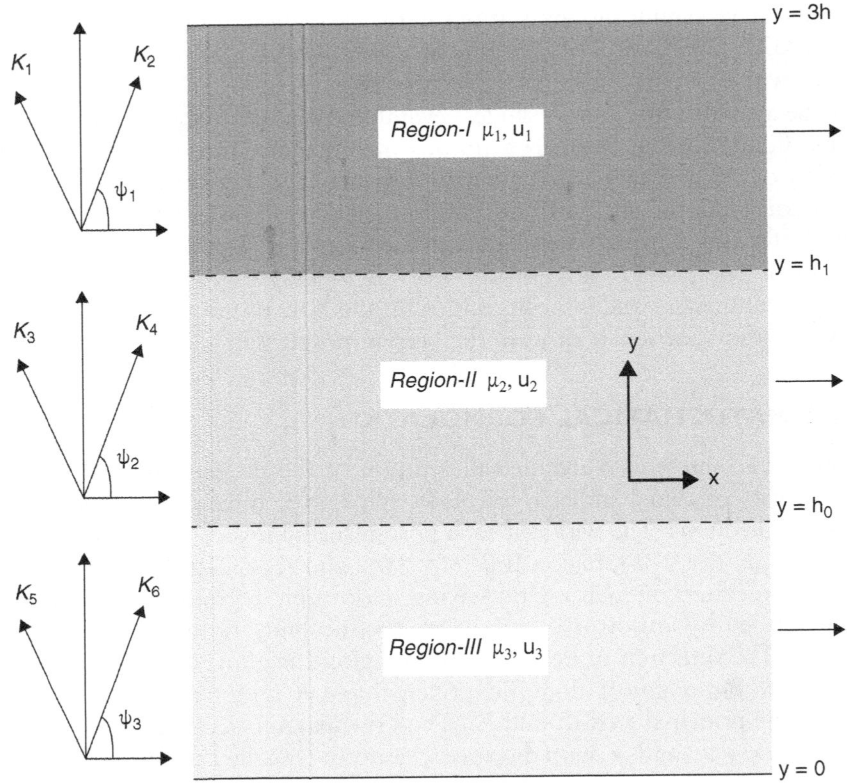

Figure 11.1 Geometry of physical model.

$$K_{II} = \begin{pmatrix} K_3 \sin^2 \psi_2 + K_4 \cos^2 \psi_2 & (K_4 - K_3)\sin \psi_2 \cos \psi_2 \\ (K_4 - K_3)\sin \psi_2 \cos \psi_2 & K_4 \sin^2 \psi_2 + K_3 \cos^2 \psi_2 \end{pmatrix}, \tag{11.4}$$

$$K_{III} = \begin{pmatrix} K_5 \sin^2 \psi_3 + K_6 \cos^2 \psi_3 & (K_6 - K_5) \sin \psi_3 \cos \psi_3 \\ (K_6 - K_5)\sin \psi_3 \cos \psi_3 & K_6 \sin^2 \psi_3 + K_5 \cos^2 \psi_3 \end{pmatrix}, \tag{11.5}$$

where i = 1, 2, 3 and i = *I*, *II*, *III* corresponds to region-I, region-II, and region-III, respectively.

The assumption of fully developed flow leads to $v_i = 0$ (velocity component in the y-direction) and from equation of continuity $u_i = u_i(y)$ (velocity component in the x-direction), where i=1, 2, 3. Based on these assumptions, the components of the mass and momentum equations can be expressed for each region as follows:

Region-I $\left(h_1 \leq y \leq 3h\right)$:

$$\frac{du_1}{dx} = 0, \tag{11.6}$$

$$au_1 = -\frac{K_1}{\mu_1}\frac{\partial p_1}{\partial x} + K_1\frac{\mu_{e1}}{\mu_1}\frac{d^2u_1}{dy^2}, \tag{11.7}$$

$$bu_1 = \frac{K_1}{\mu_1}\frac{\partial p_1}{\partial y}, \tag{11.8}$$

where $a = \sin^2\psi_1 + K_I^*\cos^2\psi_1, b = \left(1 - K_I^*\right)\sin\psi_1\cos\psi_1$.

Region-II $\left(h_0 \leq y \leq h_1\right)$:

$$\frac{du_2}{dx} = 0, \tag{11.9}$$

$$cu_2 = -\frac{K_3}{\mu_2}\frac{\partial p_2}{\partial x} + K_3\frac{\mu_{e2}}{\mu_2}\frac{d^2u_2}{dy^2}, \tag{11.10}$$

$$du_2 = \frac{K_3}{\mu_2}\frac{\partial p_2}{\partial y}, \tag{11.11}$$

where $c = \sin^2\psi_2 + K_{II}^*\cos^2\psi_2, d = \left(1 - K_{II}^*\right)\sin\psi_2\cos\psi_2$.

Region-III $\left(0 \leq y \leq h_0\right)$:

$$\frac{du_3}{dx} = 0, \tag{11.12}$$

$$eu_3 = -\frac{K_5}{\mu_3}\frac{\partial p_3}{\partial x} + K_5\frac{\mu_{e3}}{\mu_3}\frac{d^2u_3}{dy^2}, \tag{11.13}$$

$$fu_3 = \frac{K_5}{\mu_3}\frac{\partial p_3}{\partial y}, \tag{11.14}$$

where $e = \sin^2\psi_3 + K_{III}^*\cos^2\psi_3, f = \left(1 - K_{III}^*\right)\sin\psi_3\cos\psi_3$.

Here, $K_I^* = \frac{K_1}{K_2}$, $K_{II}^* = \frac{K_3}{K_4}$, and $K_{III}^* = \frac{K_5}{K_6}$ are permeability ratio of region-I, region-II, and region-III, respectively.

From Eqs. (11.6)–(11.14), we can say that the anisotropic parameters control the pressure gradient variation in x and y directions. Now, operate Eqs. (11.7) and (11.8) with $\frac{\partial}{\partial x}$ on both sides and use Eq. (11.6), we get

$$\frac{\partial}{\partial x}\left(\frac{\partial p_1}{\partial x}\right) = 0 \text{ and } \frac{\partial}{\partial x}\left(\frac{\partial p_1}{\partial y}\right) = 0, \tag{11.15}$$

that is,

$$\frac{\partial}{\partial x}\left(\frac{\partial p_1}{\partial x}\right) = 0 \text{ and } \frac{\partial}{\partial y}\left(\frac{\partial p_1}{\partial x}\right) = 0, \tag{11.16}$$

From Eq. (11.16), $\frac{\partial p_1}{\partial x} =$ constant, similarly $\frac{\partial p_2}{\partial x} =$ constant and $\frac{\partial p_3}{\partial x} =$ constant (Karmakar & Raja Shekhar, 2016). For the present problem, we assumed that $\frac{\partial p_1}{\partial x} = \frac{\partial p_2}{\partial x} = \frac{\partial p_3}{\partial x} = G$ (which is equal to constant applied pressure gradient).

11.3 SOLUTION

Introducing the following dimensionless number

$$x^* = \frac{x}{h}, y^* = \frac{y}{h}, P_0 = \frac{V_0 \mu_2}{h}, P_i^* = \frac{p_i}{P_0},\ u_i^* = \frac{u_i}{V_0}, \text{ where } i = 1, 2, \text{ and } 3.$$

And after dropping the *, the dimensionless governing equations for each region are:

Porous region-I ($\varepsilon_1 \le y \le 3$):

$$\frac{a}{Da_1} u_1 = -\frac{1}{\sigma_1}\frac{\partial p_1}{\partial x} + \lambda_1 \frac{d^2 u_1}{dy^2}, \tag{11.17}$$

$$\frac{b}{Da_1} u_1 = \frac{1}{\sigma_1}\frac{\partial p_1}{\partial y}, \tag{11.18}$$

where $\varepsilon_1 = \frac{h_1}{h}, Da_1 = \frac{K_1}{h^2}, \sigma_1 = \frac{\mu_1}{\mu_2}, \lambda_1 = \frac{\mu_{e1}}{\mu_1}$.

Porous region-II ($\varepsilon_0 \le y \le \varepsilon_1$):

$$\frac{c}{Da_2}u_2 = -\frac{\partial p_2}{\partial x} + \lambda_2 \frac{d^2u_2}{dy^2}, \tag{11.19}$$

$$\frac{d}{Da_2}u_2 = \frac{\partial p_2}{\partial y}, \tag{11.20}$$

where $\varepsilon_0 = \frac{h_0}{h},\ Da_2 = \frac{K_3}{h^2},\ \lambda_2 = \frac{\mu_{e2}}{\mu_2}$.

Porous region-III ($0 \le y \le \varepsilon_0$):

$$\frac{e}{Da_3}u_3 = -\frac{1}{\sigma_2}\frac{\partial p_3}{\partial x} + \lambda_3 \frac{d^2u_3}{dy^2}, \tag{11.21}$$

$$\frac{f}{Da_3}u_3 = \frac{1}{\sigma_2}\frac{\partial p_3}{\partial y}, \tag{11.22}$$

where $Da_3 = \frac{K_5}{h^2}, \sigma_2 = \frac{\mu_3}{\mu_2}, \lambda_3 = \frac{\mu_{e3}}{\mu_3}$.

The ratios of effective viscosity and viscosity $\lambda_1 = \frac{\mu_{e1}}{\mu_1}, \lambda_2 = \frac{\mu_{e2}}{\mu_2}$, and $\lambda_3 = \frac{\mu_{e3}}{\mu_3}$ are considered equal to unity in this problem as taken by others (Karmakar & Raja Shekhar, 2016; Kumar & Madasu, 2024)

The general solution of the Eqs. (11.17), (11.19), and (11.21) are:

$$u_1 = c_1 e^{-y\alpha} + c_2 e^{y\alpha} - \frac{G}{\sigma_1 \alpha^2}, \tag{11.23}$$

$$u_2 = c_3 e^{-y\beta} + c_4 e^{y\beta} - \frac{G}{\beta^2}, \tag{11.24}$$

$$u_3 = c_5 e^{-y\gamma} + c_6 e^{y\gamma} - \frac{G}{\sigma_2 \gamma^2}, \tag{11.25}$$

Where $\alpha = \sqrt{\frac{a}{Da_1}}, \beta = \sqrt{\frac{c}{Da_2}}, \gamma = \sqrt{\frac{e}{Da_3}}.$

We consider the no-slip boundary conditions at both plates, and continuity of shear stress and velocity at fluid–fluid interfaces. Then, the dimensionless boundary conditions are:

$$\text{At } y = 3, u_1 = 0. \tag{11.26}$$

$$\text{At } y = \varepsilon_1, u_1 = u_2, \sigma_1 \frac{du_1}{dy} = \frac{du_2}{dy} \tag{11.27}$$

$$\text{At } y = \varepsilon_0, u_2 = u_3, \frac{du_2}{dy} = \sigma_2 \frac{du_3}{dy} \tag{11.28}$$

$$\text{At } y = 0,\ u_3 = 0. \tag{11.29}$$

Now, implement the Eqs. (11.26)–(11.29), in Eqs. (11.23)–(11.25). The matrix form of a system of linear equations is

$$Mx = N, \tag{11.30}$$

where

$$M = \begin{bmatrix} e^{-3\alpha} & e^{3\alpha} & 0 & 0 & 0 & 0 \\ e^{-\varepsilon_1\alpha} & e^{\varepsilon_1\alpha} & -e^{-\varepsilon_1\beta} & -e^{\varepsilon_1\beta} & 0 & 0 \\ -\sigma_1\alpha e^{-\varepsilon_1\alpha} & \sigma_1\alpha e^{\varepsilon_1\alpha} & \beta e^{-\varepsilon_1\beta} & -\beta e^{\varepsilon_1\beta} & 0 & 0 \\ 0 & 0 & e^{-\varepsilon_0\beta} & e^{\varepsilon_0\beta} & -e^{-\varepsilon_0\gamma} & -e^{\varepsilon_0\gamma} \\ 0 & 0 & -\beta e^{-\varepsilon_0\beta} & \beta e^{\varepsilon_0\beta} & \sigma_2\gamma e^{-\varepsilon_0\gamma} & -\sigma_2\gamma e^{\varepsilon_0\gamma} \\ 0 & 0 & 0 & 0 & 1 & 1 \end{bmatrix}, x$$

$$= \begin{bmatrix} c_1 \\ c_2 \\ c_3 \\ c_4 \\ c_5 \\ c_6 \end{bmatrix}, N = \begin{bmatrix} \frac{G}{\sigma_1\alpha^2} \\ \frac{G}{\sigma_1\alpha^2} - \frac{G}{\beta^2} \\ 0 \\ \frac{G}{\beta^2} - \frac{G}{\sigma_2\gamma^2} \\ 0 \\ \frac{G}{\sigma_2\gamma^2} \end{bmatrix}$$

After solving Eq. (11.30), the arbitrary constants $c_1, c_2, c_3, c_4, c_5, c_6$ can be obtained.

The dimensionless form of flow rate throughout the channel is given by

$$Q = \int_0^{\varepsilon_0} u_3 dy + \int_{\varepsilon_0}^{\varepsilon_1} u_2 dy + \int_{\varepsilon_1}^{3} u_1 dy. \tag{11.31}$$

The dimensionless shear stress at both plates is given by

1. At upper plate: $\tau_A \big|_{y=3} = \dfrac{du_3}{dy}\Big|_{y=3}.$
2. At lower plate: $\tau_B \big|_{y=0} = \dfrac{du_1}{dy}\Big|_{y=0}.$

11.4 RESULTS AND DISCUSSION

The effect of physical parameters on fluid flow and shear stress at plates are presented graphically and discussed in this section. The fixed parametric values for all graphs are $G = -3$, $\varepsilon_0 = 1$, and $\varepsilon_1 = 2$ and values of other parameters are mentioned along with discussion. The velocity profiles demonstrating the effect of varying permeability ratios and anisotropic angles for each region are presented in Figure 11.2. The curves indicate that the velocity decreases as the permeability ratios and anisotropic angles increase across all regions. In Figure 11.3, the effect of permeability ratios $K_I^*, K_{II}^*, K_{III}^*$ of region-I, region-II, and region-III, respectively, are presented. The dash-dotted curve represents the impact of K_I^*, black curve represents the impact of K_{II}^*, and dotted curve represents the impact of K_{III}^* on fluid's velocity. For this, we consider the same anisotropic angles for all regions $(\psi_1 = \psi_2 = \psi_3 = 0)$, saturated with fluid of same viscosity $\sigma_1 = \sigma_2 = 1$, and also fixed Darcy numbers, tha tis, $Da_1 = Da_2 = Da_3 = 0.1$, which means K_1, K_3, K_5 are fixed. Since anisotropic angles are 0, then increase in permeability ratios lead to decrease in permeability along K_2, K_4, K_6. Hence, velocity in each region is promoted with a decrease in permeability ratios. To observe the impact of K_{III}^* on fluid's velocity, we consider the other regions are isotropic, that is, $K_{II}^* = K_{III}^* = 1$. It is observed that velocity achieves maximum at $K_I^* = 0.5$ rather than isotropic case ($K_I^* = 1$) and minimum at $K_I^* = 2$ in region-I. Similarly, other curves are plotted, and same pattern observed for other two regions, that is, velocity achieves maximum at $K_{II}^* = 0.5$ and minimum at $K_{II}^* = 2$ in region-II and same behaviour in region-III, maximum at $K_{II}^* = 0.5$ and minimum at K_{II}^*. When $K_I^* = 1, K_{II}^* = 1, K_{III}^* = 1$ (isotropic case), then fluid's velocity of all three regions is overlapped. At plates, velocity is zero due to no slip conditions.

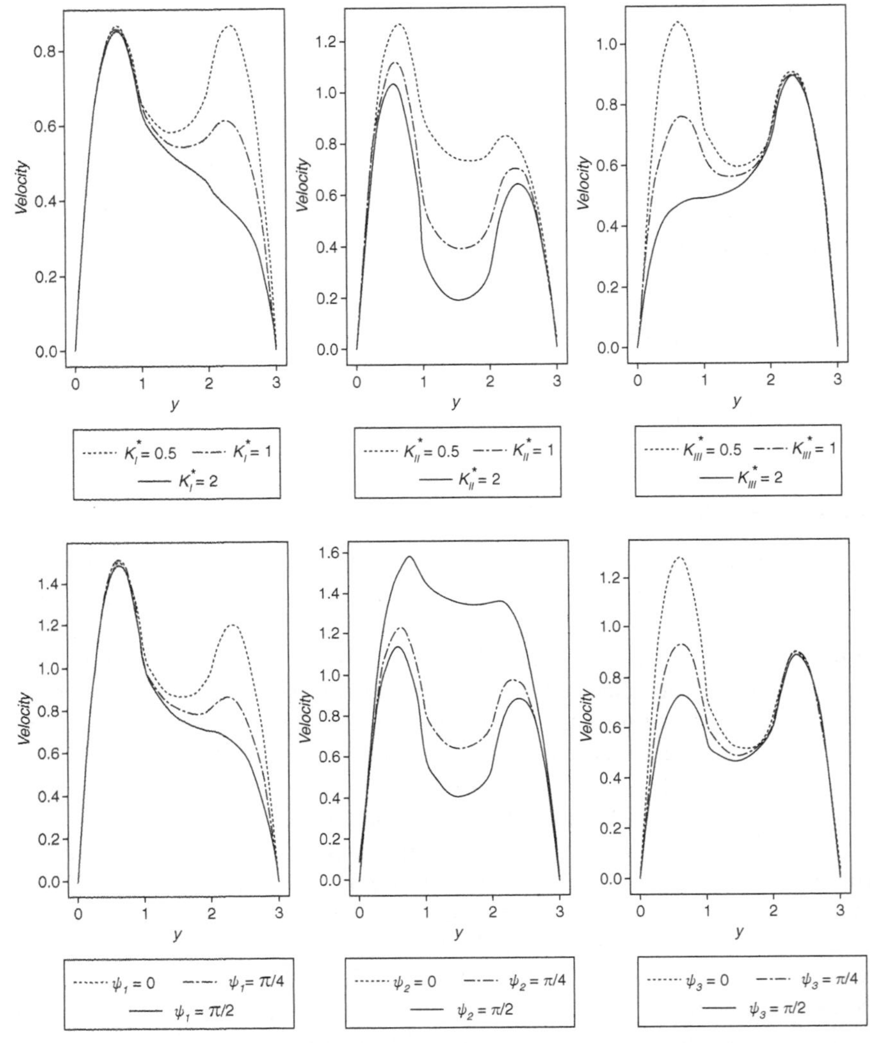

Figure 11.2 Variations in velocity profiles for fully anisotropic case when permeability ratios and anisotropic angles of each region vary.

Figure 11.4 shows the effect of anisotropic angles (ψ_1, ψ_2, ψ_3) on fluid's velocity. The dash-dotted curve represents the impact of ψ_1, black curve represents the impact of ψ_2, and dotted curve represents the impact of ψ_3 on fluid's velocity of region-I, region-II, and region-III, respectively. The values of viscosity ratios, Darcy numbers are same as in Figure 11.2 and other parameters $K_I^* = 0.25, K_{II}^* = K_{III}^* = 1$ for region-I, $K_{II}^* = 0.25, K_I^* = K_{III}^* = 1$

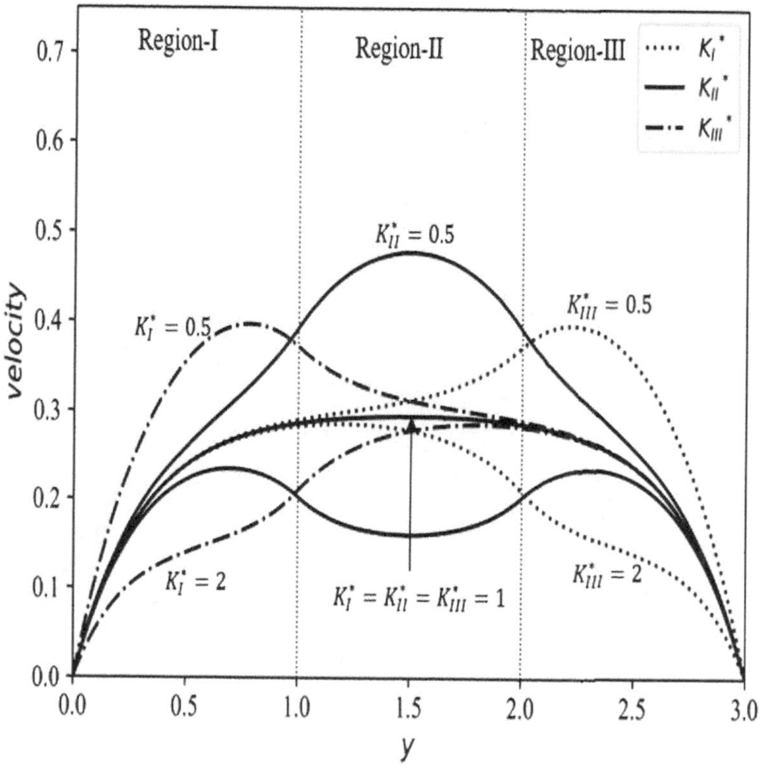

Figure 11.3 Variations in velocity profiles of all regions for different permeability ratios.

for region-II, and $K_{III}^* = 0.25, K_I^* = K_{III}^* = 1$ for region-III. Since permeability ratios are fixed, then with an increase in anisotropic angles, the resistance to fluid flow increases, causing a decrease in velocity. At $\psi_1 = \psi_2 = \psi_3 = \pi/2$, the fluid's velocity of all regions is minimum and same, and maximum velocity occurs at angle 0.

The impact of viscosity ratios on fluid's velocity are shown in Figure 11.5. Since we assumed that $\mu_1 \le \mu_2$ and $\mu_3 \le \mu_2$, then with an increase in viscosity ratios (σ_1, σ_2) velocity decreases. When $\sigma_1 = 0.3$, that is, $\mu_1 = 0.3\mu_2$ and $\sigma_2 = 1$, this implies that region-I's fluid is less viscous than region-II and viscosity of region-II and region-III are equal. Therefore, velocity reduces due to high viscosity of both region-I and region-II. Similarly, when $\sigma_2 = 0.3$ means $\mu_3 = 0.3\mu_2$, that is, viscosity of region-III is less than the viscosity of region-II; the same result of impact of σ_2 on velocity is observed, as for σ_1. The velocity curve of same viscosity $(\sigma_1 = \sigma_2)$ in all regions are same.

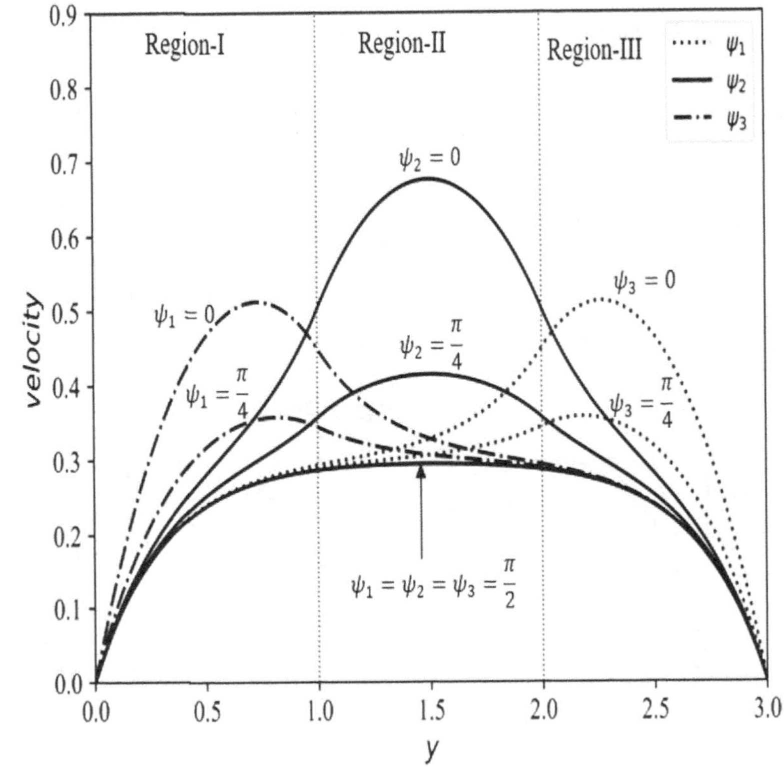

Figure 11.4 Variations in velocity profiles of all regions for different anisotropic angles.

Figure 11.6 illustrates the velocity profile for different values of viscosity ratios, and Darcy numbers $\left(Da_1 = Da_2 = Da_3 = Da\right)$. The velocity profiles enhance with an increase in Darcy number and velocity of immiscible fluids $\left(\sigma_1 \neq \sigma_2\right)$ is greater than miscible fluid $\left(\sigma_1 = \sigma_2\right)$ (Yadav et al., 2023). With an increase in Darcy number permeability along the x-direction, it results in an increase in velocity.

The impact of $(K_I^*, K_{II}^*, K_{III}^*)$, $\left(\psi_1, \psi_2, \psi_3\right)$, and $\left(\sigma_1, \sigma_2\right)$ on flow rate Q for different values of Darcy numbers are presented in Table 11.1. When $Da_1 > Da_2 < Da_3$, the flow rate decreases with increase in $(K_I^*, K_{II}^*, K_{III}^*)$, $\left(\psi_1, \psi_2, \psi_3\right)$, and $\left(\sigma_1, \sigma_2\right)$. Also, we found that the flow rate is maximum when $Da_1 > Da_2 < Da_3$, and minimum for $Da_1 > Da_2 > Da_3$, as compared to $Da_1 > Da_2 = Da_3$. The numerical values of the effect of $(K_I^*, K_{II}^*, K_{III}^*)$, $\left(\psi_1, \psi_2, \psi_3\right)$, and $\left(\sigma_1, \sigma_2\right)$ on shear stress at upper and lower plates are mentioned in Table 11.2 for different values of Darcy numbers. We can see that when $Da_1 > Da_2$, or $Da_1 = Da_2$, or $Da_1 > Da_2$, then the magnitude of

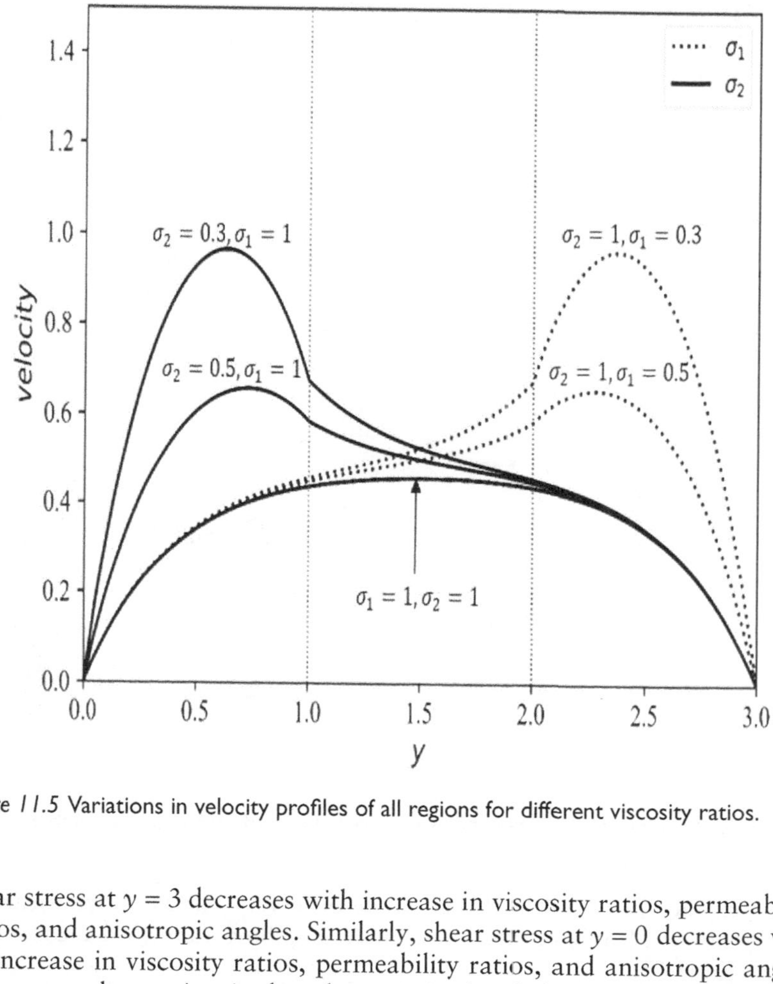

Figure 11.5 Variations in velocity profiles of all regions for different viscosity ratios.

shear stress at $y = 3$ decreases with increase in viscosity ratios, permeability ratios, and anisotropic angles. Similarly, shear stress at $y = 0$ decreases with an increase in viscosity ratios, permeability ratios, and anisotropic angles. One more observation is that the magnitude of shear stress at $y = 0$ and $y = 3$ are equal for $Da_1 < Da_2$ and $Da_2 > Da_3$; $Da_1 = Da_2$ and $Da_2 = Da_3$; and $Da_1 > Da_2$ and $Da_2 < Da_3$.

11.5 PARTICULAR CASES

Two different cases of the present problem are as follows:

1 When viscosity ratios $\sigma_1 = \sigma_2 = 1$, that is, $\mu_1 = \mu_2 = \mu_3$, then it will represent the flow of a single fluid through three adjacent layers of anisotropic porous medium. This situation could be possible in petroleum reservoirs when either gas, oil, or water is left after the

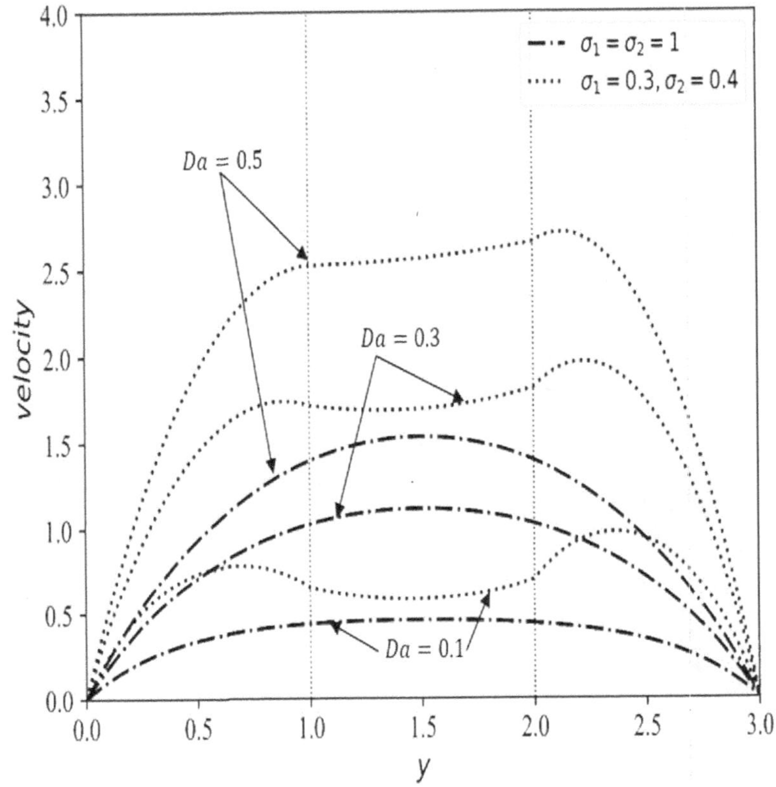

Figure 11.6 Variations in velocity profiles of all regions for different Darcy numbers.

extraction of oil. In this case, we observed that as the permeability ratio and anisotropic angle increases, flow rate Q decreases (see Table 11.3).

2 When all three regions have the same anisotropy, that is, $\psi_1 = \psi_2 = \psi_3 = \psi$, $K_I^* = K_{II}^* = K_{III}^* = K^*$, and different viscosity $\sigma_1 \neq \sigma_2$, then it will represent the flow of three immiscible fluids through an anisotropic porous medium. This type of scenario could be possible when gas, oil, and water are trapped in a single porous rock in the reservoir. The suppressing effect of different permeability ratios and anisotropic angles on velocity profile are presented in Figures 11.7 and 11.8. We observed that as the permeability ratio and anisotropic angle increase, flow rate Q decreases (see Table 11.4).

3 When $Da_1, Da_2, Da_3 \to \infty, \sigma_1 = \sigma_2 = 1$, then it will represent the plane Poiseuille flow. And expression for the same is

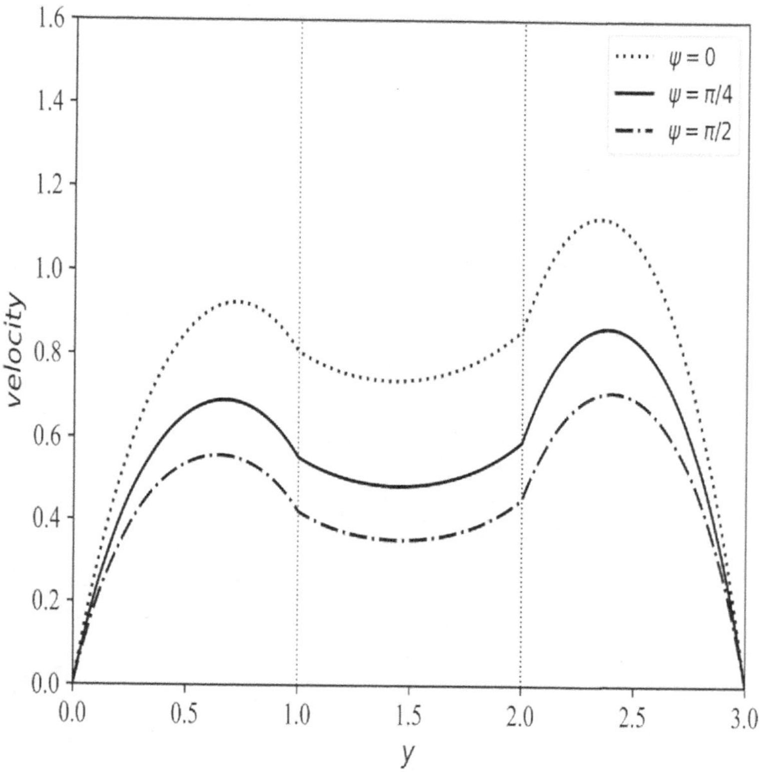

Figure 11.7 Variations in velocity profiles of all regions for different permeability ratios for single anisotropic porous medium.

$$u_{\infty} = \frac{Gy^2}{2} + c_1 y + c_2, \tag{11.32}$$

where u_{∞} is the velocity profile for high Darcy number. After implementing the no-slip boundary conditions, we get $u_{\infty} = \frac{G}{2}\left(y^2 - 2y\right)$.

11.6 CONCLUSIONS AND FUTURE SCOPE

This study explores the flow of three immiscible fluids of different viscosities through a horizontal channel filled with three layers of anisotropic porous medium of different anisotropy. The effects of various key parameters on fluid velocity and shear stress are presented. The finding shows that the permeability ratios, anisotropic angles, and viscosity ratios suppress the velocity

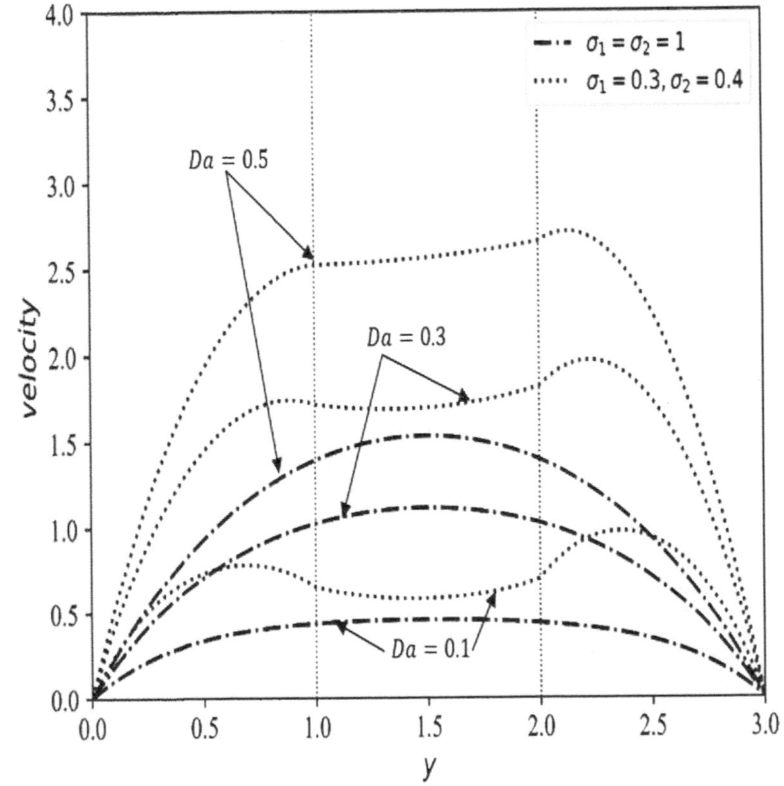

Figure 11.8 Variations in velocity profiles of all regions for different anisotropic angles for single anisotropic porous medium.

Table 11.1 Variations in flow rate Q for different values of permeability ratios, anisotropic angles, and viscosity ratios

		$Da_1 > Da_2 < Da_3$	$Da_1 > Da_2 = Da_3$	$Da_1 > Da_2 > Da_3$
K_I^*	0.5	1.11464	0.99244	0.88665
	1	1.01787	0.89670	0.79184
	2	0.89941	0.77951	0.67580
K_{II}^*	0.5	1.36339	1.19062	1.04552
	1	1.01787	0.89670	0.79184
	2	0.77016	0. 68043	0.60052
K_{III}^*	0.5	1.13029	1.01787	0.89670
	1	1.01787	0.89670	0.79184
	2	0.89670	0.79184	0.71610
ψ_1	0	1.18025	1.05735	0.95094
	$\pi/4$	1.08679	0.96489	0.85937

Table 11.1 (Cont.)

		$Da_1 > Da_2 < Da_3$	$Da_1 > Da_2 = Da_3$	$Da_1 > Da_2 > Da_3$
	$\pi/2$	1.01787	0.89670	0.79184
ψ_2	0	1.76491	1.52063	1.32147
	$\pi/4$	1.24281	1.08910	0.95875
	$\pi/2$	1.01787	0.89670	0.79184
ψ_3	0	1.21477	1.13029	1.01787
	$\pi/4$	1.09654	0.97853	0.86018
	$\pi/2$	1.01787	0.89670	0.79184
σ_1	0.3	2.00484	1.84206	1.70377
	0.7	1.48806	1.32990	1.19565
	0.9	1.37611	1.21967	1.08693
σ_2	0.3	1.98187	1.73366	1.46204
	0.7	1.49259	1.31785	1.13968
	0.9	1.38595	1.22729	1.06998

Table 11.2 Variations in τ_A and τ_B for different values of permeability ratios, anisotropic angles, and viscosity ratios

		At $y = 3$		
		$Da_1 < Da_2$	$Da_1 = Da_2$	$Da_1 > Da_2$
σ_1	0.3	−3.75824	−3.65504	−4.46359
	0.5	−2.3414	−2.26274	−2.81127
	0.9	−1.36148	−1.31001	−1.67019
K_I^*	0.5	−2.67676	−2.60451	−3.27724
	1	−2.36464	−2.31587	−3.00810
	2	−1.95374	−1.92902	−2.60451
ψ_1	0	−3.80424	−3.61399	−4.04762
	$\pi/4$	−2.87755	−2.78798	−3.43566
	$\pi/2$	−2.36464	−2.31587	−3.00810
		At $y = 0$		
		$Da_2 > Da_3$	$Da_2 = Da_3$	$Da_2 < Da_3$
σ_2	0.3	3.75824	3.65504	4.46359
	0.5	2.3414	2.26274	2.81127
	0.9	1.36148	1.31001	1.67019
K_{III}^*	0.5	2.67676	2.60451	3.27724
	1	2.36464	2.31587	3.00810
	2	1.95374	1.92902	2.60451
ψ_3	0	3.80424	3.61399	4.04762
	$\pi/4$	2.87755	2.78798	3.43566
	$\pi/2$	2.36464	2.31587	3.00810

Table 11.3 Variations in flow rate Q for different values of permeability ratios and anisotropic angles in case 1

K_I^*	*0.5*	*0.983087*	ψ_1	*0*	*1.17359*
	1	0.88281		$\pi/4$	0.98308
	2	0.77815		$\pi/2$	0.88281
K_{II}^*	0.5	0.98494	ψ_2	0	1.47513
	1	0.87485		$\pi/4$	1.06076
	2	0.74245		$\pi/2$	0.87485
K_{III}^*	0.5	1.06076	ψ_3	0	1.32973
	1	1.03240		$\pi/4$	1.13495
	2	0.98714		$\pi/2$	1.03240

Table 11.4 Variations in flow rate Q for different values of permeability ratios, anisotropic angles, and viscosity ratios in case 2

$K_I^* = K_{II}^* = K_{III}^*$	*0.5*	*1*	2
	1.46622	1.15927	0.82509
$\psi_1 = \psi_2 = \psi_3$	0	$\pi/4$	$\pi/2$
	3.39753	1.69680	1.15927
σ_1	0.3	0.5	0.9
	1.97413	1.69680	1.46209
σ_2	0.3	0.5	0.9
	1.86316	1.58696	1.35375

in the channel, whereas the Darcy number promotes the velocity. The shear stress at upper and lower plates decreases with an increase in anisotropic parameters, viscosity ratios, and Darcy numbers. In the same manner, the flow rate also behaves in a same manner. We found that if velocity increases, then shear stress at both plates and flow rate increases. The current model is valuable for studying the flow behavior of petroleum in reservoirs as well as the filtration process. A specific application of this model, which involves single fluid flow through multiple layers of anisotropic porous channels, is beneficial for filtering contaminated water. Additionally, further exploration of the unsteady two-dimensional case of this model is necessary for a better understanding of its dynamics.

REFERENCES

Allan, F. M., & Hamdan, M. H. (2002). Fluid mechanics of the interface region between two porous layers. *Applied Mathematics & Computation*, 128(1), 37–43.

Allan, F. M., Hajji, M. A., & Anwar, M. N. (2009). The characteristics of fluid flow through multilayer porous media. *Journal of Applied Mechanics*, 76, 14501–7.

Ansari, I. A., & Deo, S. (2017). Effect of magnetic field on the two immiscible viscous fluids flow in a channel filled with porous medium. *National Academy Science Letters*, 40, 211–214.

Bird, R.B., Stewart, W., & Lightfoot, E. (2002). *Transport Phenomena*. New York: John Wiley & Sons.

Degan, G., Zohoun, S., & Vasseur, P. (2002). Forced convection in horizontal porous channels with hydrodynamic anisotropy. *International Journal of Heat & Mass Transfer*, 45(15), 3181–3188.

Deo, S., & Maurya, D. K. (2022). Investigation of MHD effects on micropolar–Newtonian fluid flow through composite porous channel. *Microfluidics & Nanofluidics*, 26(8), 64.

Devakar, M., & Raje, A. (2018). A study on the unsteady flow of two immiscible micropolar & Newtonian fluids through a horizontal channel: A numerical approach. *The European Physical Journal Plus*, 133(5), 180.

Devakar, M., & Raje, A. (2019). A magnetohydrodynamic time dependent model of immiscible Newtonian & micropolar fluids through a porous channel: a numerical approach. *Journal of Applied Fluid Mechanics*, 12(2), 603–615.

Jaiswal, S., & Yadav, P. K. (2020). Flow of micropolar–Newtonian fluids through the composite porous layered channel with movable interfaces. *Arabian Journal for Science & Engineering*, 45(2), 921–934.

Jaiswal, S., & Yadav, P. K. (2023). Physics of generalized Couette flow of immiscible fluids in anisotropic porous medium. *International Journal of Modern Physics B*, 2450377.

Kamışlı, F., & Öztop, H. F. (2008). Second law analysis of the 2D laminar flow of two-immiscible, incompressible viscous fluids in a channel. *Heat & Mass Transfer*, 44, 751–761.

Karmakar, T. (2021). Physics of unsteady Couette flow in an anisotropic porous medium. *Journal of Engineering Mathematics*, 130(1), 8.

Karmakar, T., & Raja Sekhar, G. P. (2016). Effect of anisotropic permeability on fluid flow through composite porous channel. *Journal of Engineering Mathematics*, 100(1), 33.

Kumar, A., & Madasu, K. P. (2024). Annular generalized Couette flow of immiscible viscous fluids in an anisotropic porous medium. *Journal of the Brazilian Society of Mechanical Sciences and Engineering*, 46(9), 580.

Kumar Yadav, P., Jaiswal, S., Asim, T., & Mishra, R. (2018). Influence of a magnetic field on the flow of a micropolar fluid sandwiched between two Newtonian fluid layers through a porous medium. *The European Physical Journal Plus*, 133, 1–13.

Lu, D. C., Farooq, U., Hayat, T., Rashidi, M. M., & Ramzan, M. (2018). Computational analysis of three layer fluid model including a nanomaterial layer. *International Journal of Heat & Mass Transfer*, 122, 222–228.

Rajeev, A., & Mahanthesh, B. (2021). Multilayer flow & heat transport of nanoliquids with nonlinear Boussinesq approximation & viscous heating using differential transform method. *Heat Transfer*, 50(5), 4309–4327.

Tyvand, P. A., & Storesletten, L. (1991). Onset of convection in an anisotropic porous medium with oblique principal axes. *Journal of Fluid Mechanics*, 226, 371–382.

Umavathi, J. C., Chamkha, A. J., & Shekar, M. (2014). Flow & heat transfer of two micropolar fluids separated by a viscous fluid layer. *International Journal of Microscale & Nanoscale Thermal & Fluid Transport Phenomena*, 5(1), 23.

Umavathi, J. C., & Hemavathi, K. (2019). Flow & heat transfer of composite porous medium saturated with nanofluid. *Propulsion & Power Research*, 8(2), 173–181.

Umavathi, J. C., Kumar, J. P., & Chamkha, A. J. (2008). Flow & heat transfer of a micropolar fluid sandwiched between viscous fluid layers. *Canadian Journal of Physics*, 86(8), 961–973.

Umavathi, J. C., Liu, I. C., Prathap-Kumar, J., & Shaik-Meera, D. (2010). Unsteady flow & heat transfer of porous media sandwiched between viscous fluids. *Applied Mathematics & Mechanics*, 31, 1497–1516.

Umavathi, J. C., & Sutkar, P. S. (2023). Amplification of heat transfer in three immiscible fluids: micropolar nanofluid encased with porous matrix. *Journal of Nanofluids*, 12(4), 1107–1118.

Yadav, P. K., Jaiswal, S., Verma, A. K., & Chamkha, A. J. (2023). Magnetohydrodynamics of immiscible Newtonian fluids in porous regions of different variable permeability functions. *Journal of Petroleum Science & Engineering*, 220, 111113.

Zaytoon, M. A., Alderson, T. L., & Hamdan, M. H. (2016). Flow through a layered porous configuration with generalized variable permeability. *International Journal of Enhanced Research in Science, Technology & Engineering*, 5(6), 1–21.

Chapter 12

Exact solutions of micropolar stagnation point slip flow with heat transfer

Special functions categories

Waqar Khan Usafzai and Emad Hassan Aly

12.1 INTRODUCTION

The analysis of fluid dynamics near deformable surfaces, particularly stretching or shrinking sheets, has garnered sustained research attention due to its critical role in industrial processes. Applications span polymer manufacturing (e.g., extrusion and thin-film production), metallurgical operations (e.g., metal rolling), and advanced engineering domains like aerospace coatings and automated material handling. Hiemenz [1] pioneered exact solutions for stagnation-point flows within the Navier–Stokes framework. Later, Wang [2] extended this work to shrinking sheets, establishing foundational insights into boundary-layer behavior. These studies laid the groundwork for analyzing complex fluid–structure interactions, though challenges persist in extending such solutions to non-Newtonian systems.

Micropolar fluid, which is a class of complex fluids, accounts for the presence of rigid, randomly oriented microparticles, which contribute to additional rotational and couple stress, making micropolar fluid models suitable for describing the behavior of material like liquid crystal, polymeric suspensions, and blood flow. The inherent complexity of micropolar fluids, arising from their micro-rotational dynamics, expands the governing equation system. This heightened mathematical structure poses significant challenges in obtaining closed-form analytical solutions. The stagnation point flow of a micropolar fluid is studied by a numerous number of researches using either approximate or computational methods for solutions, for instance see [3–15] and the references therein. In the predominantly mentioned investigations here, they have been restricted to the non-exact solutions only. On the other hand, exact solutions, if attainable, provide deeper theoretical understanding, validate numerical and experimental solutions and serve as benchmarks for testing the reliability of approximate solutions.

Addressing the critical gap in exact solutions for nonlinear micropolar fluid systems, this work derives analytical results for stagnation-point boundary layer flow over a deformable surface, incorporating wall slip, permeability, and heat transfer effects.

DOI: 10.1201/9781003642169-12

12.2 MATHEMATICAL FORMULATION OF THE PROBLEM

This investigation focuses on the steady stagnation-point flow of an incompressible micropolar fluid interacting with a deformable boundary exhibiting velocity slip and thermal effects. The surface kinematics are characterized by the velocity profile Ux, where distinct regimes emerge: $U = 0$ indicates a fixed surface, $U > 0$ corresponds to an elongating boundary, and $U < 0$ describes a contracting configuration. The flow dynamics result from multiple potential driving mechanisms, including boundary deformation, transverse mass transfer at the wall, stagnation effects, or synergistic interactions between these factors. The velocity field components in the Cartesian directions are expressed as

$$u_w(x) = Ux, \quad v_w = v_w(x), \tag{12.1}$$

In Cartesian coordinates (x, y) with the velocity vector $\mathbf{V} = u(x,y)\mathbf{i} + v(x,y)\mathbf{j}$, the governing equations are given by (see [16–18])

$$\frac{\partial u}{\partial x} + \frac{\partial v}{\partial y} = 0, \tag{12.2}$$

$$\rho\left(u\frac{\partial u}{\partial x} + v\frac{\partial u}{\partial y}\right) = \frac{-\partial p}{\partial x} + (\mu + \kappa)\nabla^2 u + \kappa\frac{\partial G}{\partial y}, \tag{12.3}$$

$$\rho\left(u\frac{\partial v}{\partial x} + v\frac{\partial v}{\partial y}\right) = \frac{-\partial p}{\partial y} + (\mu + \kappa)\nabla^2 v - \kappa\frac{\partial G}{\partial x}, \tag{12.4}$$

$$\rho j\left(u\frac{\partial G}{\partial x} + v\frac{\partial G}{\partial y}\right) = -2\kappa G + \delta\nabla^2 G + \left(\frac{\partial u}{\partial y} - \frac{\partial v}{\partial x}\right), \tag{12.5}$$

where $\{\rho, \mu, \kappa, \alpha, j\}$ are respectively the density, the viscosity, the microrotation viscosity, the spin gradient viscosity, and the microinertia of the micropolar fluid. G is the microrotation component and it can be said that G is total spin of the fluid media and microelements for the fluid flow mechanism. In cases when the microelements are negligible, the flow field reduces to the regular viscous fluid. Therefore, the requirement that G = angular velocity be a possible solution, then we find it only if

$$\alpha = \left(\mu + \frac{\kappa}{2}\right) = \mu\left(1 + \frac{\lambda}{2}\right)j, \tag{12.6}$$

where $\lambda = \kappa/\mu$ is the micropolar parameter.

It is well-known that in the case of steady flow the axial velocity component u at the outer edge of the boundary layer coincides with that in the outer flow $u_e(x)$, and we may write

$$\frac{-1}{\rho}\frac{\partial p}{\partial x} = u_e \frac{du_e}{dx}, \tag{12.7}$$

where Equations (12.3)–(12.5) can be reduced to ordinary differential equations only when $u_e(x) = cx$, with c as a constant.

The boundary conditions are considered as (see [19–22])

$$v = v_w, \quad u = au_w(x) + D\frac{\partial u}{\partial y}, \quad G = -n\frac{\partial u}{\partial y}, \quad \text{for } y = 0, \tag{12.8}$$

$$u \to u_e(x), \quad G \to 0, \quad \text{when} \quad y \to \infty, \tag{12.9}$$

with $v_w < 0$ and $v_w > 0$ being the suction and injection at the wall, a shows the stretching or shrinking rate, D is the velocity slip, and n with the range $0 \le n \le 1$ is the concentration of microelements.

The temperature τ at the wall for the present problem is considered to be either constant $\tau_w(x) = \tau_w$ or linearly growing such that $\tau_w(x) = \tau_\infty + x\tau_0$, with τ_∞ and τ_0 as the far-field and reference temperatures, respectively. The equation governing the heat transfer is given as

$$u\frac{\partial \tau}{\partial x} + v\frac{\partial \tau}{\partial y} = \delta\frac{\partial^2 \tau}{\partial y^2}, \tag{12.10}$$

where δ is the thermal diffusivity. The thermal boundary conditions are

$$\tau = \tau_w(x), \text{ when } y = 0, \text{ and } \tau \to \tau_\infty \text{ when } y \to \infty. \tag{12.11}$$

12.3 MODEL TRANSFORMATION

To solve Equations (12.3) and (12.5) with boundary conditions (12.8) and (12.9), the similarity transformations are in the following form (Usafzai et al. [23], Usafzai and Aly [24]):

$$f(\eta) = \frac{\psi(x,\eta)}{x\sqrt{vU}}, \quad g(\eta) = \frac{G(x,\eta)}{x\sqrt{U^3 v^{-1}}}, \quad \theta(\eta) = \frac{\tau - \tau_\infty}{\tau_w - \tau_\infty}, \tag{12.12}$$

with boundary layer variable $\eta = \sqrt{U/v}y$, and the stream function ψ defined as $u = \partial\psi/\partial y$, and $v = -\partial\psi/\partial x$.

Inserting the similarity variables (12.12) into Equations (12.2)–(12.5) we obtain

$$(1+\lambda)f''' + ff'' - f'^2 + \lambda g' + d^2 = 0, \tag{12.13}$$

$$\left(1+\frac{\lambda}{2}\right)g'' + fg' - gf' - \lambda(2g + f'') = 0, \tag{12.14}$$

$$\theta'' + \Pr f\theta' = 0, \tag{12.15}$$

$$\theta'' + \Pr(f\theta' - f'\theta) = 0, \tag{12.16}$$

where Equations (12.15) and (12.16) are remodeled for constant wall temperature, and linearly growing wall temperature, respectively. The boundary conditions are transformed as

$$f(0) = s,\ f'(0) = a + bf''(0),\ f'(\infty) = d, \tag{12.17}$$

$$g(0) = -nf''(0),\ \ g(\infty) = 0, \tag{12.18}$$

$$\theta(0) = 1,\ \theta(\infty) = 0. \tag{12.19}$$

The quantities

$$\Pr = \nu/\delta,\ s = -v_w/\sqrt{\nu U},\ b = D\sqrt{U/\nu},\ d = c/U(>0), \tag{12.20}$$

represent the Prandtl number, wall mass flux with $s > 0$ and $s <$ are used for suction and injection, the slip and elasticity strength parameter.

12.4 EXACT ANALYTICAL SOLUTIONS

To solve the coupled nonlinear boundary value problem of the current type, the exponential solutions are proposed. Let us write these solutions as (see [23, 25])

$$f(\eta) = s + d\eta + \frac{a-d}{\gamma(1+b\gamma)}\left(1 - e^{-\gamma\eta}\right), \tag{12.21}$$

$$g(\eta) = \frac{n(a-d)\gamma}{1+b\gamma}e^{-\gamma\eta}, \tag{12.22}$$

where for physically meaningful solutions are $\gamma > 0$. The parameter γ emerges as a solution of the cubic equation shown below derived from substituting $f(\eta)$ and $g(\eta)$ into (12.13), with its value entirely determined by the set of parameters $\{a, b, d, s, \gamma\}$,

$$\begin{aligned}&(2+\lambda)b\gamma^3 + (2+\lambda-2abs-2abc\eta)\gamma^2 \\ &-2(s+2bc+c\eta)\gamma - 2(a+c) = 0.\end{aligned} \tag{12.23}$$

In the following, we focus on the solution of equation (23) for special cases of parametric values.

12.4.1 When *d* = 0

In this case, the cubic (12.23) reduces to

$$(2+\lambda)b\gamma^3 + (2+\lambda-2abs)\gamma^2 - 2s\gamma - 2a = 0, \tag{12.24}$$

which can be rearranged as

$$\gamma^3 + a_2\gamma^2 + a_1\gamma + a_0 = 0, \tag{12.25}$$

where

$$a_2 = \frac{(2+\lambda-2abs)}{(2+\lambda)b},\quad a_1 = \frac{-2s}{(2+\lambda)b},\quad a_0 = \frac{-2a}{(2+\lambda)b}. \tag{12.26}$$

Equation (12.25) can further be converted to an incomplete cubic equation

$$t^3 + pt + q = 0, \tag{12.27}$$

where $t = \gamma + \frac{a_2}{3}$ and

$$p = a_1 - \frac{a_2^2}{3}, q = \frac{2a_2^3}{27} - \frac{a_1a_2}{3} + a_0.$$

The three roots $\gamma_1, \gamma_2, \gamma_3$ of the (12.27) are

$$\gamma_1 = \frac{-2(1-bs)+\lambda}{3(b\lambda+2b)} + \frac{\left(\sqrt{3}\sqrt{4p^3+27q^2}-9q\right)^{\frac{1}{3}}}{2^{\frac{1}{3}}3^{\frac{2}{3}}} - \frac{\sqrt[3]{\frac{2}{3}}p}{\left(\sqrt{3}\sqrt{4p^3+27q^2}-9q\right)^{\frac{1}{3}}}, \tag{12.28a}$$

$$\gamma_2 = \frac{-2(1-bs)+\lambda}{3(b\lambda+2b)} + \frac{\left(1+i\sqrt{3}\right)p}{2^{\frac{2}{3}}\sqrt[3]{3}\left(\sqrt{3}\sqrt{4p^3+27q^2}-9q\right)^{\frac{1}{3}}} - \frac{\left(1-i\sqrt{3}\right)\left(\sqrt{3}\sqrt{4p^3+27q^2}-9q\right)^{\frac{1}{3}}}{2^{\frac{4}{3}}3^{\frac{2}{3}}}, \tag{12.28b}$$

$$\gamma_3 = \frac{-2(1-bs)+\lambda}{3(b\lambda+2b)} + \frac{\left(1-i\sqrt{3}\right)p}{2^{\frac{2}{3}}\sqrt[3]{3}\left(\sqrt{3}\sqrt{4p^3+27q^2}-9q\right)^{\frac{1}{3}}} - \frac{\left(1+i\sqrt{3}\right)\left(\sqrt{3}\sqrt{4p^3+27q^2}-9q\right)^{\frac{1}{3}}}{2^{\frac{4}{3}}3^{\frac{2}{3}}}. \tag{12.28c}$$

The corresponding constant wall temperature and linearly increasing wall temperature along with the boundary conditions under the intermediate variable $z = e^{-\gamma\eta}$ introduced in (12.15) can be rewritten as:

Constant wall temperature

$$z\theta'' + (1-\alpha+\omega z)\theta' = 0, \tag{12.29}$$

Linear growing wall temperature

$$z\theta'' + (1-\alpha+\omega z)\theta' - \omega\theta = 0, \tag{12.30}$$

with the boundary conditions

$$\theta(0) = 0, \qquad \theta(1) = 1. \tag{12.31}$$

The exact thermal solutions and thermal gradients for these two cases are given by:
Constant Wall temperature

$$\theta(\eta) = \frac{\left(\omega^{\alpha} e^{\frac{\alpha}{2}\gamma\eta} M_{\frac{\alpha}{2},\frac{\alpha}{2}+\frac{1}{2}}\left(\omega e^{-\gamma\eta}\right) + (1+\alpha)\,\omega^{\frac{3\alpha}{2}} e^{\frac{-1}{2}\omega e^{-\gamma\eta}}\right) e^{\omega-\alpha\gamma\eta-\frac{1}{2}\omega e^{-\gamma\eta}}}{\omega^{\alpha} M_{\frac{\alpha}{2},\frac{\alpha}{2}+\frac{1}{2}}(\omega)\, e^{\frac{\omega}{2}} + (1+\alpha)\,\omega^{\frac{3\alpha}{2}}}, \tag{12.32}$$

$$-\theta'(0)=\frac{-\alpha\lambda\omega^{\alpha}}{\omega^{\alpha}+e^{\omega}\left(\Gamma(\alpha+1)-\Gamma(\alpha+1,\omega)\right)}, \tag{12.32}$$

Linearly increasing wall temperature

$$\theta(\eta)=\frac{{}_1F_1\left(2;1+\alpha;\omega e^{-\gamma\eta}\right)e^{\omega-\alpha\gamma\eta-\omega e^{-\gamma\eta}}}{{}_1F_1(2;1+\alpha;\omega)}, \tag{12.33}$$

$$-\theta'(0)=\frac{-\gamma(1+2\omega)\omega^{\alpha}+\omega\gamma(3-2\alpha+2\omega)\left(\Gamma(\alpha)-\Gamma(\alpha,\omega)\right)e^{\omega}}{\omega^{\alpha}+e^{\omega}(1-\alpha+\omega)\left(\Gamma(\alpha)-\Gamma(\alpha,\omega)\right)}. \tag{12.35}$$

where $M_{k,m}(x),\Gamma(a,x)$, and ${}_1F_1(a;b;x)$ are the Whittaker, gamma and hypergoemetric functions respectively. The quantities α and ω are

$$\alpha=\frac{\Pr}{\gamma}\left(s+\frac{a}{\gamma(1+b\gamma)}\right),\omega=\frac{a\Pr}{\gamma^{2}(1+b\gamma)}.$$

12.4.2 When a = 1 and d = 1

The format of $f(\eta)$ given in (21) for $a=1$ and $d=1$ reduces to algebraic type solution

$$f(\eta)=s+\eta, \tag{12.36}$$

which holds valid for $s\in(-\infty,\infty)$, and no restriction is needed to be imposed on s. The corresponding heat transfer relations can be obtained as:
Constant wall temperature

$$\theta(\eta)=\frac{1-\mathrm{Erf}\left(\sqrt{\frac{\Pr}{2}}(s+\eta)\right)}{1-\mathrm{Erf}\left(\sqrt{\frac{\Pr}{2}}s\right)},\quad -\theta'(0)=\frac{\sqrt{\frac{2\Pr}{\pi}}e^{\frac{-Prs^{2}}{2}}}{1-\mathrm{Erf}\left(\sqrt{\frac{\Pr}{2}}s\right)} \tag{12.37}$$

Linearly increasing wall temperature

$$\theta(\eta)=\frac{e^{\frac{-1}{2}Pr(2s+\eta)\eta}\left(\sqrt{2\pi\Pr}(s+\eta)e^{\frac{1}{2}\Pr(s+\eta)^{2}}\mathrm{Erfc}\left(\sqrt{\frac{\Pr}{2}}(s+\eta)\right)-2\right)}{e^{\frac{\Pr s^{2}}{2}}\sqrt{2\pi\Pr}sErfc\left(\sqrt{\frac{\Pr}{2}}s\right)-2}, \tag{12.38}$$

$$-\theta'(0) = \frac{\sqrt{\Pr}\,\mathrm{Erfc}\left(\sqrt{\frac{\Pr}{2}}s\right)}{\sqrt{\frac{2}{\pi}}e^{\frac{-1}{2}\Pr s^2} - s\sqrt{Pr}\,\mathrm{Erfc}\left(\sqrt{\frac{\Pr}{2}}s\right)}, \tag{12.39}$$

where Erf and Erfc are the error and complementary error functions respectively.

12.4.3 When $a = -1$ and $d = -1$

In this case when $f(\eta)$ reduces to

$$f(\eta) = s - \eta, \tag{12.40}$$

also valid for all values of s. The case $a = -1$ and $d = -1$ does not produce any real solution for temperature distribution.

12.4.4 When $b \to \infty$

In the limiting case of slip, that is, when $b \to \infty$, solution $f(\eta)$ given in (12.21) becomes

$$f(\eta) = s + d\eta, \tag{12.41}$$

which is true for all values of s and d. The profiles of fixed wall temperature and increasing wall temperature with their corresponding temperature gradients are given as:

Constant wall temperature

$$\theta(\eta) = \frac{1 - \mathrm{Erf}\left(\frac{\sqrt{\Pr}(s + d\eta)}{\sqrt{2}\sqrt{d}}\right)}{1 - \mathrm{Erf}\left(\frac{\sqrt{\Pr s}}{\sqrt{2}\sqrt{d}}\right)}, \quad -\theta'(0) = \frac{-\sqrt{\frac{2}{\pi}}\sqrt{d\Pr}e^{\frac{-\Pr(s+d\eta)^2}{2d}}}{1 - \mathrm{Erf}\left(\frac{\sqrt{\Pr s}}{\sqrt{2}\sqrt{d}}\right)}, \tag{12.42}$$

Linearly increasing wall temperature

$$\theta(\eta) = \frac{\left(\sqrt{2\pi}\sqrt{\Pr}(s + d\eta)\mathrm{Erfc}\left(\frac{\sqrt{\frac{\Pr}{d}}s + d\eta}{\sqrt{2}}\right) - 2\sqrt{d}e^{\frac{-\Pr(s+d\eta)^2}{2d}}\right)e^{\frac{\Pr s^2}{2d}}}{\sqrt{2\pi}\sqrt{\Pr s}e^{\frac{\Pr s^2}{2d}}\mathrm{Erfc}\left(\frac{s\sqrt{\frac{\Pr}{d}}}{\sqrt{2}}\right) - 2\sqrt{d}}, \tag{12.43}$$

$$-\theta'(0) = \frac{d\sqrt{\pi \mathrm{Pr}}\mathrm{Erfc}\left(\frac{\sqrt{\mathrm{Pr}}}{\sqrt{2}\sqrt{d}}\right)}{\sqrt{2d}e^{\frac{-\mathrm{Pr}s^2}{2d}} - s\sqrt{\pi \mathrm{Pr}}\mathrm{Erfc}\left(\frac{\sqrt{\mathrm{Pr}}}{\sqrt{2}\sqrt{d}}\right)}. \tag{12.44}$$

12.4.5 When $d = 0$ and $\lambda = 0$

In the situation when $d = 0$ and $\lambda = 0$ along with $b \approx 0$, Equations (12.13) and (12.14) together with the boundary conditions in (12.17) become

$$f''' + ff'' - f'^2 = 0, \tag{12.45}$$

$$g'' + fg' - f'g = 0, \tag{12.46}$$

$$f(0) = s, \; f'(0) = a, \; f'(\infty) = 0, \tag{12.47}$$

$$g(0) = \frac{-1}{2}f''(0), \; g(\infty) = 0. \tag{12.48}$$

The solutions of $f(\eta)$ and $g(\eta)$ become

$$f = s + \frac{a}{\gamma}\left(1 - e^{-\gamma\eta}\right), \; g = \frac{a\gamma}{2}e^{-\gamma\eta}, \tag{12.49}$$

where γ is the quadratic root of the equation

$$\gamma^2 - s\gamma - a = 0, \tag{12.50}$$

given by

$$\gamma = \frac{1}{2}\left(s + \pm\sqrt{s^2 + 4a}\right). \tag{12.51}$$

It is worth highlighting that in the case of a regular viscous fluid ($b = 0$), the solution aligns with the findings of Yao et al. [26]. For the shrinking sheet scenario ($a = -1$), Equation (12.46) reduces to the solution derived by Fang et al. [27]. Similarly, in the stretching sheet case ($a = 1$) with $s = 0$, Equation (12.46) coincides with the classical solution reported by Crane [28]. The exact solutions of temperature profiles for both the constant and linearly growing wall cases given in Equations (12.28) and (12.29) become Constant wall temperature

$$\theta(\eta)=\frac{\left(\omega^{\alpha}e^{\frac{\alpha}{2}\gamma\eta}M_{\frac{\alpha}{2},\frac{\alpha}{2}+\frac{1}{2}}\left(\omega e^{-\gamma\eta}\right)+\left(1+\alpha\right)\omega^{\frac{3\alpha}{2}}e^{\frac{-1}{2}\omega e^{-\gamma\eta}}\right)e^{\omega-\alpha\gamma\eta-\frac{1}{2}\omega e^{-\gamma\eta}}}{\omega^{\alpha}M_{\frac{\alpha}{2},\frac{\alpha}{2}+\frac{1}{2}}\left(\omega\right)e^{\frac{\omega}{2}}+\left(1+\alpha\right)\omega^{\frac{3\alpha}{2}}}, \tag{12.52}$$

$$-\theta'(0)=\frac{-\alpha\gamma\omega^{\alpha}}{\omega^{\alpha}+e^{\omega}\left(\Gamma\left(\alpha+1\right)-\Gamma\left(\alpha+1,\omega\right)\right)}, \tag{12.53}$$

Linearly increasing wall temperature

$$\theta(\eta)=\frac{e^{\omega-\alpha\gamma\eta-\omega e^{-\gamma\eta}}\,{}_1F_1\left(2;1+\alpha;\omega e^{-\gamma\eta}\right)}{{}_1F_1\left(2;1+\alpha;\omega\right)}, \tag{12.54}$$

$$-\theta'(0)=\frac{-(1+2\omega)\,\omega^{\alpha}\gamma+\omega\gamma(3-2\alpha+2\omega)\left(\Gamma\left(\alpha\right)-\Gamma\left(\alpha,\omega\right)\right)e^{\omega}}{\omega^{\alpha}+e^{\omega}\left(1-\alpha+\omega\right)\left(\Gamma\left(\alpha\right)-\Gamma\left(\alpha,\omega\right)\right)}, \tag{12.55}$$

where α and ω can now be expressed as $\alpha=\Pr\left(a+s\gamma\right)/\gamma^{2}$ and $\omega=aPr/\gamma^{2}$.

12.4.6 When $d \neq 0, \pm 1$

For this case, Equation (12.23) is satisfied when

$$\gamma=\frac{1}{b},\quad s=\frac{\lambda+2}{2b},\quad d=\frac{-a}{3}. \tag{12.56}$$

Therefore $f\left(\eta\right)$ and $g\left(\eta\right)$ can be written as

$$f(\eta)=s+d\eta+\frac{1}{2}b(a-d)\left(1-e^{\frac{-\eta}{b}}\right),\quad g(\eta)=\frac{-(a-d)e^{\frac{-\eta}{b}}}{2b}. \tag{12.57}$$

with the corresponding constant wall temperature and heat gradient given by

$$\theta(\eta)=1-\frac{\int_{0}^{\eta}e^{\frac{-1}{2}\Pr\left((2s+b(a-d))x+dx^{2}+b^{2}(a-d)e^{\frac{-x}{b}}\right)}dx}{\int_{0}^{\infty}e^{\frac{-1}{2}\Pr\left((2s+b(a-d))x+dx^{2}+b^{2}(a-d)e^{\frac{-x}{b}}\right)}dx}, \tag{12.58}$$

$$-\theta'(0) = \frac{e^{\frac{-\mathrm{Pr}}{2}b^2(a-d)}}{\int_0^\infty e^{\frac{-1}{2}\mathrm{Pr}\left((2s+b(a-d))x+dx^2+b^2(a-d)e^{\frac{-x}{b}}\right)}dx}. \tag{12.59}$$

12.5 RESULTS AND DISCUSSION

The analytical investigation conducted for the micropolar slip flow in this analysis can be validated by comparing the current results with the existing results as shown in Table 12.1–12.4, which is clear demonstration on the reliability of our results.

The γ-solution curves showing unique and dual solutions for the stretching and shrinking sheets, respectively, are plotted in Figures 12.1(a) and (b). It can be seen that the general rule for the stretching sheet exhibiting a unique solution can be observed for this highly complex and nonlinear micropolar stagnation point flow system for $b = 1, \lambda = 1$. We observed that this unique solution extends for all the values of wall transpiration, that is, $s \in R$. Also, the dual solutions profile for the shrinking sheet is observed only for the suction case ($s > 0$), when $b = 1$ and $\lambda = 1$.

Table 12.1 Values of $-\theta'(0)$ for varying Pr showcasing the constant temperature case when $s = 0, d = 0.5$

	Stretching sheet ($a > 0$)	
Pr	*Recent results*	*Turkyilmazoglu and Pop [25]*
0.5	0.490387078	0.49038708
1	0.715435623	0.71543562
2	1.039158868	1.03915887
5	1.688620272	1.68862027
10	2.424446386	2.42444639

Table 12.2 Values of $-\theta'(0)$ for varying Pr showcasing the constant temperature case when $s = 0, d = 0.5$

	Shrinking sheet ($a < 0$)	
Pr	*Recent results*	*Turkyilmazoglu and Pop [25]*
0.5	0.132656811	0.13265677
1	0.112315831	0.11231583
2	0.065974764	0.06597476
5	0.008721676	0.00872168
10	0.000196175	0.00019617

Table 12.3 Variation in $f''(0)$ for stretching sheet ($a > 0$) at $b = 0 = s, d = 1$

a	Ishak et al. [7]	Present results		
		$\lambda = 0$	$\lambda = 1$	$\lambda = 2$
0	1.232588	1.232588	1.006408	0.871609
0.1	1.146561	1.146561	0.936166	0.810770
0.2	1.051130	1.051130	0.858246	0.743283
0.5	0.713295	0.713295	0.582404	0.504384
1	0	0	0	0
2	-1.887306	-1.887306	-1.540979	-1.334530
3	-4.276540	-4.276540	-3.491781	-3.023974
5	-10.264749	-10.264749	-8.381133	-7.258275

Table 12.4 Wall shear stress values $f''(0)$ for shrinking sheet configurations ($a < 0$) with $b = 0 = s$, and elasticity parameter $d = 1$

a	Wang [2]	Current analysis		
		$\lambda = 0$	$\lambda = 1$	$\lambda = 0$
-0.25	1.402241	1.402241	1.144934	0.991612
-0.50	1.495670	1.495670	1.221231	1.057764
-0.75	1.489298	1.489299	1.216066	1.053489
-1.00	1.328817	1.328820	1.085196	0.940886
-1.15	1.082230	1.082245	0.884456	0.769305
-1.2465	0.554300	0.586975	0.515377	0.489453

The effect of mass transpiration on both stretching sheet and shrinking sheet configurations when $d = 0$ are shown in Figures 12.2(a) and (b). The results reveal that an inverse relationship between the boundary layer thickness and the wall mass transpiration parameter s, with this trend being consistently observed across the tested range $s = -2$ to 2). In Figure 12.2(b), the profile of shrinking sheet boundary layer is depicted for the suction case only where the boundary layer is depressed when s decreases for a particular set of parametric values $b = 0.3, \lambda = 0.1$.

The thermal boundary layers for constant and linearly increasing wall temperatures are illustrated in Figures 12.3(a) and (b). For a fixed wall temperature with the parameter set $a = 1$, $b = 1$, $\Pr = 1$, $\lambda = 1$, and $s = \{-1, -0.5, 0, 0.5, 1\}$, the temperature profiles decrease while attaining larger values along the η-axis. In contrast, for a linearly growing wall temperature, the profiles corresponding to smaller wall mass flux values $s = \{-0.7, -0.3, 0, 0.3, 0.7\}$ asymptotically approach zero for smaller values of η.

In order to analyze the results, numerical computations are also carried out giving rise to the dual solutions for $f''(0), g'(0)$ and $\theta'(0)$ as shown in Figures 12.4. The existence of critical numbers a_c in these graphs for

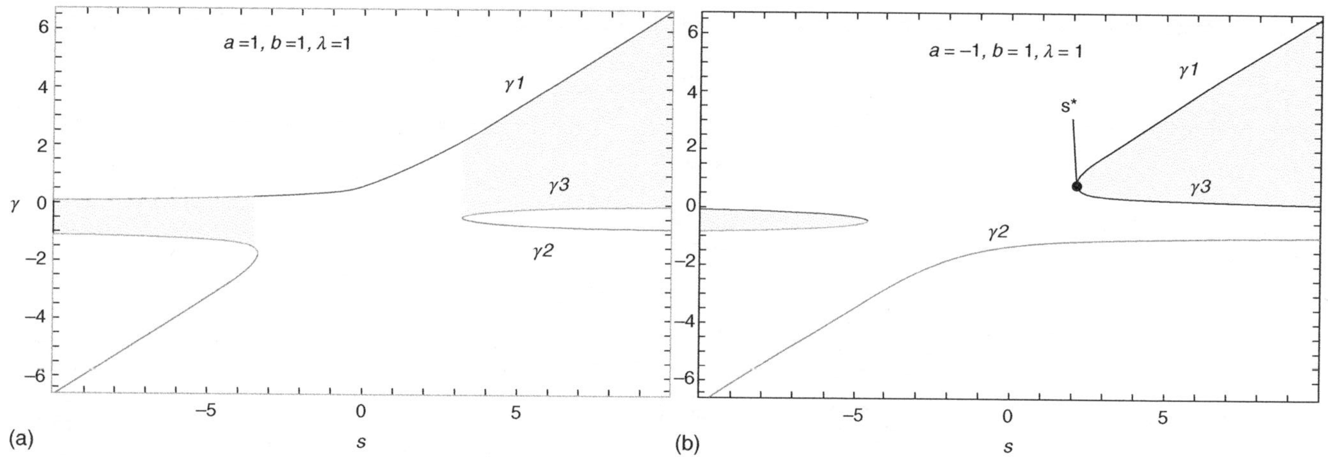

Figure 12.1 Dependence of the characteristic parameter γ on the mass transpiration coefficient s for: (a) sheet elongation (positive s) and (b) sheet contraction (negative s).

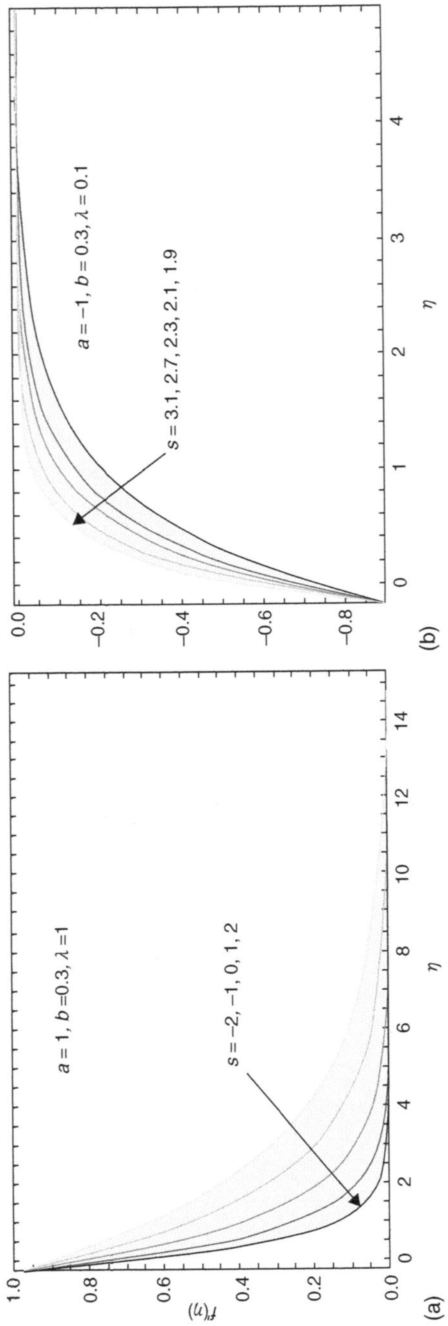

Figure 12.2 Dimensionless velocity distribution $f'(\eta)$ across the boundary layer for (a) stretching and (b) shrinking surface cases.

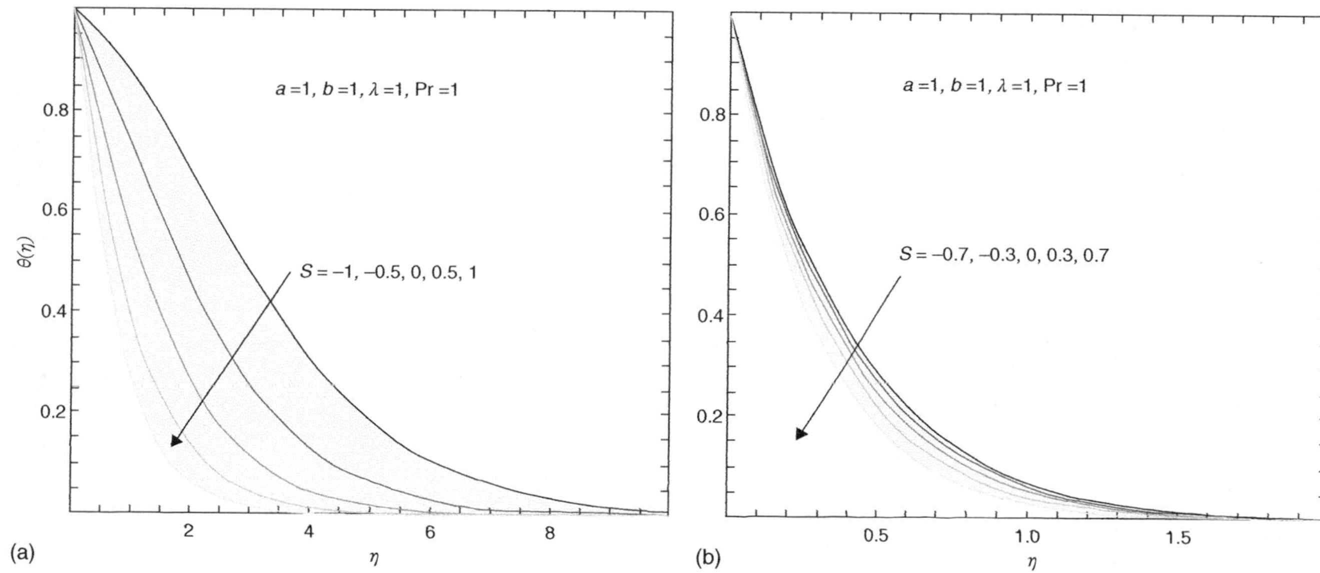

Figure 12.3 Dimensionless temperature $\theta(\eta)$ profiles demonstrating the effect of transpiration rate *s* on: (a) isothermal wall and (b) linearly heated wall boundary conditions.

the skin friction, couple stress and reduced Nusselt number show the behavior of dual solutions in the positive and negative a-region. It can be seen that $f''(0)$, is increasing for $a < a_c$, and a decreasing when $a > a_c$ where $a_c = -1$, as given in Figure 12.4(a). It also shows that as λ increases, $f''(0)$ decreases. The couple stress profile given in Figure 12.4(b) for $a_c = -0.5$ decreases with increasing λ. The sinusoidal behavior of thermal boundary layer for the linearly growing wall temperature with two critical points $a_{c_1} = -2.1596$, and $a_{c_2} = -1.7919$ for $\lambda = 0$ and $\lambda = 0.3$ are given in Figure 12.4(c). It is observed that the solutions for $\lambda = 0, 0.3$ decrease in regions $a < a_{c_1} = -2.1596$ and $a > a_{c_2} = -1.7919$, whereas solutions increase for $a_{c_1} = -2.1596 \le a_c \le a_{c_2} = -1.7919$. In Figure 12.4(d), the case of constant wall thermal gradient is depicted for $\lambda = 0$ and 0.1.

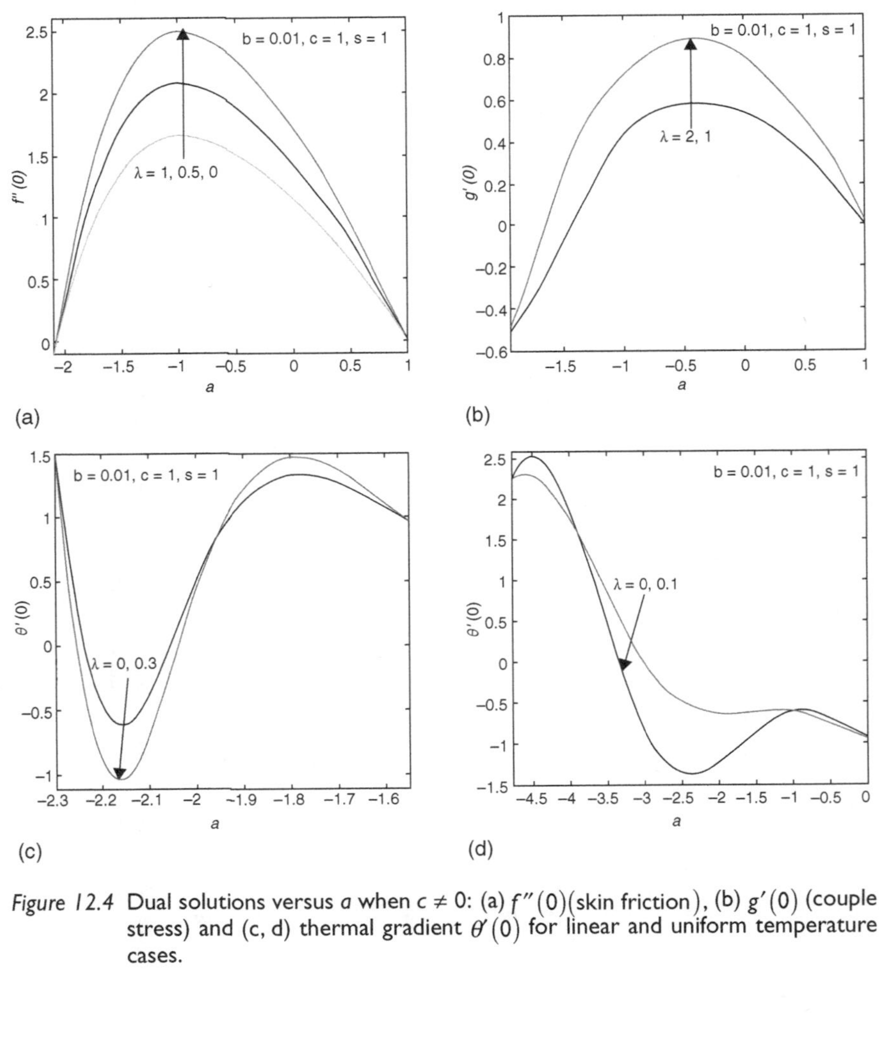

Figure 12.4 Dual solutions versus a when $c \neq 0$: (a) $f''(0)$(skin friction), (b) $g'(0)$ (couple stress) and (c, d) thermal gradient $\theta'(0)$ for linear and uniform temperature cases.

ACKNOWLEDGMENT

The first author thanks the financial sponsorship by Nanjing Institute of Technology for Young Scholars Research Support Program under Grant No. YKJ202126.

REFERENCES

[1] K. Hiemenz. Die Grenzschicht an einem in den gleichformingenFlussigkeits-stromeingetauchtengradenKreiszylinder, *Dinglers Polytech. J.* 326 (1911) 321–324.

[2] C.Y. Wang, Stagnation flow towards a shrinking sheet, *International Journal of Non-Linear Mechanics* 43 (2008) 377–382.

[3] R. Nazar, N. Amin, D. Filip, I. Pop, Stagnation point flow of a micropolar fluid towards a stretching sheet. *International Journal of Non-Linear Mechanics*, 39 (2004) 1227–1235.

[4] Y.Y. Lok, N. Amin, D. Campean, I. Pop, Steady mixed convection flow of a micropolar fluid near the stagnation point on a vertical surface, *International Journal of Numerical Methods for Heat Fluid Flow*, 15 (2005) 654–670.

[5] Y.Y. Lok, N. Amin, I. Pop, Unsteady mixed convection flow of a micropolar fluid near the stagnation point on a vertical surface, *International Journal of Thermal Sciences* 45 (2006) 1149–1157.

[6] T. Hayat, T. Javed, Z. Abbas, MHD flow of a micropolar fluid near a stagnation-point towards a non-linear stretching surface, *Nonlinear Analysis: Real World Applications* 10 (2009) 1514–1526.

[7] A. Ishak, Y.Y. Lok, I. Pop, Stagnation-point flow over a shrinking sheet in a micropolar fluid, *Chemical Engineering Communications* 197 (2010) 1417–1427.

[8] N.A. Yacob, A. Ishak, I. Pop, Melting heat transfer in boundary layer stagnation-point flow towards a stretching/shrinking sheet in a micropolar fluid, *Computers & Fluids* 47(2011) 16–21.

[9] M.A. Mahmoud, S.E. Waheed, MHD stagnation point flow of a micropolar fluid towards a moving surface with radiation, *Meccanica* 47(2012) 1119–1130.

[10] H. Rosali, A. Ishak, I. Pop, Micropolar fluid flow towards a stretching/shrinking sheet in a porous medium with suction, *International Communications in Heat and Mass Transfer* 39 (2012) 826–829.

[11] M.A. Abbas, N. Faraz, Y.Q. Bai, Y. Khan, Analytical study of the non orthogonal stagnation point flow of a micro polar fluid, *Journal of King Saud University-Science* 29 (2017) 126–132.

[12] A. Kumar, V. Sugunamma, N. Sandeep, Impact of non-linear radiation on MHD non-aligned stagnation point flow of micropolar fluid over a convective surface, *Journal of Non-Equilibrium Thermodynamics* 43(2018) 327–345.

[13] S. Anuradha, R. Punithavalli, MHD boundary layer flow of a steady micropolar fluid along a stretching sheet with binary chemical reaction, *International Journal of Applied Engineering Research* 14 (2019) 440–446.

[14] M.S. Kausar, A. Hussanan, M. Waqas, M. Mamat, Boundary layer flow of micropolar nanofluid towards a permeable stretching sheet in the presence of porous medium with thermal radiation and viscous dissipation, *Chinese Journal of Physics* 78 (2022) 435–452.

[15] M.N. Sadiq, B. Sarwar, M. Sajid, N. Ali, Heat transfer in unsteady separated stagnation point flow of a micro-polar fluid: Cattaneo–Christov model, *Journal of Thermal Analysis and Calorimetry* 147 (2022) 5199–5209.

[16] W.K. Usafzai, E.H. Aly, M.M. Tharwat, A.M. Mahros, Modeling of micropolar nanofluid flow over flat surface with slip velocity and heat transfer: Exact solution, *Alexandria Engineering Journal* 75 (2023) 313–323.

[17] W.K. Usafzai, E.H. Aly, Hiemenz flow with heat transfer in a slip condition micropolar fluid model: Exact solution, *International Communications in Heat and Mass Transfer* 144 (2023) 106775.

[18] W.K. Usafzai, E.H Aly, S.A. Shad, Hiemenz flow for a micropolar nanofluid with bidirectional flexible surface and heat transfer, *Thermal Science and Engineering Progress* 45 (2023) 102119.

[19] W.K. Usafzai, E.H. Aly, I. Pop, Hiemenz stagnation point flow of a second order micropolar slip flow with heat transfer, *International Journal of Numerical Methods for Heat Fluid Flow* 34 (2024) 1277–1296.

[20] W.K. Usafzai, E.H. Aly, S. Kim, Exact analytical solutions for micropolar magnetohydrodynamic flow: Insights into velocity slip and heat transfer characteristics, *Physics of Fluids* 36 (2024) 032914.

[21] W.K. Usafzai, E. H. Aly, I. Pop, Multiple exact solutions in tri-hybrid nanofluid flow: a study of elastic surface effects. *International Journal of Numerical Methods for Heat & Fluid Flow* 35 (2025) 277–299. https://doi.org/10.1108/HFF-08-2024-0610

[22] W.K. Usafzai, N.S. Wahid, N.M. Arifin, E.H. Aly, MHD flow and heat transfer of nanotriple (Cu-AlO-Ag): exact solutions, *Chinese Journal of Physics* 93 (2025) 56–74.

[23] W.K. Usafzai, E.H. Aly, A.S. Alshomrani, M. Zakaullah, Multiple solutions for nanofluids flow and heat transfer in porous medium with velocity slip and temperature jump, *International Communications in Heat and Mass Transfer* 131 (2022) 105831.

[24] W.K. Usafzai, E.H. Aly, Multiple exact solutions for micropolar slip flow and heat transfer of a bidirectional moving plate, *Thermal Science and Engineering Progress* 37 (2023) 101584.

[25] M. Turkyilmazoglu, I. Pop, Exact analytical solutions for the flow and heat transfer near the stagnation point on a stretching/shrinking sheet in a Jeffrey fluid, *International Journal of Heat and Mass Transfer* 57 (2013) 82–88.

[26] S. Yao, T. Fang, Y. Zhong, Heat transfer of a generalized stretching/shrinking wall problem with convective boundary conditions, *Communications in Nonlinear Science and Numerical Simulation* 16 (2011) 752–760.

[27] T. Fang, S. Yao, J. Zhang, A. Aziz, Viscous flow over a shrinking sheet with a second order slip flow model, *Communications in Nonlinear Science and Numerical Simulation* 15 (2010) 1831–1842.

[28] L.J. Crane, Flow past a stretching plate, *Journal of Applied Mathematics and Physics (ZAMP)* 21 (1970) 645–647.

Chapter 13

Application of Jacobi algorithm to physical problems

Prasanna Bansode

13.1 INTRODUCTION

Eigenvalue problems are fundamental in both theoretical and applied physics, appearing in contexts such as vibrational analysis, stability problems, and quantum mechanics. Many physical systems governed by differential equations can be reduced, through discretization techniques, to algebraic eigenvalue problems. Efficient and reliable numerical algorithms are therefore essential for solving these equations, particularly when analytical solutions are unavailable or impractical.

The Jacobi eigenvalue algorithm, a classical method for diagonalizing symmetric matrices, offers a robust and conceptually straightforward approach. Despite its relatively slow convergence compared to modern alternatives, the Jacobi algorithm remains useful due to its guaranteed stability and simplicity in implementation. These characteristics make it particularly effective for moderate-sized systems and for educational purposes where interpretability is prioritized.

In computational physics, Jacobi's algorithm is frequently applied to model systems such as the classical wave equation and the Schrödinger equation. These systems can be rescaled into dimensionless forms, allowing the transformation of differential equations into matrix equations that are solvable via linear algebra techniques. This transformation provides a pathway to compute eigenfrequencies in mechanical systems or energy levels in quantum systems.

This chapter presents a numerical study of eigenvalue problems arising from classical and quantum mechanical systems using the Jacobi algorithm. Benchmarking against analytical solutions enables verification of accuracy, while variations in system parameters provide insight into the algorithm's performance. The study demonstrates the algorithm's flexibility and its relevance in solving diverse physical problems.

DOI: 10.1201/9781003642169-13

13.2 RELATED WORK

Eigenvalue problems play a central role in computational physics and engineering. Numerous numerical techniques have been developed to compute eigenvalues and eigenvectors of symmetric matrices, which are often encountered in physical systems. One of the earliest and most notable algorithms for this purpose is the Jacobi eigenvalue algorithm, introduced in the mid-twentieth century and praised for its conceptual simplicity and guaranteed convergence for symmetric matrices [1].

In structural mechanics, particularly in problems involving vibrations and buckling of beams and membranes, finite difference methods are commonly used to discretize differential operators, transforming partial differential equations into matrix eigenvalue problems. The Jacobi algorithm has been applied successfully in this context, although it is less efficient than modern algorithms like the QR method or divide-and-conquer methods [2]. Nevertheless, its pedagogical value and robustness make it a viable choice for solving problems like the buckling beam equation discussed in this work.

In quantum mechanics, especially in the discretization of the Schr¨odinger equation for systems such as the harmonic oscillator or two-electron atoms, Jacobi's method provides a straightforward approach to finding the energy eigenvalues and eigenfunctions [3]. While alternative methods such as shooting or variational methods are often employed in quantum mechanical computations, the Jacobi method stands out due to its direct applicability to the discretized Hamiltonian matrix [4].

The use of Jacobi's algorithm in educational contexts is also well-documented. It offers a valuable demonstration of iterative diagonalization and is particularly effective in helping students understand the nature of orthogonal transformations and the diagonalization process [5]. More recent work has also explored the optimization of Jacobi's algorithm through parallel computing approaches to address its slow convergence rate [6].

Additionally, comparisons between Jacobi's method and other eigensolvers reveal important trade-offs. While the Jacobi algorithm is computationally expensive for large scale problems, its ability to preserve orthogonality exactly and its simplicity in implementation make it attractive for smaller-scale systems or as a baseline for performance evaluation [7].

In summary, although not the most efficient algorithm in the modern computational landscape, the Jacobi method continues to be relevant due to its reliability, accuracy for small to mid-sized systems, and educational utility. This chapter builds on this tradition by applying Jacobi's method to a series of classical and quantum mechanical problems to demonstrate its applicability and performance characteristics in physical simulations.

13.3 METHOD

13.3.1 Classical wave-the buckling beam problem

The first example of an eigenvalue problem I present is the classical wave equation for a buckling beam in one dimension

$$\gamma[2]u(x)x = -F(x)u(x) \tag{13.1}$$

Here, $u(x)$ represents the vertical displacement of the beam along the y-direction. I let $x \in [0, L]$ for a beam length L. The constant γ is a material-dependent parameter specifying the beam rigidity, and F is the force applied at the interval $(0, L)$. Next, the Dirichlet boundary conditions are imposed so that $u(0) = u(L) = 0$. I consider the three parameters F, L, and γ to be known. In order to make the equation more convenient to handle, I scale the integration variable to the beam length as

$$\rho = \frac{x}{L} \tag{13.2}$$

So that the new dimensionless integration variable $\rho \in [0,1]$. I can now rewrite(13.1) as

$$[2]u(\rho)\rho = -\frac{FL^2}{\gamma}u(\rho) = -\lambda u(\rho) \tag{13.3}$$

where I define λ as $\frac{FL^2}{\gamma}$. In order to solve (13.3) for the λ's numerically the discretization $\rho \to \rho_i = \rho_0 + ih$ is introduced, where with $i = 0, 1, 2, 3........, n-1$. The equation written on discrete form is then

$$[2]u_i x \approx -\frac{u_{i+1} - 2u_i + u_{i-1}}{h^2} = \lambda u_i \tag{13.4}$$

where the step length $h = \rho_n - \rho_0$, $\rho_0 = \rho_{\text{min}} = 0$ and $\rho_n = \rho_{\text{max}} = 1$ are the boundaries and n is the grid size. Here $u_{i\pm1}$ denotes $u(\rho_i \pm h)$.

The above discrete equation can easily be formulated as a matrix equation

$$A\vec{u} = \lambda\vec{u} \tag{13.5}$$

by introducing the vector $\vec{u}^T = [u_1, u_2.........., u_{n-1}]$ and the matrix

$$A = \begin{bmatrix} d & a & 0 & 0 & \dots & 0 & 0 \\ a & d & a & 0 & \dots & 0 & 0 \\ 0 & a & d & a & 0 & \dots & 0 \\ \dots & \dots & \dots & \dots & \dots & \dots & \dots \\ 0 & \dots & \dots & \dots & a & d & a \\ 0 & \dots & \dots & \dots & \dots & a & d \end{bmatrix} \tag{13.6}$$

where $a = -\frac{1}{h^2}$ and $d = \frac{2}{h^2}$. The problem is now transformed into a linear algebra eigenvalue problem. Since the analytical eigenvalues

$$\lambda_i = d + 2a\cos\left(\frac{j\pi}{n+1}\right), \qquad j = 1,2,3........,n \tag{13.7}$$

are known I can later compare these to those found numerically.

13.3.2 Eigenvalue solver: Jacobi's algorithm

The eigenvalue solver I will show applies the Jacobi algorithm, which converges quite stably yet very slowly. The essence of the algorithm is that I can rotate the column space of A around different axes in the n dimensional space using similarity transforms, obtaining a diagonal matrix D being similar to A. Due to the similarity between D and A, the diagonal elements $d_{ii} = \lambda_i$, being the eigenvalues of D, are also eigenvalues of A.

The similarity transforms used in the Jacobi algorithm arebased on a sequence of orthogonal transforms. In an orthogonal transformation, an orthogonal basis $F = \{\vec{v}_i \mid i = 1,2....,n\}$, keeps its orthogonality. Meaning that the basis vectors in the new basis $G = \{\vec{w}_i \mid i = 1,2....,n\}$ where $\vec{w}_i = U\vec{v}_i$ form a new orthogonal basis for the same space as before the transformation if U is orthogonal,that is, $UU^T = U^TU = I_n$ giving $U^{-1} = U^T$. Therefore, the inner product of basis vectors in G gives

$$\vec{w}_i \cdot \vec{w}_j = w_i^T w_j = (Uv_i)^T Uv_j \tag{13.8}$$

$$= v_i^T U^T U v_j = v_i^T v_j = \delta_{ij} \tag{13.9}$$

due to the orthogonality of the basis F.

When transforming A into a diagonal matrix D, I need to eliminate the off-diagonal elements of A. A single similarity transforms B of the matrix A is written as

$$B = S^T AS \tag{13.10}$$

for a rotational matrix

$$S = \begin{bmatrix} 1 & 0 & 0 & 0 & \cdots & 0 \\ 0 & 1 & 0 & 0 & \cdots & 0 \\ \cdots & \cdots & \cdots & \cdot\cdot & \cdots & \cdot\cdot \\ 0 & \cos\theta & \cdots & \cdots & \sin\theta & 0 \\ \cdots & \cdots & \cdots & \cdots & \cdots & \cdots \\ 0 & -\sin\theta & \cdots & \cdots & \cos\theta & 0 \\ \cdots & \cdots & \cdots & \cdots & \cdots & \cdots \\ 0 & 0 & 0 & 0 & 1 & 0 \\ 0 & 0 & 0 & 0 & 0 & 1 \end{bmatrix} \tag{13.11}$$

rotating the column space basis an angle θ in the $\rho_k - \rho_l -$ plane, since $S_{kk} = S_{ll} = \cos\theta = c$, $S_{kl} = \sin\theta = s = -S_{lk}$. Since the rotational matrix S is orthogonal, the matrix B is similar to A.

Multiplying out the above similarity transform I get

$$b_{ii} = a_{ii}, \quad i \neq k, i \neq l \tag{13.12}$$

$$b_{ik} = a_{ik}c - a_{il}s, \quad i \neq k, i \neq l \tag{13.13}$$

$$b_{il} = a_{il}c + a_{ik}s, \quad i \neq k, i \neq l \tag{13.14}$$

$$b_{kk} = a_{kk}c^2 - 2a_{kl}cs + a_{ll}s^2 \tag{13.15}$$

$$b_{ll} = a_{ll}c^2 + 2a_{kl}cs + a_{kk}s^2 \tag{13.16}$$

$$b_{kl} = \left(a_{kk} - a_{ll}\right)cs + a_{kl}\left(c^2 - s^2\right) \tag{13.17}$$

Since I want to zero out the off-diagonals of A, I require that (13.17) is zeroed out, that is, $b_{kl} = b_{lk} = 0$. Defining

$$\cot(2\theta) = \tau = \frac{a_{ll} - a_{kk}}{2a_{kl}} \tag{13.18}$$

we get a quadratic equation $t^2 + 2\tau t - 1 = 0$, where $t = \tan\theta$, following the requirement $b_{kl} = 0$. From this quadratic equation, I get that

$$t = -\tau \pm \sqrt{1 + \tau^2} \tag{13.19}$$

$$c = \frac{1}{\sqrt{1+t^2}} \tag{13.20}$$

$$s = tc \tag{13.21}$$

This way, I can tune the angle θ in order to zero out b_{kl} and then find the remaining elements of B. The way t is defined is numerically unstable. If $\tau \gg 1$ a computer will let $1+\tau^2 = \tau^2$ resulting in $t = -\tau \pm \tau$,which is not correct. Using some simple algebra I rewrite $t = -\frac{1}{-\tau \mp \sqrt{1+\tau^2}}$, which is more numerically stable. Let $t = \frac{1}{\tau + \sqrt{1+\tau^2}}$ whenever $\tau > 0$ and let $t = -\frac{1}{-\tau + \sqrt{1+\tau^2}}$ when $\tau < 0$, thus circumventing subtraction of almost equal numbers in the denominator which can resulting in division by zero.

As shown by [8, Ch. 7.4] for a two-matrix without loss of generality (w. l. o. g.), the off-diagonal norm of a similarity transform B is

$$Off(B)^2 = \sum_{i \neq j} \left| b_{ij}^2 \right| = \|A\|_F^2 - 2a_{kl} \tag{13.22}$$

being reduced in size for each transform. To determine which elements a_{kl} should be eliminated each time I perform a similarity transform, I look at (13.22); if I choose a_{kl} to be the greatest off-diagonal element of A I reduce $Off(B)^2$ by the largest possible value for each iteration.

Since I want to end up with a diagonal matrix D, I need to make the off-diagonal norm approach zero through multiple similarity transforms. In practice, this corresponds to iterate over similarity transforms, until

$$Off(A) \geq \max_{i \neq j} \left(a_{ij}^2 \right) = a_{kl}^2 \leq \varepsilon \tag{13.23}$$

for a tolerance ε typically $\sim 10^{-8}$. I can thus use $\max_{i \neq j} \left(a_{ij}^2 \right) \leq \varepsilon$ as a stopping criterion for the while loop performing the similarity transforms, instead of calculating $Off(A)$ which can be time-consuming.

After having looped over the similarity transforms, the remaining matrix should be diagonally within the tolerance, making it easy to extract the eigenvalues λ_i. To find the corresponding eigenvectors, I rotate an arbitrary orthogonal basis for $\Re^n$ for each loop iteration using the corresponding rotation S. I then contain the input basis vectors in a matrix E, for example, the identity matrix, and rotate them one rotation by multiplying SE. This is also done in the loop so that

$$e_{ik} = ce_{ik} - se_{il} \tag{13.24}$$

$$e_{il} = ce_{il} + se_{ik} \tag{13.25}$$

As discussed by [8, p. 217] the Jacobi algorithm usually requires between $3n^2 - 5n^2$ rotations, and each rotation requires $\sim 4n$ operations, resulting in $13n^3 - 20n^3$ operations to zero out the off-diagonal elements. Since I only consider sparse matrices in my case, I expect $\sim n$ rotations needed to diagonalize.

The Jacobi algorithm can now be used to solve eigenvalue problems for symmetrical matrices like the buckling beam wave or quantum mechanical issues, as shown in the following subsection.

13.3.3 Schrödinger's equation:one electron in a harmonic oscillator potential

Next, I show how to apply the Jacobi algorithm to find the energy eigenvalues of the Hamiltonian operator for an electron in a harmonic oscillator potential $V(r)$ in three dimensions. For simplicity, I only consider the radial Schrödinger equation

$$-\frac{h^2}{2m}\left(\frac{1}{r^2}rr^2r - \frac{l(l+1)}{r^2}\right)R(r) + V(r)R(r) = ER(r) \tag{13.26}$$

where $V(r) = \frac{1}{2}kr^2$ with $k = m\omega^2$, and the energies E of the electron in the harmonic oscillator in three dimensions are given as

$$E_{nl} = h\omega\left(2n + l + \frac{3}{2}\right) \tag{13.27}$$

where n = 0,1,2,... and l = 0,1,2,..., representing the principal and orbital momentum quantum numbers. I let the radial distance $r \in [0,\infty)$. Next I let $R(r) = \frac{u(r)}{r}$ so that

$$-\frac{h^2}{2m}[2]ru(r) + \left(V(r) + \frac{l(l+1)}{r^2}\frac{h^2}{2m}\right)u(r) = Eu(r) \tag{13.28}$$

In order to reflect reality, I must require $u(0) = u(\infty) = 0$, as the wave function of the electron must fall off to zero when going to infinity.

To ensure safer handling of the equation, I introduce the dimension less radial distance $\rho = \frac{r}{\alpha}$, where α is a parameter with dimension length. Inserting this into (13.28), I obtain

$$-\frac{h^2}{2m\alpha^2}[2]\rho u(\rho) + \left(V(\rho) + \frac{l(l+1)}{\rho^2}\frac{h^2}{2m\alpha^2} \right) u(\rho) = Eu(\rho) \tag{13.29}$$

To simplify things further, I only consider the case where $l = 0$. Inserting the dimension less potential $V(\rho) = \frac{k}{2}\alpha^2\rho^2 u(\rho)$ into (13.29) I can rewrite the equation as

$$-\frac{h^2}{2m\alpha^2}[2]\rho u(\rho) + \frac{k}{2}\alpha^2\rho^2 u(\rho) = Eu(\rho) \tag{13.30}$$

by multiplying both sides by $\frac{2m\alpha^2}{h^2}$. In order to simplify things even further, I introduce a natural length scale so that

$$\alpha = \left(\frac{h^2}{mk} \right)^{1/4} \tag{13.31}$$

being set by the mass of the electron m and the strength potential k and I let the frequency $\omega = 1$. The energy eigenvalues are then defined as

$$\lambda = \frac{2m\alpha^2}{h^2}E \tag{13.32}$$

The Schrödinger equation is now in a dimensionless form

$$-[2]\rho u(\rho) + \rho^2 u(\rho) = \lambda u(\rho) \tag{13.33}$$

where $V(\rho) = \rho^2$ represents the effective potential. To solve for the eigenvalues numerically, I discretize the equation so that

$$-\frac{u_{i+1} - 2u_i + u_{i-1}}{h^2} + \rho_i^2 u_i \tag{13.34}$$

$$= -\frac{u_{i+1} - 2u_i + u_{i-1}}{h^2} + V_i u_i = \lambda u_i \tag{13.35}$$

letting the effective potential $V_i = \rho_i^2$. The step size h has the same form as shown earlier only for a different ρ_n. Since I need the wave function $u(\rho)$ to tend to zero at infinity, $\rho \to \infty$, I must set $\rho_{max} = \rho_n = \infty$. However, representing infinity on a computer is not possible, meaning I need to find a suitably large number to approximate infinity. The approximation of infinity will also be dependent on the matrix size, meaning I have to balance my choice of parameters ρ_{max} and n in order to achieve the wanted precision in the eigenvalues. From this discretization, I can write the equation on the same matrix form as (13.5), this time using diagonal elements $d_i = \frac{2}{h^2} + V_i$ and similarly the secondary diagonals $e_i = -\frac{1}{h^2}$. Having the quantum mechanical problem in matrix form, I can use Jacobi's algorithm to solve it and then compare the numerical eigenvalues to the analytical solutions $\lambda = 3, 7, 11, 15........$

13.3.4 Schrödinger's equation: two electrons in a harmonic oscillator potential

Another problem from quantum mechanics worth considering is the Coulomb interaction between two electrons in a three-dimensional harmonic oscillator potential, which has analytical solutions usable for comparison. When adding another electron to the earlier single-electron case, I also need to introduce the Coulomb interaction, resulting in the Schrödinger equation on the following form

$$\left(-\frac{h^2}{2m}[2]r_1 - \frac{h^2}{2m}[2]r_2 + \frac{1}{2}kr_1^2 + \frac{1}{2}kr_2^2 + \frac{\beta e^2}{|\vec{r}_1 - \vec{r}_2|}\right) u(r_1, r_2) = E^{(2)}u(r_1, r_2) \tag{13.36}$$

Here the Coulomb potential $V_{coul} = \frac{\beta e^2}{|\vec{r}_1 - \vec{r}_2|}$ for $\beta e^2 = 1.44\ eVnm$, the energy of the two-electron system is $E^{(2)}$ and the positions of the electrons are given as r_1 and r_2.

It is now convenient to introduce the relative distance $\vec{r} = |\vec{r}_1 - \vec{r}_2|$ and the center of-mass $\vec{R} = \frac{1}{2}(\vec{r}_1 - \vec{r}_2)$ in order to split the Schrödinger equation in two by separation of variables, $u(r) = \psi(r)\phi(r)$. I now only consider the equation for $\psi(r)$ as the center of-mass equation is simply another harmonic oscillator previously shown how to solve. I introduce a convenient scaling method to make the equation dimensionless so that it can be solved using

Jacobi's algorithm. As this is completely analogous to the previously shown single-electron case, I jump ahead to the final form of the dimensionless equation (see Appendix for a detailed description of scaling and separation of variables):

$$-[2]\rho\psi(\rho) + \omega_r^2 \rho^2 \psi(\rho) + \frac{1}{\rho} = \lambda\psi(\rho) \tag{13.37}$$

The frequency ω_r is a parameter directly linked to the strength of the harmonic oscillator potential. The discretization and reformulation as a matrix problem, in this case, are done completely analogously to the single electron case, this time using diagonal elements $d_i = \frac{2}{h^2} + V_i$ for $V_i = \omega_r^2 \rho^2 + \frac{1}{\rho}$.

We are only interested in the ground state of the system; thus, I let $l = 0$. The system's ground state eigenvalues, found through the Jacobi algorithm, are then compared to the analytical solutions, and the eigenstates are plotted for different strengths ω_r of the harmonic oscillator potential.

13.4 RESULTS

The buckling beam equation, given by (13.5) was solved using the Jacobi algorithm as described in Section 13.3.2. Plotting the relative error between the first three of the numerical eigenvalues and analytical eigenvalues given by (13.7) as a function of increasing the grid size n produced the result shown in Figure 13.1. The number of similarity transforms needed is plotted as a function of matrix dimensions n for the buckling beam problem. In this plot, I compare it to the analytical function $f(n) = \frac{3}{2}n^2$ as well. This is shown in Figure 13.2. The CPU time for a complete run-through of the Jacobi algorithm is also plotted as a function of matrix dimensions n. This is shown in Figure 13.3.

For the harmonic oscillator potential described in Section 13.3.3 the first four numerical eigenvalues were compared with the known analytical eigenvalues for this problem, namely 3, 7, 11, and 15. Figure 13.4 shows the absolute error of the first four eigenvalues compared with their analytical solutions as a function of the dimensionless maximum distance ρ_{max}. This result was achieved using a matrix of dimensions $n = 400$ for each ρ_{max}.

For the two electrons in a harmonic oscillator potential problem, as described in Section 13.3.4, the first eigenstate is plotted as a function of ρ for varying degrees of the potential strength ω_r. This is shown in Figure 13.5. The ground state eigenvalues λ_1 for selected values for the potential strength ω_r are tabulated in Table 13.1.

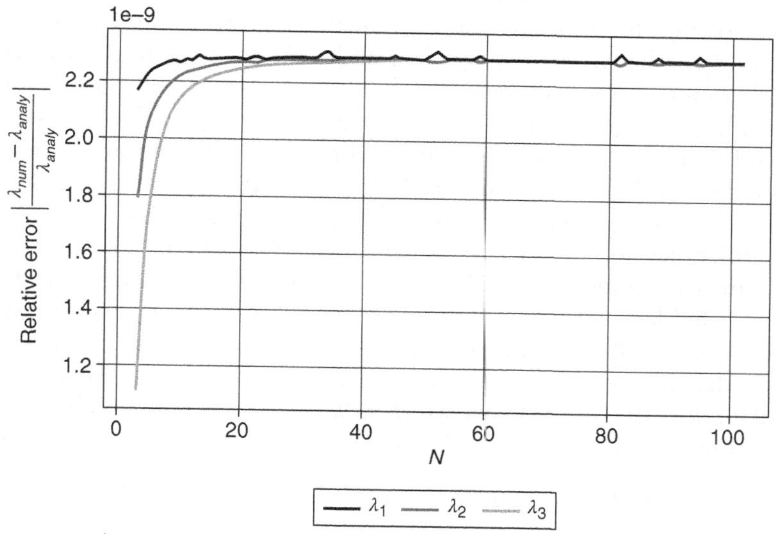

Figure 13.1 The relative error between the numerical and analytical eigen values, given by Equation (13.7), as a function of n for the buckling beam problem given by Equation (13.5). Here n starts at $n = 3$.

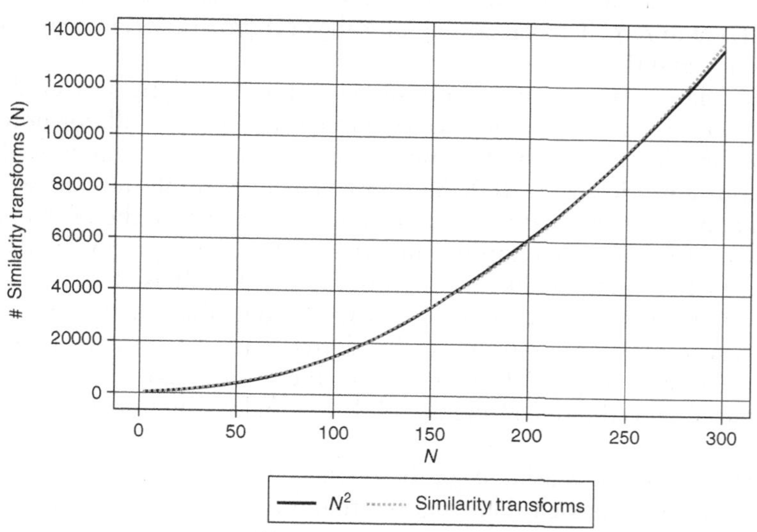

Figure 13.2 The number of similarity transforms conducted before the off-diagonal elements are set below a tolerance when solving the buckling beam problem using the Jacobi algorithm.

13.5 DISCUSSION

When solving the buckling beam problem using the Jacobi algorithm, the relative error between the analytical and numerical eigenvalues increases rapidly before it quickly settles on a constant relative error for the three eigenvalues considered. This is seen in Figure 13.1. In the figure, n starts at $n = 3$. For a 2×2 matrix, the numerical result should reproduce the exact solution to machine precision. It is worth noting that the difference between the constant settling value and the initial error is relatively small. I have set a tolerance for treating the off-diagonal elements as zero to $\varepsilon = 10^{-8}$. The similarity transforms ends when the off-diagonal elements are below this threshold, which is constant for every n. In order to further test this method, one could study the settling value as a function of ε. It is also worth noting that I have only learned this behavior up to n=100. This behavior could change after further increasing n. When plotting the number of similarity transforms, as shown in Figure 13.2, one can see that they increase exponentially as the grid size n increases. Here, I compare it to $f(n) = \frac{3}{2}n^2$ as it has a similar lope. This strongly indicates that the number of similarity transforms is proportional to n^2, as the theory suggests. The CPU time shown in Figure 13.3 shows the same behavior as the number of similarity transforms plots. For a dense matrix, the CPU time should behave proportionally to n^3. I have a tridiagonal matrix with a large number of values already set to zero. This can account for the deviation between the measured CPU time and the theoretical.

The relative error between the numerical and analytical ground state energy eigen values of a single electron in a harmonic oscillator potential as seen in Figure 13.4 shows how the error behaves as a function of ρ_{max} at a constant grid size n=400. All four eigenvalues result in relative errors starting out quite high at between ten and 100. However, the relative error quickly decreases when approaching $\rho_{max} \in (4,6)$. Each graph has its minimum in this interval, but not at the same value. It is thus impossible to minimize the relative error for all four eigenvalues at the same time. One has to make a compromise (see dotted line). I found the best compromise to beat $\rho_{max} \approx 5.2$. At this value, the fourth relative error graph is minimized, and the remaining are all below the wanted precision of four digits.

When further increasing ρ_{max}, the relative errors start to grow again, as their minima are passed. This suggests that for a grid size of $n = 400$, a sufficient approximation for an infinite radial distance is $\rho_{max} = 5.2$, as it is the value at which the relative error is at its smallest and at the same time below the set tolerance. Since the minima of the relative errors move upwards for increasing ρ_{max}, I cannot represent more than four eigenvalues to the wanted precision. In order to find more eigenvalues to the set tolerance I would

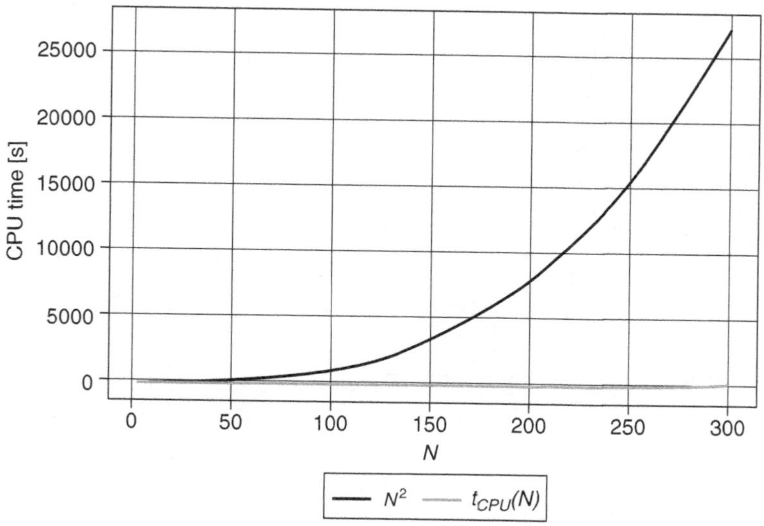

Figure 13.3 The CPUtime as a function of matrix dimension n when solving the buckling beam problem with the Jacobi algorithm.

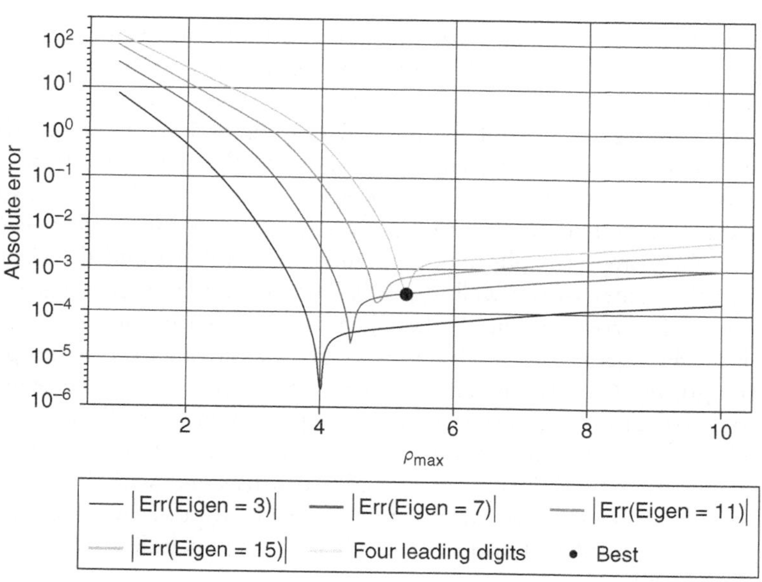

Figure 13.4 The absolute error of the first four eigenvalues for the harmonic oscillator potential described in Section 13.3.3 as a function of varying the dimensionless maximum distance ρ_{max}.

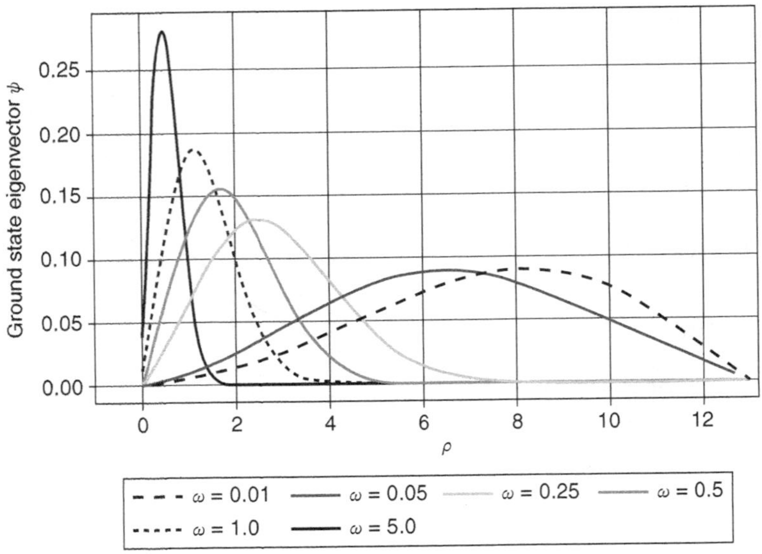

Figure 13.5 The eigenvector for the ground state in the two-electron harmonic oscillator potential as described in Section 13.3.4 for varying ω.

thus have to increase the grid size n. Thus a further improvement would be to grid over both n and ρ_{max} to produce a contour plot of the relative error, illustrating the best combination of the two variables.

For the two-electron problem, as shown in Figure 13.5, I can clearly see that when increasing the potential strength ω_r, the wave function peaks closer to the origin while spreading out more for lower ω_r. This is consistent with expectations from quantum mechanics, as increasing ω_r decreases the width of the potential well, making it more probable to find the electrons closer to the center of the well at $\rho = 0$. The ground state energies λ_1 seen in Table 13.1 corresponding to the eigenstates shown in Figure 13.5 illustrate how the energy of the electrons behaves as a function of the potential strength ω_r. Clearly, the energy eigenstates increase in value when the potential strength is increased. Thus, a steeper potential well increases the system's ground-state energy. Furthermore, since I find that my numerical eigenvalues for $\omega_r = 0.25$ and $\omega_r = 0.05$ are consistent with the analytical ones found by [9] (being about the same up to a factor of 2 due to a different scaling; $\lambda_1^{\omega_r=0.25} = 0.6250$ and $\lambda_1^{\omega_r=0.05} = 0.175$), I conclude that my results are quite sound.

Table 13.1 Potential strength and corresponding ground state energy eigenvalues for two-electron harmonic oscillator problem(Gridsize: $n = 300$)

ω_r	λ_1
0.01	0.420
0.05	0.475
0.25	1.250
0.50	2.230
1.00	2.058
5.00	17.448

13.6 CONCLUSION

We have implemented the Jacobi algorithm for finding eigenvalues and applied it to three physical problems. The numerical solutions provided by the algorithm were in agreement with values calculated analytically within an acceptable tolerance. The performance of the algorithm differed from what one would expect, which was reasonable due to the fact that the problems at hand provided tridiagonal matrices as opposed to dense matrices. For the harmonic oscillator potential problem my results were made with a constant n. A natural next step would be to make a grid over different combinations of n and ρ_{max} to find the optimal ρ_{max} for a given matrix dimension.

REFERENCES

[1] G. H. Golub and C. F. Van Loan, *Matrix Computations*. JHU Press, 2013.

[2] W. H. Press, S. A.Teukolsky, W. T.Vetterling, and B. P.Flannery, *Numerical Recipes*, 3rd Edition: *The Art of Scientific Computing*. Cambridge University Press, 2007.

[3] D. J. Griffiths and D. F.Schroeter, *Introduction to Quantum Mechanics*. Cambridge University Press, 2018. 11

[4] L. N. Trefethen and D. Bau, *Numerical Linear Algebra*. SIAM, 1997.

[5] M. H. Jensen, *Computational Physics: Lecture Notes Fall 2015*, 2015, available athttps://github.com/CompPhysics/ComputationalPhysics.

[6] Z. Bai, J. Demmel, J. Dongarra, A. Ruhe, and H. van der Vorst, *Templates for the Solution of Algebraic Eigenvalue Problems: A Practical Guide*. SIAM, 2000.

[7] Y. Saad, *Numerical Methods for Large Eigenvalue Problems*. SIAM, 2011.

[8] M. H. Jensen, "Computational physics, lecture notes fall 2015,"2015, https://github.com/CompPhysics/ComputationalPhysics/blob/master/doc/Lectures/lectures2015.pdf, Retrieved September6, 2019.
[9] Taut, M., Two electrons in an external oscillator potential: Particular analytic solutions of a Coulomb correlation problem. *Physical Review A*, 48(5), 3561, 1993.
[10] Department of Physics,"Project 2, deadline September 30 computational physics ifys3150/fys4150," Norway, Sep. 2019, https://github.com/Comp Physics/ Computational Physics/tree/master/doc/Projects/2019/Project2/pdf, Retrieved September 27, 2019.

APPENDIX

Consider the two-electron Schrödinger equation

$$\left(-\frac{h^2}{2m}[2]r_1 - \frac{h^2}{2m}[2]r_2 + \frac{1}{2}kr_1^2 + \frac{1}{2}kr_2^2 + \frac{\beta e^2}{|\vec{r}_1 - \vec{r}_2|}\right) u(r_1, r_2) = E^{(2)}u(r_1, r_2) \tag{13.38}$$

The next step in transforming the equation to dimensionless form is to introduce the relative distance $\vec{r} = (\vec{r}_1 - \vec{r}_2)$ and the center-of-mass $\vec{R} = \frac{1}{2}(\vec{r}_1 - \vec{r}_2)$ enabling to me write the Schrödinger equation as

$$\left(-\frac{h^2}{m}[2]r - \frac{h^2}{4m}[2]R + \frac{1}{4}kr^2 + kR^2 + \frac{\beta e^2}{r}\right)u(r, R) = E^{(2)}u(r, R) \tag{13.39}$$

Since the relative distance and the center-of-mass are two independent degrees of freedom, I can use the separation of variables $u(r) = \psi(r)\phi(R)$ to isolate the Schrödinger equation for the relative distance. Since the center-of-mass equation is just a regular harmonic oscillator potential, similar to the one I solved earlier, I neglected it from here on. The energy of the two-electron system can then be separated too, as $E^{(2)} = E_r + E_R$, where E_r is the energy of the relative distance equation and E_R is the harmonic oscillator energy of the center-of-mass frame. The Schrödinger equation for the relative distance

$$\left(-\frac{h^2}{m}[2]r + \frac{1}{4}kr^2 + \frac{\beta e^2}{r}\right)\psi(r) = E_r \psi(r) \tag{13.40}$$

can, similarly to the single-electron case, be made dimensionless by introducing the dimensionless variable $\rho = \frac{r}{\alpha}$.

This then gives the equation

$$-[2]\rho\psi(\rho) + \frac{1}{4}\frac{mk}{\hbar^2}\alpha^4\rho^2\psi(\rho) + \frac{m\alpha\beta e^2}{\rho\hbar^2}\psi(\rho) = \frac{m\alpha^2}{\hbar^2}E_r\psi(\rho) \tag{13.41}$$

In order to write it in a similar form to (13.33), I define the frequency

$$\omega_r^2 = \frac{1}{4}\frac{mk}{\hbar^2}\alpha^4 \tag{13.42}$$

where I require

$$\frac{m\alpha\beta e^2}{\hbar^2} = 1 \tag{13.43}$$

introducing a new natural lengths scale

$$\alpha = \frac{\hbar^2}{m\beta e^2} \tag{13.44}$$

Finally, I let the energy eigenvalues be rewritten in dimensionless form:

$$\lambda = \frac{m\alpha^2}{\hbar^2}E \tag{13.45}$$

resulting in the final form of the Schrödinger equation,

$$-[2]\rho\psi(\rho) + \omega_r^2\rho^2\psi(\rho) + \frac{1}{\rho} = \lambda\psi(\rho) \tag{13.46}$$

where ω_r is a parameter characterizing the harmonic oscillator.

Index

Adiabatic, 156
Aerospace, 3, 14, 15
Agglomeration, 138
Algebraic solution, 217
Analysis, 37
Analytical solutions, 211, 214, 228
Angular velocity, 212
Anisotropic angle, 190, 191, 193, 199–208
Anisotropic parameters, 191, 193, 196, 208
Anisotropic permeability, 193, 209
Anisotropic porous channel, 190, 208
ANOVA, 138, 143, 147, 150
Apparent viscosity, 20
Arbitrary constants, 199
Archimedes, 18
Arrhenius-type equations, 5
Average squared residual, 122

Base fluid, 108, 115–116, 129
Bejan number, 27
Bingham model, 28
Bioconvection Lewis number, 156, 166
Biot number, 60, 103, 110
Bi-viscous Bingham model, 28
Blood flow, 33
Boundary conditions (BC), 8, 9, 11, 17, 31, 119, 139, 140, 142, 198, 205, 213, 216, 219, 225, 228
Boundary layer, 37, 39, 41, 47, 49, 51, 52, 54, 56, 57, 81, 137, 139, 140, 211, 213, 222, 224, 226, 227, 228
Boundary layer theory, 19
Brinkman, 193
Brownian motion, 26, 37, 59, 79
Brownian parameter, 137, 143
Buoyancy-driven, 155
Buoyancy force, 86

Cardiovascular modeling, 33
Carreau model, 26
Cartesian coordinates, 212
Casson fluid, 59
Casson model, 30
Cattaneo–Christov model, 80
Cavity, 174
Central composite design (CCD), 150
Chemical reactions, 137, 139, 143, 151
Coarse, 161
Coefficient of drag, 107, 117
Coefficient of kinematic viscosity, 19
Cogley radiation, 80
Complex fluids, 211
Compressible flows, 5, 7
COMSOL multiphysics, 160, 174
Concentration, 60
Concentration Buoyancy, 147
Concentration contours, 165
Concentration slip length, 59
Concentration slip parameter, 59
Conduction, 1–3, 9, 11–13, 15
Consistency coefficient, 24
Consistency index, 4
Constant wall temperature, 216–220
Continuity, 194, 198
Contour plots, 147
Convection, 1–3, 9, 11, 13, 60, 78
Convective coefficient, 2, 3, 9
Convective conditions, 143, 151
Convective transport, 164, 166
Convergence criterion, 143
Coordinate systems, 14
Couette–Poiseuille, 191
Couple stress, 211, 216
Critical, 211, 222, 226
Cross viscosity model, 26

Darcy–Forchheimer, 107, 109–111, 132
Darcy friction factor, 11
Darcy law, 60
Darcy number, 166, 190, 199, 200, 202, 204, 205, 208
Darcy parameter, 98

Darcy permeability (K), 107, 112–113, 117
Deborah term, 80
Decrease, 199, 201–204, 208
Deformable surface, 211
Deformation, 49
Degree Of Freedom, 150
Degree of shear thinning, 27
Density, 5, 7, 11, 59, 107, 112–113, 115–118
Dilatant fluid, 30
Dimensionless, 196, 199
Dimensionless parameters, 141, 156
Dirichlet, 4, 5, 15
DNS, 6, 14
Drag coefficient, 65, 83
Drilling fluids, 25
Dual solutions, 221, 222, 226
Dufour–Soret effects, 61
Dufour diffusivity, 59
Dufour number, 59
Dynamic viscosity coefficient, 19

Eckert number (Ec), 78, 117, 120–121, 126–128, 143
Effective viscosity, 190, 193, 197
Eigenvectors, 230, 234
Elasticity, 86
Elasticity parameter, 222
Electrical conductivity, 59, 78, 107
Electrically conducting, 37, 40–41, 58
Ellis model, 27
Enclosures, 154
Energy levels, 229, 235–238, 242
Engine oil (EO), 110, 115–116
Error (relative/absolute), 238–241
Error function, 217
Exact solutions, 211, 213, 215, 217, 219, 221, 223, 225, 227, 228
Exponential heat source/sink parameter, 98
Extending sheet, 137, 138
Extra coarse, 161
Extremely coarse, 161
Eyring–Powell (E-P), 109

Finite difference methods, 230–231
Finite element method, 14
Finite volume method, 14
Flow behavior index, 25
Flow curve, 19
Flow path, 84
Flow rate, 199, 202, 204, 206, 208
Fluid(s), 4, 5, 7, 11, 15, 211, 212, 219, 227, 228
Fluid parameter, 41, 44–45
Force, 231
Forced convection, 79
Forchheimer, 154
Four parameter model, 26
Fourier's law, 1
Free convection, 78
Fully developed flow, 194
F-Value, 150

Gamma function, 217
Gaseous phase, 18
Governing equation, 193, 196
Gradient(s), 49, 52–53, 212, 216, 218, 220, 226
Graphene oxide, 102
Grashof number, 79
Gravitational acceleration, 7
Grid size, 231, 237–243
Ground state, 238, 240, 242
Gyrotaxis, 155

Hall effects, 108
Hamiltonian, 230, 235, 238
Harmonic oscillator, 230, 235–238, 241–243
Hartmann number, 26, 166, 171
Heat and Mass Flux, 139
Heat capacity, 99
Heat flux, 80
Heat generation, 174
Heat generation rate, 78
Heat sink, 62
Heat source, 60, 78
Heat transfer (HT), 1–3, 8–11, 13–15, 78, 110–111, 118, 128–129, 131, 143, 211, 213, 217, 227, 228
Herschel Bulkley model, 29
Heterogeneous, 164
Hiemenz flow, 211, 228
Homogeneous, 38, 57
HPC, 15
HVAC, 3, 6, 10, 12, 13
Hybrid nanofluid (HNF), 78, 108–111, 115, 129, 131, 155
Hydrodynamic lubrication, 22
Hypergeometric functions, 217

Identity tensor, 7
Immiscible fluids, 191–193, 202, 204, 205, 209, 210
Implementation, 229–230, 234, 243
Incompressibility, 160
Incompressible, 193, 209
Incompressible flow, 62
Increase, 199, 201–204, 208
Inertia coefficient parameter, 107, 117, 120–121, 123
Influence, 39–41, 45, 51, 53, 55–56
Initial conditions, 8, 9
Injection, 213, 214

Insulators, 1
Interfaces, 191, 198, 208, 209
Isothermal walls, 159
Isotherms, 162
Isotropic, 191, 192, 199
Issac Newton, 18
Iterative diagonalization, 230

Jacobi algorithm, 229–243
Jeffry fluid, 138
Joule heating, 37, 62, 79, 156, 172, 173

Keller box, 40, 57
Kinematic viscosity, 59, 107, 116–118
Kinetic energy, 6, 8, 71, 86, 100
Knudsen number, 5

Laminar flow, 6, 11, 13
Laplace's equation, 9
Layer, 191–193, 203, 205, 208–210
LES, 14
Lid-driven, 174
Linear algebra, 229, 231–232
Linearly growing wall temperature, 213, 216–218, 220
Lorentz force (LF), 38, 40–41, 47, 57, 66, 84, 100, 125, 129–131, 138, 143, 166

Magnetic field (MF), 59, 78, 110, 125–126, 129–131, 166
Magnetic parameter, 78
Magnetohydrodynamics (MHD), 59, 78, 108, 110, 129–131, 154
Magnetorheological fluid modeling, 33
Mass transfer, 38, 40–41, 52, 55–57, 61, 137, 173
Mass transpiration, 222
MATLAB, 137
Matrix (symmetric, tridiagonal, dense), 229–235, 237–243
Maximum, 199, 201, 202
Maxwell fluid, 60, 80
Maxwell hybrid nanofluid, 93
Mechanical systems, 229, 230, 235
Mesh, 160
Mesh independence, 160
Mesh refinement, 161
MHD, 177
Microbial migration, 156
Microchannel, 137
Microorganiasm, 61
Microorganism diffusivity, 156
Microorganism transport, 166
Micropolar, 192, 209, 210
Micropolar fluid, 60, 211, 212, 227, 228
Micropolar parameter, 212
Microrotation, 212
Minimum, 199, 201, 202
Mixed convection, 61, 79
Modified Hartman Number, 143, 147
Momentum equations, 194
Momentum quantum numbers, 235
Momentum slip length, 59
Momentum slip parameter, 59
Motile microorganism, 155
Multivariate classic model, 147

Nanofluid (NF), 78, 108, 110, 129, 137–139, 141, 143, 151
Nanomaterials, 90
Nanoparticles (NP), 108–109, 115
Natural convection, 60
Navier–Stokes equations, 5–7, 13
Navier–Stokes framework, 211
Neumann, 8, 9
Newton's law of viscosity, 4
Newtonian, 190–193, 209, 210
Newtonian fluid, 19, 156
Newtonian nanofluid, 37
Non-diagonal, 191
Nonlinear, 211, 214, 221, 227, 228
Nonlinear stretching surface, 139
Non-Newtonian (N-N), 108, 110–111, 124, 191, 192
Non-Newtonian systems, 211
Non-Newtonian fluid(s), 20, 60, 80
Non-Newtonian liquids, 137
Non-uniform heat source, 81
Non-uniform inertia coefficient for porous media, 107, 117
Nth Order Chemical Reaction, 139, 143, 151
Numerical solutions, 229–232, 237–243
Nusselt number, 2, 65, 79, 97, 149, 159, 171, 179, 226

Objectives, 41
Off-diagonal elements, 232–234, 239–240
Oobleck, 33
Orthogonality/orthogonal transforms, 230, 232, 234
Oxygen concentration, 166
Oxygen transport, 156
Oxytaxis, 164, 166

Parallel plates, 191, 192
Parameter, 212, 214, 215, 222, 223
Parameters (system/material), 229, 231–232, 237–238, 242
Parametric analysis, 160
Particular cases, 193, 203
PCMs, 12
PDEs, 8, 9, 13
Performance, 229–230, 243
Permeability, 59, 78, 190–193, 196, 199–201, 203–208

Permeability ratio, 193, 196, 199–201, 203–208
Permeability tensor, 190, 193
Permeable surface, 138
Petroleum reservoirs, 191, 193, 203
Physics based models, 31
Plane Poiseuille flow, 204
Porosity, 59, 78
Porous matrix, 191, 210
Porous medium, 166, 191–193, 203–206, 209, 210, 227, 228
Potential (effective, oscillator, Coulomb), 235–238, 242–243
Power law model, 25
Prandtl fluid, 37
Prandtl number, 59, 78, 166, 183, 214
Pressure, 5–7, 10, 11, 190, 193, 196
Pressure gradient, 193, 196
Principal axis, 193
Pseudoplasticity, 25
P-Value, 150

Quantum mechanics, 229–230, 235–238, 242
Quantum numbers, 235–236
Quickly, 47

Radiation, 1–3, 15, 60, 78
RANS, 6, 14
Rayleigh number, 162
Ree–Eyring (RE), 108–111
Relative error, 238–240
Relaxation time, 59, 78
Reliability, 230
Residual norms, 160
Response surface methodology, 138, 139
Reynolds number (Re), 6, 11, 15, 65, 79, 107, 117, 121, 123, 156
Rheological modelling, 23
Riga plate, 109, 138, 139
Rigidity, 231
Rivlin–Erickson, 42
Robin, 8, 9
R-Squared, 150
Runge–Kutta (R-K) method, 137, 139

Scaling (variables, parameters), 229, 231, 236–238
Schmidt Number, 143, 151
Schrodinger equation, 229–230, 235–238
Second-Grade (SG) fluid, 137–151
Separation of variables, 237–238
Shape factor (SF), 109, 115
Shear rate, 148–151
Shear stress, 19, 198, 199, 202, 203, 205, 208
Shear thickening fluid, 30
Shear thinning fluid, 25
Sherwood number, 65, 160, 173
Shooting technique, 142
Shrinking sheet, 211, 221, 222, 227, 228
Similarity transformations, 141, 213
Similarity transforms, 232–234, 239–240
Skin friction, 226
Skin friction coefficient, 118, 129–131, 139
Skin friction, Sherwood number, 179
Slip, 198, 199, 205
Slip velocity, 96
Solid matrix, 191
Solutal profile, 143
Soret diffusivity, 59
Soret number, 59
Specific heat (C_p), 1, 59, 78, 107, 115–117
Spin gradient viscosity, 212
Squeezing effect, 60
Stability, 6, 10, 15, 229
Stagnation-point, 38, 58
Stagnation-point flow, 211, 212
States of matter, 18
Statistical approach, 147
Steady, 193
Stefan–Boltzmann constant, 3
Stream function, 213
Streamline patterns, 162
Stress tensor, 7
Stretching sheet, 211, 212, 221, 222
Stretching surface, 59
Suction, 213, 214, 222
Surface plots, 149
Symmetric matrices, 229–230, 232, 235
System of linear equations, 198

Temperature, 1–3, 5, 8–13, 59, 213, 214, 216–221, 225, 226
Ternary nanofluid, 79
Thermal boundary layer, 81, 222
Thermal conductivity (TC), 1, 8, 9, 12, 59, 78, 108, 111, 129, 131
Thermal diffusivity, 59, 78, 166, 213
Thermal energy, 86
Thermal extension coefficient, 78
Thermal gradient, 216, 226
Thermal radiation, 94, 110, 131, 138
Thermal slip length, 59
Thermal slip parameter, 59
Thermodynamics, 8, 12
Thermophoresis, 59
Thermophoresis number, 78
Thermophoretic effect, 37
Thermophoretic parameters, 143
Three parameter model, 27, 29
Time, 1, 4, 6, 9, 10, 13–15
Time-dependent fluids, 20
Time-independent fluids, 20

Tolerance, 234, 240
Transpiration, 221, 222, 223, 225
Tridiagonal matrix, 240, 243
Turbulence theory, 19
Turbulent flow, 6, 10, 11, 13–15
Two parameter model, 30

Unique, 221
Unsteady, 4, 13, 14, 40, 56–57, 79

Variable thermal conductivity parameter (ϵ), 107, 120–121, 128
Velocity, 190, 192, 193, 198–206, 208
Velocity boundary layer, 81
Velocity distribution, 224
Velocity gradient, 4, 143
Velocity profile, 143, 151, 192, 199–206
Velocity slip, 212, 213, 228
Vibrations, 229–231
Viscoelastic fluid, 61
Viscoelastic parameter, 143, 147, 151
Viscoplastic fluid, 28
Viscosity, 4–7, 19, 98, 190–193, 197, 199–208, 212
Viscosity ratio, 190, 193, 199–203, 205–208
Viscous, 191, 192, 201, 209, 210
Viscous dissipation, 37, 137, 143, 151
Viscous flow, 61
Von Kármán similarity, 116

Wall mass flux, 214, 222
Wall mass transpiration, 222
Water, 37
Wave equation/wave function, 229, 231, 235–237, 242
Weissenberg number (We), 107, 117–118, 120–121, 129–131
Whittaker function, 217
Williamson fluid, 79
Williamson term, 78

Yield stress, 20